The Future of Healthcare: Biomedical Technology and Integrated Artificial Intelligence

The Future of Healthcare: Biomedical Technology and Integrated Artificial Intelligence

Guest Editors

**Juvenal Rodriguez-Resendiz
Gerardo I. Pérez-Soto
Karla Anhel Camarillo-Gómez
Saul Tovar-Arriaga**

Basel • Beijing • Wuhan • Barcelona • Belgrade • Novi Sad • Cluj • Manchester

Guest Editors

Juvenal Rodriguez-Resendiz
Engineering Faculty
UAQ
Queretaro
Mexico

Gerardo I. Pérez-Soto
Engineering Faculty
UAQ
Queretaro
Mexico

Karla Anhel Camarillo-Gómez
Department of Mechanical
Engineering
Tecnológico Nacional
de México en Celaya
Celaya
Mexico

Saul Tovar-Arriaga
Engineering Faculty
UAQ
Queretaro
Mexico

Editorial Office
MDPI AG
Grosspeteranlage 5
4052 Basel, Switzerland

This is a reprint of the Special Issue, published open access by the journal *Technologies* (ISSN 2227-7080), freely accessible at: www.mdpi.com/journal/technologies/special_issues/H45GFE5110.

For citation purposes, cite each article independently as indicated on the article page online and using the guide below:

Lastname, A.A.; Lastname, B.B. Article Title. *Journal Name* **Year**, *Volume Number*, Page Range.

ISBN 978-3-7258-2670-4 (Hbk)
ISBN 978-3-7258-2669-8 (PDF)
https://doi.org/10.3390/books978-3-7258-2669-8

© 2024 by the authors. Articles in this book are Open Access and distributed under the Creative Commons Attribution (CC BY) license. The book as a whole is distributed by MDPI under the terms and conditions of the Creative Commons Attribution-NonCommercial-NoDerivs (CC BY-NC-ND) license (https://creativecommons.org/licenses/by-nc-nd/4.0/).

Contents

About the Editors vii

Saul Tovar-Arriaga, Gerardo Israel Pérez-Soto, Karla Anhel Camarillo-Gómez, Marcos Aviles and Juvenal Rodríguez-Reséndiz
Perspectives, Challenges, and the Future of Biomedical Technology and Artificial Intelligence
Reprinted from: *Technologies* **2024**, *12*, 212, https://doi.org/10.3390/technologies12110212 1

Minjae Kim and Sunghoi Hong
Integrating Artificial Intelligence to Biomedical Science: New Applications for Innovative Stem Cell Research and Drug Development
Reprinted from: *Technologies* **2024**, *12*, 95, https://doi.org/10.3390/technologies12070095 10

Maria Carolina Avelar, Patricia Almeida, Brigida Monica Faria and Luis Paulo Reis
Applications of Brain Wave Classification for Controlling an Intelligent Wheelchair
Reprinted from: *Technologies* **2024**, *12*, 80, https://doi.org/10.3390/technologies12060080 30

Mohammad Asif Hasan, Fariha Haque, Saifur Rahman Sabuj, Hasan Sarker, Md. Omaer Faruq Goni and Fahmida Rahman et al.
An End-to-End Lightweight Multi-Scale CNN for the Classification of Lung and Colon Cancer with XAI Integration
Reprinted from: *Technologies* **2024**, *12*, 56, https://doi.org/10.3390/technologies12040056 56

Tanvi Chandel, Victor Miranda, Andrew Lowe and Tet Chuan Lee
Blood Pressure Measurement Device Accuracy Evaluation: Statistical Considerations with an Implementation in R
Reprinted from: *Technologies* **2024**, *12*, 44, https://doi.org/10.3390/technologies12040044 80

Pedro Moltó-Balado, Silvia Reverté-Villarroya, Victor Alonso-Barberán, Cinta Monclús-Arasa, Maria Teresa Balado-Albiol and Josep Clua-Queralt et al.
Machine Learning Approaches to Predict Major Adverse Cardiovascular Events in Atrial Fibrillation
Reprinted from: *Technologies* **2024**, *12*, 13, https://doi.org/10.3390/technologies12020013 97

Alejandro Villanueva Cerón, Eduardo López Domínguez, Saúl Domínguez Isidro, María Auxilio Medina Nieto, Jorge De La Calleja and Saúl Eduardo Pomares Hernández
Level of Technological Maturity of Telemonitoring Systems Focused on Patients with Chronic Kidney Disease Undergoing Peritoneal Dialysis Treatment: A Systematic Literature Review
Reprinted from: *Technologies* **2023**, *11*, 129, https://doi.org/10.3390/technologies11050129 111

Edgar Rafael Ponce de Leon-Sanchez, Jorge Domingo Mendiola-Santibañez, Omar Arturo Dominguez-Ramirez, Ana Marcela Herrera-Navarro, Alberto Vazquez-Cervantes and Hugo Jimenez-Hernandez et al.
Fuzzy Logic System for Classifying Multiple Sclerosis Patients as High, Medium, or Low Responders to Interferon-Beta
Reprinted from: *Technologies* **2023**, *11*, 109, https://doi.org/10.3390/technologies11040109 122

Rafael Ortiz-Feregrino, Saul Tovar-Arriaga, Jesus Carlos Pedraza-Ortega and Juvenal Rodriguez-Resendiz
Segmentation of Retinal Blood Vessels Using Focal Attention ConvolutionBlocks in a UNET
Reprinted from: *Technologies* **2023**, *11*, 97, https://doi.org/10.3390/technologies11040097 142

María Gonzalez-Moreno, Carlos Monfort-Vinuesa, Antonio Piñas-Mesa and Esther Rincon
Digital Technologies to Provide Humanization in the Education of the Healthcare Workforce: A Systematic Review
Reprinted from: *Technologies* **2023**, *11*, 88, https://doi.org/10.3390/technologies11040088 .... **156**

Marcos Aviles, Juvenal Rodríguez-Reséndiz and Danjela Ibrahimi
Optimizing EMG Classification through Metaheuristic Algorithms
Reprinted from: *Technologies* **2023**, *11*, 87, https://doi.org/10.3390/technologies11040087 .... **174**

Reem Alotaibi and Felwa Abukhodair
Radiation Dose Tracking in Computed Tomography Using Data Visualization
Reprinted from: *Technologies* **2023**, *11*, 74, https://doi.org/10.3390/technologies11030074 .... **196**

About the Editors

Juvenal Rodriguez-Resendiz

Juvenal Rodriguez-Resendiz was with West Virginia University as a Visiting Professor in 2012. Currently, he is the coordinator of the Engineering Ph.D. at Querétaro State University (UAQ), in México. There, he has taught more than 200 digital signal processing and research methodology courses. He belongs to the Mexican Academy of Sciences. He has higher membership among researchers, called SNII level 3, given by the Mexican government. He has developed more than 50 industrial projects by linking UAQ and the government. His team has published 200 scientific papers. He patented more than 10 innovations. He has won national and international prizes because of his academic and innovation developments. He has been the advisor of more than 200 theses of undergraduate, master, and doctoral grades. He has been invited to give 30 conferences around the world. Because of his research, he collaborates with 30 national and international institutions.

Gerardo I. Pérez-Soto

Gerardo I. Pérez-Soto received a Ph.D. degree from the Unviersidad de Guanajuato. In 2013, he joined the Universidad Autonoma de Queretato (UAQ) as an Assistant Professor. He is currently a Professor of the Facultad de Ingenieria, UAQ. His current research field focuses on theoretical kinematics, humanoid robots, mobile robots, and assembly automatization and applications of mechanical engineering on industrial processes. He is a member of the Mexican Association on Robotics and Industry. He received several fellowships and awards, including the A. T. Yang Memorial Award in Theoretical Kinematics as the Best Paper by the American Society of Mechanical Engineering in 2007 and 2011. He is with the National System of Researchers, CONACyT, Mexico.

Karla Anhel Camarillo-Gómez

Karla Anhel Camarillo-Gómez received a Ph.D. degree from the Tecnologico Nacional de Mexico en la Laguna. In 2009, she joined the Tecnologico Nacional de Mexico en Celaya as a Professor, where she is currently a Professor and the Head of the research projects with the Department of Mechanical Engineering. Her current research field focuses on the modeling and control of robots, control of nonlinear systems, stability analysis of nonlinear systems, development of rehabilitation systems, humanoid robots, and assembly automatization and applications of vision control on industrial processes. She is a member of the Mexican Association on Robotics and Industry (AMRob) and the Co-Chair of HuroCup of Federation International of Robo-Sports Association (FIRA). She received several fellowships and awards for robotics from AMRob and FIRA.

Saul Tovar-Arriaga

Saúl Tovar-Arriaga obtained his B.Sc. in Electronics Engineering at the Technology Institute of Queretaro, his M. Sc. in Mechatronics at the University of Siegen, Germany, and his Dr. rer. hum. biol. (doctor of science in human biology) at the University of Erlangen-Nuremberg, Germany. He is currently full professor at the Autonomous University of Queretaro. His research interests include automatic illness diagnosis, surgical robotics, and machine learning applications.

Editorial

Perspectives, Challenges, and the Future of Biomedical Technology and Artificial Intelligence

Saul Tovar-Arriaga [1,†], Gerardo Israel Pérez-Soto [1,†], Karla Anhel Camarillo-Gómez [2,†], Marcos Aviles [1,*,†] and Juvenal Rodríguez-Reséndiz [1,*,†]

1. Facultad de Ingeniería, Universidad Autónoma de Querétaro, Querétaro 76010, Mexico; saul.tovar@uaq.mx (S.T.-A.); israel.perez@uaq.mx (G.I.P.-S.)
2. Departamento de Ingeniería Mecánica, Tecnológico Nacional de México en Celaya, Celaya 38010, Mexico; karla.camarillo@itcelaya.edu.mx
* Correspondence: marcosaviles@ieee.org (M.A.); juvenal@uaq.edu.mx (J.R.-R.)
† These authors contributed equally to this work.

Citation: Tovar-Arriaga, S.; Pérez-Soto, G.I.; Camarillo-Gómez, K.A.; Aviles, M.; Rodríguez-Reséndiz, J. Perspectives, Challenges, and the Future of Biomedical Technology and Artificial Intelligence. *Technologies* **2024**, *12*, 212. https://doi.org/10.3390/technologies12110212

Received: 12 October 2024
Accepted: 16 October 2024
Published: 24 October 2024

Copyright: © 2024 by the authors. Licensee MDPI, Basel, Switzerland. This article is an open access article distributed under the terms and conditions of the Creative Commons Attribution (CC BY) license (https://creativecommons.org/licenses/by/4.0/).

1. Introduction

Biomedical technologies are the compound of engineering principles and technologies used to diagnose, treat, monitor, and prevent illness. Advances in science and engineering continuously improve them, and artificial intelligence (AI) is not getting behind. In the last few years, artificial intelligence has grown faster and has the potential to revolutionize the human experience [1]. With recent advances in generative models such as GPT-4 [2], advancements in biomedical technologies are expected to be realized at a faster pace [3]. In this manuscript, we aim to contribute to the growth of biomedical technologies boosted by AI.

2. The Challenges of Modern Healthcare

Each era in the history of medicine has had its own challenges. Modern medicine faces many challenges as well, such as economic constraints, a growing population, and increased life expectancy, to name just a few. Moreover, there is an alarming increase in cancer [4], cardiovascular [5], and neurodegenerative illnesses [6], many of which could be controlled if they are diagnosed on time. On the other hand, advancements in many areas of medicine have resulted in a high level of specialization [7]. Diagnosing rare illnesses requires years of study. Furthermore, some diseases have only recently been discovered, and it is documented that AI agents have diagnosed diseases that only a few specialists are able to find [8,9]. Although this advancement is generally positive, it conveys the need for costly equipment and the need to analyze complex data.

3. The Transition from Medicine 2.0 to Medicine 3.0

There is an increasing interest in going to the next medicine paradigm, known as Medicine 3.0 [10]. Medicine 1.0 refers to the medicine that our ancestors used, which in many cases worked, but they did not know how. In Medicine 2.0, the mechanics of the treatments are better understood, but it is more oriented to reactive care when the illness is already manifested. On the other hand, Medicine 3.0 is proactive, meaning it focuses more on prevention and quality of life. It considers the patient as unique, so the care should be done accordingly. We are transitioning from Medicine 2.0 to Medicine 3.0, but making this possible will require all the high-tech we have at our disposal.

4. The Uses of AI in Prevention and Treatment

Artificial Intelligence provides methods for analyzing large quantities of data in short periods of time. Since AI systems do not need to rest, they can support a large number of patients at any time. They can help orient patients in deciding how to get medical services or make clear the steps to take to get the prescribed drugs. Internists can also use the technologies to facilitate repetitive, time-consuming tasks such as requesting patient data or writing summaries of cases [3]. Some of the tasks that AI agents can do include classification, regression, prediction, counting, answers to prompts, generating images, sounds, etc. These capabilities can be used for many biomedical technologies, including medical imaging, diagnosis, administration of patient records, design of prosthetic, enhanced vaccine development, acceleration of drug discovery, and so on.

5. Diagnosis and Preventive Care

AI has been enhancing the area of medical imaging, which is crucial for diagnosis. The most well-known imaging techniques include magnetic resonance imaging, computed tomography (CT), and X-rays, but there are other techniques such as ultrasound, PET, and retinograph, to name a few. Different sources of data have also been explored, like signals obtained from biosensors such as BCIs (brain-computer interfaces), text (e.g., Twitter), voice, sounds of organs, wearable devices, etc.

6. Embedded Artificial Intelligence

Embedded artificial intelligence has emerged as a revolutionary innovation in healthcare and medicine, allowing the direct integration of AI algorithms into medical devices and monitoring systems [11]. This technology facilitates real-time data processing, which is crucial for applications requiring immediate responses, such as critical patient monitoring, early detection of abnormalities, and personalized treatment delivery [12].

Medical devices equipped with embedded artificial intelligence can analyze physiological data locally without transmitting information to remote servers, thereby reducing latency and improving system efficiency [13]. For example, wearable devices can monitor vital signs such as heart rate, blood pressure, and glucose levels, providing instant alerts to the patient and medical staff in case abnormal values are detected [14].

Furthermore, embedded AI in point-of-care diagnostic systems has improved diagnostic accuracy, especially in resource-limited settings [15]. These systems can assist healthcare professionals in interpreting medical images using deep learning algorithms trained to recognize patterns associated with various pathologies [16].

However, implementing embedded AI in medical devices presents significant challenges. Limitations in computational resources, such as processing and memory, require the design of efficient algorithms optimized for specific hardware [17]. Standardization and regulation are also critical aspects to consider. Medical devices with embedded AI must comply with strict regulations to ensure their efficacy and safety. Ethical frameworks that address issues related to automated decision-making and liability in the event of errors or malfunctions need to be developed [18].

7. AI in Prosthetic

Artificial intelligence has revolutionized the field of prostheses, significantly improving the functionality and control of artificial limbs [19]. Advanced machine learning algorithms allow modern prostheses to interpret biosignals and convert them into precise movements, providing users with a more natural experience [20]. For example, brain-machine interfaces allow prostheses to be controlled directly by EEG signals, increasing the efficiency and usability of these devices [21]. Furthermore, AI facilitates the continuous adaptation of the prosthesis to the user, learning and adjusting to individual movement patterns to improve performance over time [22]. However, despite these advances, sensory integration and haptic feedback challenges still need to be addressed to achieve a more intuitive interaction between the user and the prosthesis [23].

8. Automate Administrative

Artificial intelligence can transform administrative tasks in the medical field. Functions that could be automated include medical record management, appointment scheduling, billing, and patient tracking [3]. This automation will not only improve efficiency and reduce errors but also allow healthcare professionals to spend more time on direct patient care, thus optimizing the quality of medical care [24].

9. Accelerating Drug Discovery

There are a variety of difficulties throughout the entire drug discovery process: time, prior validation, and robust financial and scientific investment. As an alternative method in drug discovery and development, artificial intelligence allows more effective approaches, overcoming obstacles of traditional methods [25–27]. When artificial intelligence is applied correctly, there is a considerable advantage in laboratory methods, data generation, and computational algorithms, enabling optimal decision-making even when working with incomplete information [28,29]. However, there is significant concern regarding the ethical implications of applying artificial intelligence in pharmacological research, particularly in safeguarding data privacy and security, obtaining informed consent, and ensuring human oversight in decision-making processes, where some authors indicate robust frameworks and regulations are needed [27–31].

10. Surgical Robotics

AI-driven robotic surgery is reshaping the field of surgery by equipping surgeons with real-time advanced data such as force feedback and tactile information, enhancing the identification of surgical margins, and even automating certain parts of surgical procedures [32]. In recent years, one of the main problems is the assistance in real-time during surgery [33], as well as the importance of levels of autonomy and ethical and legal considerations related to advances in surgical robotics with AI [34–36]. This remains an open and constantly evolving topic.

11. Personalized Medicine

Also known as precision medicine, this approach to healthcare focuses on every patient's specific characteristics [37]. Not all people are the same; therefore, treatment should also be according to the individual's needs. To accomplish this, personalized medicine considers genetic data [38], medical records, environment, and lifestyle [39]. Advances in biomedical technology and AI will be crucial in making personalized medicine possible.

12. Editorial Remarks

This Editorial refers to the Special Issue "The Future of Healthcare: Biomedical Technology and Integrated Artificial Intelligence," which aims to showcase innovations using artificial intelligence as a main topic for solving problems in biomedical technology through developing technologies for health and quality of life. Twenty-eight manuscripts were submitted for consideration. All were rigorously peer-reviewed by specialists in their respective areas of expertise. Eleven papers were finally accepted for publication, eight of which are original articles and three are reviews. The list of the final published articles is presented next:

1. Alotaibi, R.; Abukhodair, F. Radiation dose tracking in computed tomography using data visualization. *Technologies* **2023**, *11*, 74.
2. Aviles, M.; Rodríguez-Reséndiz, J.; Ibrahimi, D. Optimizing EMG classification through metaheuristic algorithms. *Technologies* **2023**, *11*, 87.
3. Gonzalez-Moreno, M.; Monfort-Vinuesa, C.; Piñas-Mesa, A.; Rincon, E. Digital technologies to provide humanization in the education of the healthcare workforce: A systematic review. *Technologies* **2023**, *11*, 88.
4. Ortiz-Feregrino, R.; Tovar-Arriaga, S.; Pedraza-Ortega, J.C.; Rodríguez-Reséndiz, J. Segmentation of retinal blood vessels using focal attention convolution blocks in a UNET. *Technologies* **2023**, *11*, 97.
5. de Leon-Sanchez, E.R.P.; Mendiola-Santibáñez, J.D.; Dominguez-Ramirez, O.A.; Herrera-Navarro, A.M.; Vazquez-Cervantes, A.; Jimenez-Hernandez, H.; Senties-Madrid, H.

Fuzzy logic system for classifying multiple sclerosis patients as high, medium, or low responders to interferon-beta. *Technologies* **2023**, *11*, 109.

6. Cerón, A.V.; Domínguez, E.L.; Isidro, S.D.; Nieto, M.A.M.; De La Calleja, J.; Hernández, S.E.P. Level of technological maturity of telemonitoring systems focused on patients with chronic kidney disease undergoing peritoneal dialysis treatment: A systematic literature review. *Technologies* **2023**, *11*, 129.
7. Moltó-Balado, P.; Reverté-Villarroya, S.; Alonso-Barberán, V.; Monclús-Arasa, C.; Balado-Albiol, M.T.; Clua-Queralt, J.; Clua-Espuny, J.-L. Machine learning approaches to predict Major Adverse Cardiovascular Events in atrial fibrillation. *Technologies* **2024**, *12*, 13.
8. Chandel, T.; Miranda, V.; Lowe, A.; Lee, T.C. Blood pressure measurement device accuracy evaluation: Statistical considerations with an implementation in R. *Technologies* **2024**, *12*, 44.
9. Hasan, M.A.; Haque, F.; Sabuj, S.R.; Sarker, H.; Goni, M.O.F.; Rahman, F.; Rashid, M.M. An end-to-end lightweight multi-scale CNN for the classification of lung and colon cancer with XAI integration. *Technologies* **2024**, *12*, 56.
10. Avelar, M.C.; Almeida, P.; Faria, B.M.; Reis, L.P. Applications of brain wave classification for controlling an intelligent wheelchair. *Technologies* **2024**, *12*, 80.
11. Kim, M.; Hong, S. Integrating artificial intelligence to biomedical science: New applications for innovative stem cell research and drug development. *Technologies* **2024**, *12*, 95.

The contributions of the listed articles are summarized in the following lines:

- The study presented in [40] highlights the urgent need to improve radiation dose monitoring in patients undergoing CT scans due to the increase in their use and the risks of overexposure, such as the increased risk of developing cancer. The main challenge lies in the variability of factors influencing the dose received, such as patient characteristics, equipment, and procedure. Current solutions are static, and integration difficulties are present due to the heterogeneity of hospital information systems, limiting the accuracy of user queries. The study proposes a visual analysis approach using Tableau software. It allows automated data cleaning and organization in an interactive dashboard, with multiple simultaneous filters to facilitate its exploration and manipulation. The results, evaluated by experts, show a significant improvement in the radiation dose monitoring process, with a 100% success rate, increasing user satisfaction and providing a better understanding of the analysis. The tool enables individual and group monitoring of patients and procedures, supporting the justification and optimization of these procedures through accurate and easy-to-interpret data. The work contributes to a flexible, interactive, and effective solution for monitoring radiation doses in CT, benefiting health providers, regulators, researchers, and patients by facilitating decision-making, optimizing resources, and improving the quality of radiation-related data.
- On the other hand [41] proposes a metaheuristic-based approach for hyperparameter optimization in a multilayer perceptron (MLP) to improve electromyography signal (EMG) classification, focusing on optimizing the number of neurons, learning rate, epochs, and training batches using the Particle Swarm Optimization (PSO) and Gray Wolf Optimizer (GWO) algorithms. The results show that optimizing these hyperparameters significantly improves the performance of the MLP, achieving an accuracy of 93% in the validation phase. However, it is acknowledged that using a limited database might have affected the performance, so future research with more extensive databases and data augmentation techniques is suggested. The study highlights the effectiveness of the PSO and GWO algorithms in hyperparameter optimization, avoiding manual tuning and reducing model complexity. Although potential limitations, such as stagnation in local optima, are identified, the proposed approach is a promising strategy to improve EMG signal classification, with potential application in other signal processing problems.

- The article [42] analyses the lack of university educational programs combining humanization in healthcare with digital technologies for health sciences students. A systematic review of the literature identified six studies involving 295 students, mostly nursing students, over the last ten years. Only one of the studies integrated digital strategies to teach humanization skills, and another measured the level of humanization after training. The results highlight that, although humanization in care is recognized as essential, no standardized and empirically validated university curricula combine these skills with digital technology. The authors propose a training program based on the HUMAS model, focused on developing skills such as sociability, emotional understanding, and self-efficacy, using narrative methodologies, mindfulness, and digital health technologies such as virtual reality. The importance of designing programs that prepare future health professionals to incorporate humanistic skills in their clinical practice is emphasized, especially in an increasingly digitalized medical environment. Despite the potential benefits, the lack of studies with more diverse groups, including medical students, and the scarcity of digital humanization strategies stand out as critical limitations.
- The work [43] contributed to retinal vessel segmentation, which is essential in diagnosing several illnesses, such as hypertensive retinopathy, diabetic retinopathy, and macular edema. Although there are many methods for segmentation, the authors explore the use of visual transformers, which have been successful in other applications but have the disadvantage of large computational processing. To deal with this constraint, the authors adapted the attention module of visual transformers and integrated it into a convolutional neural network (CNN) based on UNET network, achieving superior performance compared to other models.
- In [44], the authors introduce a fuzzy logic-based system, supported by the knowledge of a neurology expert, to classify patients with relapsing-remitting multiple sclerosis into three categories: high, medium, and low response to interferon-beta treatment. The system showed 100% efficiency compared to a hierarchical clustering method, which only achieved 52%. In addition, a predictive model was developed using biomarkers associated with interferon-beta response to identify suitable candidates for treatment, reaching a test accuracy of 80%. The predictive model includes data normalization steps, principal component analysis compression, and an MLP learning algorithm, which optimizes patient classification and reduces processing time. The results suggest that this approach can help avoid ineffective therapies and improve patient selection for this treatment. Despite its promising results, the study points out limitations such as the small size of the test samples, which restricted cross-validation iterations. The authors highlight the importance of continuing research with other biomarkers and exploring more advanced predictive models, such as evolutionary or deep learning algorithms, to improve performance in predicting responses to treatment.
- The manuscript [45] analyzes fourteen works that propose telemonitoring systems focused on patients with chronic kidney disease (CKD) undergoing peritoneal dialysis (PD) to determine their Technology Readiness Level (TRL). From these works, eight were classified within TRL 9, two within TRL 7, three within TRL 6, and one within TRL 4. Also, the implementations of telemonitoring systems that reached the highest level of TRL correspond to studies developed with the use of proprietary devices and services of international companies specialized in telemedicine treatment of CKD with some limitations regarding their status as proprietary systems incompatible with other devices or systems. Their main limitation is that they are oriented only to treating patients in automated peritoneal dialysis, which limits the care of patients undergoing continuous ambulatory peritoneal dialysis. So, this paper contributes as a reference for researchers and technologists focused on developing telemonitoring systems for patients with CKD undergoing PD.

- In [46], researchers implemented five machine learning techniques to obtain predictors of major adverse cardiovascular events (MACE) in atrial fibrillation (AF) patients. They used two-thirds of the data for training, employing diverse approaches and optimizing to minimize prediction errors, while the remaining third was reserved for testing and validation. The features influencing predictions included the Charlson comorbidity index, diabetes mellitus, cancer, cognitive impairment, vascular disease, chronic obstructive pulmonary disease, the Wells scales, and CHA2DS2-VASc, with specific associations identified. The contribution of the manuscript is that the AdaBoost model was the most effective in predicting MACE in patients with newly diagnosed AF, with an accuracy of 99.99%, a recall of 100%, and an F1 score of 99.97%. Also, it contributes to the optimization of treatment decisions concerning the burden of AF according to the associated risks of thromboembolism and ischemic events.
- The article [47] proposes a methodology for evaluating the accuracy blood pressure (BP) measurement, which expands the method developed by the Committee of the US Association for the Advancement of Medical Instrumentation (AAMI) with the purpose of reducing the sample size stipulated in the International Standard ISO 81060-2. This methodology is based on statistical consideration with an implementation in R and can be used for the early evaluation of experimental devices, showing the potential effects of employing different sample sizes for validating a BP measurement device. Furthermore, it compares previous studies that investigated novel BP measurement methods with different sample sizes and assesses their adherence to the current standard.
- The manuscript [48] present a deep CNN model for detecting lung and colon cancer (LCC). The proposed model achieved an accuracy of 99.20% for the overall LCC class classification and is appropriate for real-time applications, such as mobile or Internet of Medical Things devices, because it has fewer computationally expensive parameters (1.1 million) than existing models. Integrating explainable artificial intelligence (XAI) algorithms, such as Grad-CAM and SHAP, enhances the model's interpretability by providing diverse and complementary insights into feature importance, enabling a more comprehensive understanding of the model's decision-making process.
- The work in [49] contributes to the development of a brain-computer interface (BCI) designed to control a smart wheelchair through motor imagery. Two data sets were used: the first from the IV BCI competition (A) and the second obtained in the laboratory with the Emotiv EPOC device (B). The results indicate that data set A, acquired under controlled conditions and with mobile electrodes, presented a better performance with an F1 score of 0.797 and a false positive rate of 0.150. On the other hand, data set B, obtained with Emotiv EPOC, showed a lower performance due to problems with the fixed placement of the electrodes and noise in the signal, although some subjects achieved good scores. Various feature extraction techniques were evaluated regarding the methodologies, highlighting the Filter Bank Common Spatial Pattern method, whose second version produced the best results. Although the current results are unsuitable for real-time applications, the study validates the concept and the developed architecture, proposing for future work the improvement in noise removal, the use of non-linear classifiers, and the expansion of the data set to increase the generalization of the model.
- Finally, ref. [50] highlight that cultivating and differentiating stem cells, as well as demonstrating their efficacy, is a time-intensive and complex process. Their review explores the applications and advancements of AI technology in drug development, regenerative medicine, and stem cell research. They specifically focus on CNN-based models from the literature, which are used to analyze stem cell images, predict cell types, and evaluate differentiation efficiency. This comprehensive review provides valuable insights into the current state of the field and underscores the growing role of AI in both present and future stem cell research.

13. Conclusions

In this editorial article, we present an overview of biomedical technology and integrated artificial intelligence. We state the importance of these multifaceted technologies and articulate some topics that could be addressed to advance the field. We describe published articles and summarize each contribution. As expected, we received articles from diverse research fields and nationalities. Since health concerns us all, advancing biomedical research and changing the paradigm from Medicine 2.0 to Medicine 3.0 will require the scientific community's work worldwide.

Author Contributions: Conceptualization, K.A.C.-G.; methodology, S.T.-A.; writing—original draft preparation, K.A.C.-G. and M.A; writing—review and editing, M.A., G.I.P.-S., K.A.C.-G. and J.R.-R.; supervision, J.R.-R. All authors have read and agreed to the published version of the manuscript.

Funding: This research received no external funding.

Institutional Review Board Statement: Not applicable.

Data Availability Statement: Not applicable.

Conflicts of Interest: The authors declare no conflicts of interest.

References

1. Kissinger, H.A.; Schmidt, E.; Huttenlocher, D. *The Age of AI: And Our Human Future*; Little, Brown and Company: Hachette, UK, 2021.
2. Achiam, J.; Adler, S.; Agarwal, S.; Ahmad, L.; Akkaya, I.; Aleman, F.L.; Almeida, D.; Altenschmidt, J.; Altman, S.; Anadkat, S.; et al. Gpt-4 technical report. *arXiv* **2023**, arXiv:2303.08774.
3. Lee, P.; Bubeck, S.; Petro, J. Benefits, limits, and risks of GPT-4 as an AI chatbot for medicine. *N. Engl. J. Med.* **2023**, *388*, 1233–1239. [CrossRef] [PubMed]
4. WHO. *Global Cancer Burden Growing, Amidst Mounting Need for Services*; Comunicado de Prensa; WHO: Geneva, Switzerland, 2024.
5. Mensah, G.A.; Fuster, V.; Murray, C.J.; Roth, G.A.; Global Burden of Cardiovascular Diseases and Risks Collaborators. Global burden of cardiovascular diseases and risks, 1990–2022. *J. Am. Coll. Cardiol.* **2023**, *82*, 2350–2473. [CrossRef] [PubMed]
6. Steinmetz, J.D.; Seeher, K.M.; Schiess, N.; Nichols, E.; Cao, B.; Servili, C.; Cavallera, V.; Cousin, E.; Hagins, H.; Moberg, M.E.; et al. Global, regional, and national burden of disorders affecting the nervous system, 1990–2021: A systematic analysis for the Global Burden of Disease Study 2021. *Lancet Neurol.* **2024**, *23*, 344–381. [CrossRef]
7. Marques, L.; Costa, B.; Pereira, M.; Silva, A.; Santos, J.; Saldanha, L.; Silva, I.; Magalhães, P.; Schmidt, S.; Vale, N. Advancing precision medicine: A review of innovative in silico approaches for drug development, clinical pharmacology and personalized healthcare. *Pharmaceutics* **2024**, *16*, 332. [CrossRef]
8. Gertz, R.J.; Dratsch, T.; Bunck, A.C.; Lennartz, S.; Iuga, A.I.; Hellmich, M.G.; Persigehl, T.; Pennig, L.; Gietzen, C.H.; Fervers, P.; et al. Potential of GPT-4 for detecting errors in radiology reports: Implications for reporting accuracy. *Radiology* **2024**, *311*, e232714. [CrossRef]
9. Ito, N.; Kadomatsu, S.; Fujisawa, M.; Fukaguchi, K.; Ishizawa, R.; Kanda, N.; Kasugai, D.; Nakajima, M.; Goto, T.; Tsugawa, Y. The accuracy and potential racial and ethnic biases of GPT-4 in the diagnosis and triage of health conditions: Evaluation study. *JMIR Med. Educ.* **2023**, *9*, e47532. [CrossRef]
10. Attia, P.; Gifford, B. *Outlive*; Harmony Books: New York, NY, USA, 2023.
11. Patel, S.; Park, H.; Bonato, P.; Chan, L.; Rodgers, M. A review of wearable sensors and systems with application in rehabilitation. *J. Neuroeng. Rehabil.* **2012**, *9*, 21. [CrossRef]
12. Chen, J.; Ran, X. Deep learning with edge computing: A review. *Proc. IEEE Inst. Electr. Electron. Eng.* **2019**, *107*, 1655–1674. [CrossRef]
13. Chen, M.; Hao, Y.; Hwang, K.; Wang, L.; Wang, L. Disease prediction by machine learning over big data from healthcare communities. *IEEE Access* **2017**, *5*, 8869–8879. [CrossRef]
14. Baig, M.M.; GholamHosseini, H.; Connolly, M.J. Mobile healthcare applications: System design review, critical issues and challenges. *Australas. Phys. Eng. Sci. Med.* **2015**, *38*, 23–38. [CrossRef] [PubMed]
15. Esteva, A.; Kuprel, B.; Novoa, R.A.; Ko, J.; Swetter, S.M.; Blau, H.M.; Thrun, S. Dermatologist-level classification of skin cancer with deep neural networks. *Nature* **2017**, *542*, 115–118. [CrossRef] [PubMed]
16. Litjens, G.; Kooi, T.; Bejnordi, B.E.; Setio, A.A.A.; Ciompi, F.; Ghafoorian, M.; van der Laak, J.A.W.M.; van Ginneken, B.; Sánchez, C.I. A survey on deep learning in medical image analysis. *Med. Image Anal.* **2017**, *42*, 60–88. [CrossRef] [PubMed]
17. Haris, J.; Gibson, P.; Cano, J.; Agostini, N.B.; Kaeli, D. SECDA: Efficient hardware/software co-design of FPGA-based DNN accelerators for edge inference. *arXiv* **2021**, arXiv:2110.00478.

18. Vayena, E.; Blasimme, A.; Cohen, I.G. Machine learning in medicine: Addressing ethical challenges. *PLoS Med.* **2018**, *15*, e1002689. [CrossRef]
19. Farina, D.; Aszmann, O. Bionic limbs: Clinical reality and academic promises. *Sci. Transl. Med.* **2014**, *6*, 257ps12. [CrossRef]
20. Micera, S.; Carpaneto, J.; Raspopovic, S. Control of hand prostheses using peripheral information. *IEEE Rev. Biomed. Eng.* **2010**, *3*, 48–68. [CrossRef]
21. Collinger, J.L.; Wodlinger, B.; Downey, J.E.; Wang, W.; Tyler-Kabara, E.C.; Weber, D.J.; McMorland, A.J.C.; Velliste, M.; Boninger, M.L.; Schwartz, A.B. High-performance neuroprosthetic control by an individual with tetraplegia. *Lancet* **2013**, *381*, 557–564. [CrossRef]
22. Kuiken, T.A.; Li, G.; Lock, B.A.; Lipschutz, R.D.; Miller, L.A.; Stubblefield, K.A.; Englehart, K.B. Targeted muscle reinnervation for real-time myoelectric control of multifunction artificial arms. *JAMA* **2009**, *301*, 619–628. [CrossRef]
23. Clemente, F.; Valle, G.; Controzzi, M.; Strauss, I.; Iberite, F.; Stieglitz, T.; Granata, G.; Rossini, P.M.; Petrini, F.; Micera, S.; et al. Intraneural sensory feedback restores grip force control and motor coordination while using a prosthetic hand. *J. Neural Eng.* **2019**, *16*, 026034. [CrossRef]
24. Lee, P.; Goldberg, C.; Kohane, I. *The AI Revolution in Medicine: GPT-4 and Beyond*; Pearson: Upper Saddle River, NJ, USA, 2023.
25. Vijayan, R.S.K.; Kihlberg, J.; Cross, J.B.; Poongavanam, V. Enhancing preclinical drug discovery with artificial intelligence. *Drug Discov. Today* **2022**, *27*, 967–984. [CrossRef] [PubMed]
26. Tripathi, A.; Misra, K.; Dhanuka, R.; Singh, J.P. Artificial intelligence in accelerating drug discovery and development. *Recent Pat. Biotechnol.* **2023**, *17*, 9–23. [CrossRef] [PubMed]
27. Tiwari, P.C.; Pal, R.; Chaudhary, M.J.; Nath, R. Artificial intelligence revolutionizing drug development: Exploring opportunities and challenges. *Drug Dev. Res.* **2023**, *84*, 1652–1663. [CrossRef] [PubMed]
28. Santa Maria, J.P., Jr.; Wang, Y.; Camargo, L.M. Perspective on the challenges and opportunities of accelerating drug discovery with artificial intelligence. *Front. Bioinform.* **2023**, *3*, 1121591. [CrossRef]
29. Mak, K.K.; Wong, Y.H.; Pichika, M.R. Artificial intelligence in drug discovery and development. In *Drug Discovery and Evaluation: Safety and Pharmacokinetic Assays*; Springer International Publishing: Cham, Switzerland, 2023; pp. 1–38.
30. Singh, S.; Kumar, R.; Payra, S.; Singh, S.K. Artificial intelligence and machine learning in pharmacological research: Bridging the gap between data and drug discovery. *Cureus* **2023**, *15*, e44359. [CrossRef]
31. Visan, A.I.; Negut, I. Integrating artificial intelligence for drug discovery in the context of revolutionizing drug delivery. *Life* **2024**, *14*, 233. [CrossRef]
32. Knudsen, J.E.; Ghaffar, U.; Ma, R.; Hung, A.J. Clinical applications of artificial intelligence in robotic surgery. *J. Robot. Surg.* **2024**, *18*, 102. [CrossRef]
33. Panesar, S.; Cagle, Y.; Chander, D.; Morey, J.; Fernandez-Miranda, J.; Kliot, M. Artificial intelligence and the future of surgical robotics. *Ann. Surg.* **2019**, *270*, 223–226. [CrossRef]
34. Yang, G.Z.; Cambias, J.; Cleary, K.; Daimler, E.; Drake, J.; Dupont, P.E.; Hata, N.; Kazanzides, P.; Martel, S.; Patel, R.V.; et al. Medical robotics-Regulatory, ethical, and legal considerations for increasing levels of autonomy. *Sci. Robot.* **2017**, *2*, eaam8638. [CrossRef]
35. Laterza, V.; Marchegiani, F.; Aisoni, F.; Ammendola, M.; Schena, C.A.; Lavazza, L.; Ravaioli, C.; Carra, M.C.; Costa, V.; De Franceschi, A.; et al. Smart operating room in digestive surgery: A narrative review. *Healthcare* **2024**, *12*, 1530. [CrossRef]
36. Finocchiaro, M.; Banfi, T.; Donaire, S.; Arezzo, A.; Guarner-Argente, C.; Menciassi, A.; Casals, A.; Ciuti, G.; Hernansanz, A. A framework for the evaluation of human machine interfaces of robot-assisted colonoscopy. *IEEE Trans. Biomed. Eng.* **2024**, *71*, 410–422. [CrossRef] [PubMed]
37. Suwinski, P.; Ong, C.; Ling, M.H.; Poh, Y.M.; Khan, A.M.; Ong, H.S. Advancing personalized medicine through the application of whole exome sequencing and big data analytics. *Front. Genet.* **2019**, *10*, 49. [CrossRef] [PubMed]
38. Filipp, F.V. Opportunities for artificial intelligence in advancing precision medicine. *Curr. Genet. Med. Rep.* **2019**, *7*, 208–213. [CrossRef] [PubMed]
39. Johnson, K.B.; Wei, W.Q.; Weeraratne, D.; Frisse, M.E.; Misulis, K.; Rhee, K.; Zhao, J.; Snowdon, J.L. Precision medicine, AI, and the future of personalized health care. *Clin. Transl. Sci.* **2021**, *14*, 86–93. [CrossRef] [PubMed]
40. Alotaibi, R.; Abukhodair, F. Radiation dose tracking in computed tomography using data visualization. *Technologies* **2023**, *11*, 74. [CrossRef]
41. Aviles, M.; Rodríguez-Reséndiz, J.; Ibrahimi, D. Optimizing EMG classification through metaheuristic algorithms. *Technologies* **2023**, *11*, 87. [CrossRef]
42. Gonzalez-Moreno, M.; Monfort-Vinuesa, C.; Piñas-Mesa, A.; Rincon, E. Digital technologies to provide humanization in the education of the healthcare workforce: A systematic review. *Technologies* **2023**, *11*, 88. [CrossRef]
43. Ortiz-Feregrino, R.; Tovar-Arriaga, S.; Pedraza-Ortega, J.C.; Rodriguez-Resendiz, J. Segmentation of retinal blood vessels using focal attention convolution blocks in a UNET. *Technologies* **2023**, *11*, 97. [CrossRef]
44. Ponce de Leon-Sanchez, E.R.; Mendiola-Santibañez, J.D.; Dominguez-Ramirez, O.A.; Herrera-Navarro, A.M.; Vazquez-Cervantes, A.; Jimenez-Hernandez, H.; Senties-Madrid, H. Fuzzy logic system for classifying multiple sclerosis patients as high, medium, or low responders to interferon-beta. *Technologies* **2023**, *11*, 109. [CrossRef]

45. Villanueva Cerón, A.; López Domínguez, E.; Domínguez Isidro, S.; Medina Nieto, M.A.; De La Calleja, J.; Pomares Hernández, S.E. Level of technological maturity of telemonitoring systems focused on patients with chronic kidney disease undergoing peritoneal dialysis treatment: A systematic literature review. *Technologies* **2023**, *11*, 129. [CrossRef]
46. Moltó-Balado, P.; Reverté-Villarroya, S.; Alonso-Barberán, V.; Monclús-Arasa, C.; Balado-Albiol, M.T.; Clua-Queralt, J.; Clua-Espuny, J.L. Machine learning approaches to predict Major Adverse Cardiovascular Events in atrial fibrillation. *Technologies* **2024**, *12*, 13. [CrossRef]
47. Chandel, T.; Miranda, V.; Lowe, A.; Lee, T.C. Blood pressure measurement device accuracy evaluation: Statistical considerations with an implementation in R. *Technologies* **2024**, *12*, 44. [CrossRef]
48. Hasan, M.A.; Haque, F.; Sabuj, S.R.; Sarker, H.; Goni, M.O.F.; Rahman, F.; Rashid, M.M. An end-to-end lightweight multi-scale CNN for the classification of lung and colon cancer with XAI integration. *Technologies* **2024**, *12*, 56. [CrossRef]
49. Avelar, M.C.; Almeida, P.; Faria, B.M.; Reis, L.P. Applications of brain wave classification for controlling an intelligent wheelchair. *Technologies* **2024**, *12*, 80. [CrossRef]
50. Kim, M.; Hong, S. Integrating artificial intelligence to biomedical science: New applications for innovative stem cell research and drug development. *Technologies* **2024**, *12*, 95. [CrossRef]

Disclaimer/Publisher's Note: The statements, opinions and data contained in all publications are solely those of the individual author(s) and contributor(s) and not of MDPI and/or the editor(s). MDPI and/or the editor(s) disclaim responsibility for any injury to people or property resulting from any ideas, methods, instructions or products referred to in the content.

Review

Integrating Artificial Intelligence to Biomedical Science: New Applications for Innovative Stem Cell Research and Drug Development

Minjae Kim [1,2] and Sunghoi Hong [1,3,*]

1. Laboratory of Stem Cell and NeuroRegeneration, School of Biosystems and Biomedical Sciences, Seoul 02841, Republic of Korea; vscorea89@korea.ac.kr
2. Department of Biomedical Engineering, College of Health Science, Korea University, Seoul 02841, Republic of Korea
3. BK21 FOUR R&E Center for Precision Public Health, Graduate School of Korea University, Seoul 02841, Republic of Korea
* Correspondence: shong21@korea.ac.kr

Citation: Kim, M.; Hong, S. Integrating Artificial Intelligence to Biomedical Science: New Applications for Innovative Stem Cell Research and Drug Development. *Technologies* **2024**, *12*, 95. https://doi.org/10.3390/technologies12070095

Academic Editors: R. Simon Sherratt, Juvenal Rodriguez-Resendiz, Gerardo I. Pérez-Soto, Karla Anhel Camarillo-Gómez and Saul Tovar-Arriaga

Received: 18 March 2024
Revised: 21 June 2024
Accepted: 22 June 2024
Published: 26 June 2024

Copyright: © 2024 by the authors. Licensee MDPI, Basel, Switzerland. This article is an open access article distributed under the terms and conditions of the Creative Commons Attribution (CC BY) license (https://creativecommons.org/licenses/by/4.0/).

Abstract: Artificial intelligence (AI) is rapidly advancing, aiming to mimic human cognitive abilities, and is addressing complex medical challenges in the field of biological science. Over the past decade, AI has experienced exponential growth and proven its effectiveness in processing massive datasets and optimizing decision-making. The main content of this review paper emphasizes the active utilization of AI in the field of stem cells. Stem cell therapies use diverse stem cells for drug development, disease modeling, and medical treatment research. However, cultivating and differentiating stem cells, along with demonstrating cell efficacy, require significant time and labor. In this review paper, convolutional neural networks (CNNs) are widely used to overcome these limitations by analyzing stem cell images, predicting cell types and differentiation efficiency, and enhancing therapeutic outcomes. In the biomedical sciences field, AI algorithms are used to automatically screen large compound databases, identify potential molecular structures and characteristics, and evaluate the efficacy and safety of candidate drugs for specific diseases. Also, AI aids in predicting disease occurrence by analyzing patients' genetic data, medical images, and physiological signals, facilitating early diagnosis. The stem cell field also actively utilizes AI. Artificial intelligence has the potential to make significant advances in disease risk prediction, diagnosis, prognosis, and treatment and to reshape the future of healthcare. This review summarizes the applications and advancements of AI technology in fields such as drug development, regenerative medicine, and stem cell research.

Keywords: artificial intelligence; image analysis; drug discovery; stem cell therapy; convolutional neural network

1. Introduction

In the fields of bioscience and healthcare, numerous companies and researchers are dedicated to developing novel treatments, including new medicines through drug discovery, new gene therapies, and new stem cell therapies, with the aim of accurately diagnosing and treating diseases. However, these endeavors demand diverse technologies and specialized knowledge, and each area is constrained by significant time and cost limitations. To overcome these challenges, approaches such as virtual screening (VS) and molecular docking have been employed in drug development. Yet, these computational methods exhibit inaccuracies and inefficiencies, contributing to an average cost of $2.8 billion and a 15-year timeline for drug discovery [1,2]. Additionally, the complexities of aspects such as chemical structures and drug–protein interactions pose challenges, making it difficult to handle big data manually [3]. Consequently, the need to develop new methods for handling these time- and cost-intensive tasks has been recognized. Figure 1 illustrates the relationship between artificial intelligence (AI), deep learning (DL), and machine learning (ML). Artificial intelligence is widely utilized across various industries. In bioscience, there is a significant emphasis on overcoming the challenges of drug design and discovery [4], where AI is employed in processes such as drug target prediction, bioavailability prediction, and de novo drug design [5,6]. Major pharmaceutical companies, including Bayer, Roche, and Pfizer, have initiated collaborations with information technology (IT) companies to develop AI-based methodologies for drug design [1].

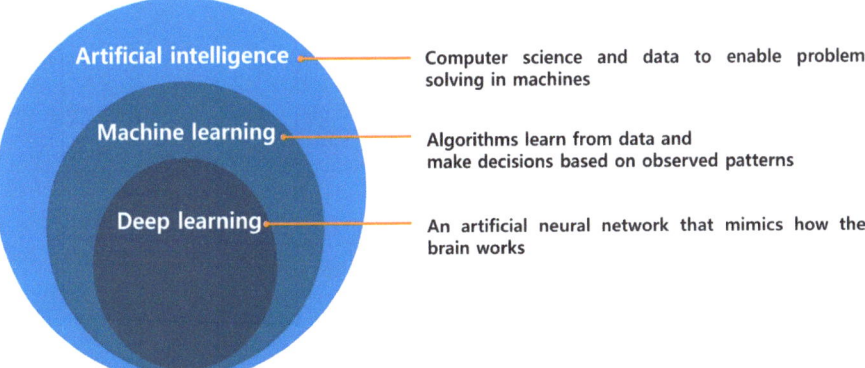

Figure 1. Artificial Intelligence encompasses a broad spectrum, with Machine Learning being one of its subsets. Within Machine Learning, the category of Deep Learning is a subset specifically within the domain of Artificial Neural Networks.

A linkage between stem cell biology and AI research has firmly established itself as a revolutionizing approach in biological studies, offering tremendous potential for elucidating the characteristics of stem cells and their applications in the field of therapeutics. Stem cells, with their remarkable abilities for self-renewal and differentiation into specific cell types, play a central role in regenerative medicine and developmental biology [7,8]. Their extraordinary potential to regenerate damaged tissues and organs has positioned them as key elements in exploring novel therapeutic interventions, ranging from cancer treatments to those addressing neurodegenerative diseases [9–11].

While numerous researchers strive to develop successful and safe stem cell therapies, experiments involving the stable cultivation of stem cells and induced differentiation into desired cell types are extremely time-consuming and labor-intensive [12]. To overcome these limitations, various global companies and research institutes are employing AI technology to investigate the safe cultivation of stem cells and the efficient development of stem cell therapies. Artificial intelligence possesses exceptional analytical and pattern recognition capabilities for processing vast and complex datasets, enabling unprecedented efficiency in exploring the intricacies of stem cell biology by analyzing and predicting high-dimensional cellular imaging data [13].

Among the various algorithms contributing to stem cell research, convolutional neural networks (CNNs) are seen as essential deep learning models for analyzing and identifying image patterns. Recent advancements in CNNs allow researchers to not only discern the types of stem and cancer cells from unlabeled images but also predict the efficiency of differentiation into desired cell types and the level of genetic safety. The advancements in artificial intelligence and deep learning present excellent opportunities for efficient drug development and early disease diagnosis and will affect diverse research areas involving stem cells. Ultimately, these technologies are poised to have a significant impact on humanity.

This review paper will explore how AI technology is utilized in the field of stem cells; introduce current cases of its application; and suggest future directions for AI development. Additionally, it will briefly present ways AI is being applied in regenerative medicine and drug development.

2. AI Technology in Regenerative Medicine

Stem cells are characterized as undifferentiated cells possessing the capability for both self-renewal division and differentiation into various cell types within an organism [14,15]. Pluripotent stem cells (Figure 2), such as embryonic stem cells (ESCs), originate from the inner cell mass of the blastocyst and have the potential to differentiate into all cell types of the embryo and adult, excluding the embryo itself. However, their study has raised ethical concerns due to embryo destruction [16]. Induced pluripotent stem cells (iPSCs), generated from artificially reprogrammed adult somatic cells, share similar functional properties with pluripotent stem cells [17], providing a valuable alternative to ESCs for drug development, disease modeling, and regenerative medicine without significant ethical concern [15,18].

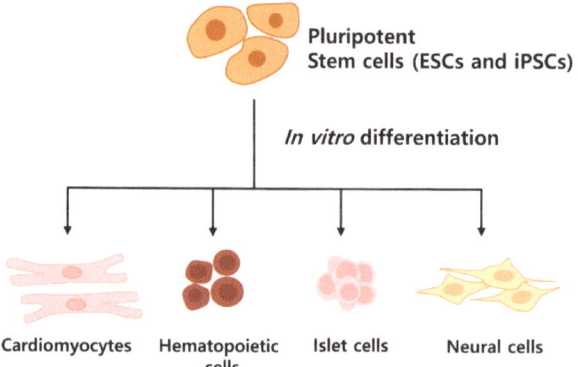

Figure 2. Differentiation of Pluripotent Stem Cells. Pluripotent stem cells can be used to obtain desired cells through differentiation. They have come to represent a crucial turning point in regenerative medicine, with potential applications in disease treatment.

Regenerative medicine, involving advanced technologies such as stem-cell-based therapies, gene therapy, and tissue engineering, aims to restore or replace damaged tissues and organs [19,20]. Therefore, regenerative medicine holds significant promise for patients afflicted with challenging diseases like heart conditions, diabetes, and neurological disorders [21,22]. However, the development of regenerative therapies necessitates the analysis of complex and extensive data, a task where the capabilities of AI technology can be applied. The remainder of this section aims to explore the utilization of AI in the field of regenerative medicine, discussing outcomes and proposing new research directions, while specific examples of AI's use in regenerative medicine are discussed in the next section. This comprehensive perspective seeks to underscore the potential growth of AI in the field.

Cell-based therapy is a promising field in regenerative medicine that utilizes stem cells to treat damaged tissues and organs. Stem cells are considered ideal candidates for repairing damaged tissues and organs [23]. Cell therapy also holds the potential to address chronic diseases without current treatment options [24]. However, there are drawbacks, such as the time- and cost-intensive cultivation of stem cells in a stable manner and the challenges of inducing specific cell differentiation. Moreover, despite ongoing clinical trials, a fully developed cure through stem cell therapy is yet to be realized [25]. Many scientists are turning to AI to analyze extensive datasets and assist in identifying optimal cells for specific patients. One of the key advantages of employing AI in stem cell therapy is its capability to predict the most effective cell types by analyzing patients' genetic information and medical records. This not only aids in identifying conditions for the differentiation of specific cells and optimizing cell cultures but also allows the prediction of cell types based solely on cell morphology. In this context, various algorithms are utilized, and we aim to explore the algorithms predominantly used and provide illustrative examples.

In DL, CNN is the most famous and commonly used algorithm. Its advantage is that it automatically identifies the characteristics of the data without human intervention. The CNN algorithm is widely applied in various fields such as computer vision, voice processing, and facial recognition [26].

The CNN model structure commonly used in the field of biosciences is illustrated in Figure 3. The main framework of the CNN structure consists of convolution layers, max-pooling layers, and softmax layers [27]. Initially, the input data passes through a series of convolution layers paired with max-pooling layers to extract features from the data. Subsequently, the classification task is carried out through fully connected layers.

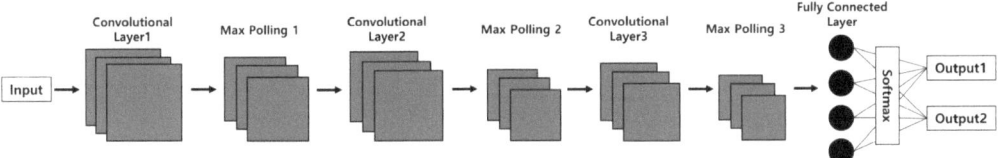

Figure 3. The common convolutional neural network model used in biosciences. The CNN model consists of three convolutional layers extracting features from input images and three max-pooling layers reducing the feature map size followed by a fully connected layer for image classification. The number of convolutional layers and max pooling layers can be customized based on the model's application.

The convolution layers play a crucial role in generating feature maps of various sizes, which are then reduced through pooling layers before proceeding to the next layer. The fully connected layers and softmax function map the extracted features into a final output, such as classification. The initial layers serve to recognize the basic structures of the images, while neurons in deeper layers are specialized in identifying more complex structures [28,29].

2.1. Potential of CNNs in Cell-Image-Based Classification

Microscopes are the most important tools in the field of medicine, allowing the close observation of cell shapes and the detection of abnormal cells. It is important in biology for researchers to maintain cell cultures safely and detect the desired cell morphologies; however, this is time-consuming and prone to errors. Deep learning can overcome these limitations by efficiently analyzing vast amounts of data. Molecular biology is one significant field where deep learning technology can be effectively applied as it deals with the unique shapes of each cell, proven through various experiments [30]. Figure 4 illustrates the capability of CNNs to discern subtle changes in cell morphology induced by a cell culture medium that are imperceptible to the human eye. Worth exploring is whether it is possible to classify the desired neuronal cells solely based on cell images by comparing them with the morphology of existing stem cells. Such cell-image-based classification aims to ascertain whether early cell differentiation can be recognized or if cell types can be identified in early differentiation. If it is successful, utilizing only the images of early differentiating cells could significantly reduce the overall time and cost involved in differentiation experiments. The results predicted by CNNs can be validated by staining stem cells and the desired differentiated cells and observing them through a confocal microscope while also assessing stem cell and differentiation-related gene expression through real-time PCR experiments.

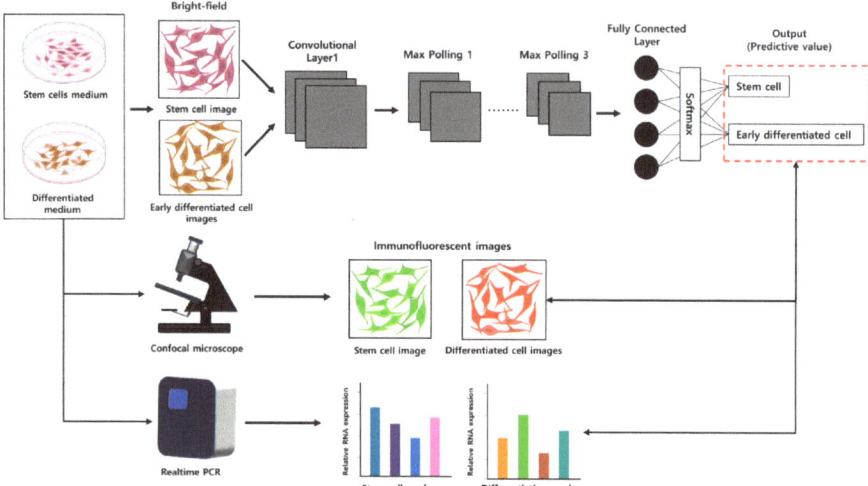

Figure 4. The overall experimental process using CNNs for assessing cell differentiation. The illustration demonstrates the CNN model's ability to distinguish subtle differences in morphology between stem cells and early differentiating cells. The accuracy of CNN predictions can be confirmed by staining both stem cells and differentiated cells, observing them with a confocal microscope, and simultaneously evaluating gene expression related to stem cell differentiation using real-time PCR.

2.2. Applications of CNNs in Stem Cell Culture and Differentiation

Pluripotent stem cells play a significant role in regenerative medicine, disease modeling, and drug testing due to their capacity to differentiate into various cell types within an organism [31,32]. Among the pluripotent stem cells, ESCs and iPSCs are two distinct forms: ESCs are derived during the early phases of embryo development, while iPSCs are generated by reprogramming genes, a process that reverses terminally differentiated somatic cells into a pluripotent state [14,20]. iPSC-derived cells offer a targeted examination of cellular physiology, rendering them valuable for activities such as drug screening, disease analysis, and regenerative medicine. Furthermore, the utilization of mature endothelial cells, derived from iPSCs via differentiation, holds potential for disease modeling and organ development [33]. However, even with well-trained researchers adept at consistently cultivating iPSCs and conducting differentiation experiments into desired cell types, significant time and budget allocation are required for this research. Therefore, trained AI is essential to effectively assist experimenters during the iPSC production phase. Table 1 presents examples of CNN applications in the field of biology and identifies the various cell types and/or disease categories tested in each application. The rest of this section details some of these studies and the major advancement in this field.

Table 1. Applications of CNNs in the stem cell field.

Cell Types	Groups	Algorithm Types	Dataset	Reference
Endothelial cells	CD31-stained cells Unstained cells	CNN	800 images	[34]
Human Embryonic stem cells (hESCs)	Cell cluster Debris Unattached cells Attached cells Dynamically blebbing cells Apoptically blebbing cells	CNN	83,000 images	[35]

Table 1. Cont.

Cell Types	Groups	Algorithm Types	Dataset	Reference
Human-induced pluripotent stem-cell-derived cardiomyocytes (hiPS-CMs)	Normal hiPSC-CM images Abnormal hiPSC-CM images	CNN	18,000 images	[36]
hiPSCs	Reprogramming hiPSCs groups (4 classes) Reprogrammed hiPSCs groups (1 class) Human CD34+ cells group (1 class)	CNN	Total 4020 images	[37]
hiPSCs	Region with no cells Region with differentiated cells Region with possibly reprogramming and reprogrammed hiPSCs	CNN	555 images	[38]
Cancer stem-like cells	Single cells (0, 2, 4, and 14 days)	CNN	1710 single cells	[39]
Mouse ESCs	Retinal organoid Non-retinal organoid	CNN-ResNet50v2	1209 images	[40]
Mesenchymal stem cells (MSCs)	Species Body weight Tissue Cell number and concentration Defect area and depth Type of cartilage damage	Artificial neural network (ANN)	15 clinical trials 29 animal models (1 goat, 6 pigs, 2 dogs, 9 rabbits, 9 rats, and 2 mice)	[41]
hiPS-CMs	Cardiac cell images	CNN	2500 quantitative phase images	[42]
A549, GM12878, MCF7 cells	Transcription factors (DNA binding motifs)	CNN	53 transcription factors	[43]
Human keratinocyte stem cells	Keratinocyte nuclei	CNN	15,040 images	[44]
Bone-marrow-derived mesenchymal stem cells (BMSCs)	10 ng/mL (BMP-12/GDF-7) for 12, 24, and 48 h 50 ng/mL (BMP-12/GDF-7) for 12, 24, and 48 h	CNN	Immunofluorescence staining images	[45]
MSCs	Single-cell RNA sequencing	CNN	RNA sequencing dataset	[46]
ESCs Neuronal progenitor cells (NPCs)	CG methylation	CNN-epiNet	CG methylation in mouse oocytes	[47]
Hematopoietic tumor cells (HTCs)	Acute myeloid leukemia Chronic myeloid leukemia B-cell acute lymphoblastic leukemia Myeloma	CNN	Ten hematopoietic tumor cell lines imaging	[48]
Neural stem cells (NSCs)	Differentiated at 1, 2, 4, and 6 days	Google Inception	Imaging with bright field and flow cytometry	[49]
MSCs	High and low multilineage differentiating stress-enduring (MUSE) cell markers	CNN-DenseNet121	6120 cell images	[50]

Table 1. Cont.

Cell Types	Groups	Algorithm Types	Dataset	Reference
hMSCs	Control group Osteogenic differentiation group Adipogenic differentiation group Osteogenic + Adipogenic differentiation group	CNN-Resnet50	2336 images (Images taken after 1, 2, 3, 5, 7, 10, and 13 days of differentiation)	[51]
hiPSCs	Images during the reprogramming process for 10 days	CNN	3000 images	[52]
Rat rBMSCs	Osteogenic differentiation at 0, 1, 4, and 7 days	Osteogenic CNN	2916 single-cell images	[53]
Human nasal turbinate stem cells (hNTSCs)	Multipotent cell images Non-multipotent cell images	CNN-DenseNet121	1254 multipotent cell images 596 non-multipotent cell images	[54]
Cancer stem cell (CSC)	CSC in images of 1-day culture CSC in images of 2-day culture	CNN	2000 images	[55]
hPSCs	Early differentiation group Late differentiation group	CNN	1331 images	[56]
hESCs	High or low pluripotency status	CNN	269 images of hPSC colonies	[57]
Hematopoietic stem cells (HSCs)	Grade I–IV Acute Graft-Versus-Host Disease (aGVHD)	CNN	18,763 patients between 16 and 80 years of age	[58]
Stem cells	Colony groups	Triplet-net CNN	Colonies images	[59]
Senescent MSCs	Senescent cells Non-senescent cells	CNN	93,907 senescent cells 46,118 non-senescent cells	[60]

The DNN model introduced by Christiansen et al. [61] can predict cell types and the location of cell nuclei from microscopic images without conducting cell immunology experiments. Segmentation based on CNNs can classify images pixel by pixel, assigning each pixel to a particular object category. Additionally, the CNN model enables the detection of object boundaries and categorization within boundary-delineated areas. Consequently, semantic segmentation finds extensive application in fields like cell biology and medicine, not only for identifying the cell's position but also for determining its categories [30].

Kim et al. [31] investigated whether a CNN algorithm, Resnet50, could distinguish subtle changes in the shapes of stem cells, including ESCs and iPSCs, under different culture conditions. These conditions included a medium containing leukemia inhibitory factor (LIF) to maintain pluripotency, a medium without LIF, and a medium with insulin/transferrin/selenium (ITS) to induce differentiation. Data obtained from transmitted light microscopy capturing changes in cell morphologies over a 24-h period were utilized, and the algorithm demonstrated an accuracy of over 95% in identifying the culture conditions and cell types solely based on cell morphology.

Edlund et al. [62] developed the LIVECell system, which can accurately classify eight different types of cells using phase-contrast microscopy images. Even without molecular labeling, this enables the visualization of not only cell types but also intracellular components, along with their localization and types [63,64]. The process of stem cell differentiation was also analyzed using AI by securing image data with a microscope. The differentiation of C2C12 cells and hematopoietic stem cells was conducted with high accuracy. Furthermore, the utilization of an RNN designed to analyze time-series data enabled accurate predictions of hematopoietic stem cell differentiation from microscopic images [65,66].

The CNN model can be employed in unlabeled cell classification systems, as demonstrated by Ugawa et al. [67], who created a ghost cell measurement system capable of identifying undifferentiated human iPSCs, iPSC-derived differentiated cells, neuroendothelial cells (NECs), and hepatic endothelial cells (HECs), and categorizing surrounding white blood cell types. Additionally, CNNs are capable of categorizing cardiac tissue contractility, molecular images, and cell morphologies [68–70].

Though theoretically straightforward, stem cell therapy becomes highly challenging if the cells are not stable or homogeneous. Moreover, conventional testing methods may incur more errors than accurate predictions [71]. Scientists believe that various artificial intelligence techniques, such as ML-SVM and DL-CNN, can assist in addressing these complexities and limitations, potentially serving as the key to perfecting the formula for stem cell therapy [71]. Fan et al. [69] derived iPSCs from human urine cells and utilized a CNN for colony recognition and a semi-supervised segmentation method, two essential aspects of machine learning, to understand visual information from limited labeled data, enabling the detection of colony positions and boundaries.

A vector-based CNN (V-CNN) model was utilized to identify iPSC colonies using phase-contrast images [72]. The input data were healthy and unhealthy iPSC colonies, and the training results of an SVM classifier model and the V-CNN model were compared. The results showed that the V-CNN model could detect the quality of iPSC colonies with an accuracy of 95.5%, while the SVM classifier exhibited an accuracy of 75.2%. Thus, the V-CNN model greatly outperformed the SVM classifier.

A CNN model was also employed to precisely distinguish between stable PSCs and those undergoing early differentiation [32]. Images were acquired at multiple time points within the 24-h period following the onset of differentiation and used to train the CNN model. The training results showed that the CNN model could identify differentiating PSCs with high accuracy just 20 min after the onset of differentiation, distinguishing between undifferentiated and differentiating cells with over 99% accuracy within 24 h.

Recently, an interesting paper focused on discerning morphological distinctions between cells derived from Parkinson's disease patients and healthy individuals [73]. Parkinson's disease has a variety of causes and progression patterns, with significant differences among patients. The team divided patients' cells into five classes, comprising a group of healthy control neurons and four different disease subtype neuron groups. Then, the group trained the model using microscopy image data from all groups obtained after multidimensional fluorescent labeling. This model achieved an accuracy of 82% using tabular data based on the fluorescent cell imagery, while utilizing the microscopy images themselves as the input data resulted in an accuracy of 95%. The model demonstrated consistent accuracy across all Parkinson's disease subtypes used in the experiment, suggesting that beyond diagnosis and drug discovery, AI technology could be used to directly identify the pathological mechanisms of Parkinson's disease.

3. AI Technology in Medical Image Analysis and New Drug Development

3.1. Image Analysis

Medical imaging, crucial for clinical diagnosis and disease treatment, plays a pivotal role in healthcare by generating visual data of the human body. Artificial intelligence has emerged as a leading technology for the analysis of medical imaging and big data [74].

In the realm of medical imaging analysis for disease diagnosis, CNN stands out as the most successful and commonly used deep learning model. This model's widespread adoption in medical image research is primarily attributed to its remarkable performance, made possible by its utilization of graphics processing units (GPUs). Numerous studies have actively employed CNNs to predict and diagnose various diseases, including research projects utilizing MRI images to predict Alzheimer's disease, analyzing CT images to identify pancreatic cancer, and focusing on the early diagnosis of breast cancer [75–77].

The application of CNNs plays a crucial role in predicting and treating a diverse range of diseases. They facilitate efficient diagnosis and the prescription of treatments by analyzing image data through segmentation and classification tasks, enabling clinicians to make effective diagnostic and treatment decisions based on a detailed analysis of visual information [28]. Table 2 elucidates instances of CNN utilization within the biomedical domain, serving as exemplary cases for research and clinical applications.

Table 2. Applications of CNNs in the biomedical field.

Disease Types	Groups	Algorithm Types	Dataset	Reference
Alzheimer's Disease (AD)	AD Mild Cognitive Impairment (MCI) Negative Control	Deep 3D CNN	AD Neuroimaging Initiative MRI images	[78]
Abnormal breast	Normal breast Abnormal breast	CNN	209 normal mammogram images 113 abnormal mammogram images	[79]
Spinal Muscular Atrophy (SMA)	SMA subjects Healthy subjects	CNN	Cell images	[80]
Parkinson	MR Imaging	CNN-Alexnet	MR imaging	[81]
Skin lesion	Dermoscopy images	CNN	2750 skin lesion imaging	[82]
Amyotrophic Lateral Sclerosis (ALS)	ALS subjects Healthy subjects	CNN-VGG16	iPSC cell image	[83]
Lumbar degenerative disease Osteoporosis	Negative group Positive group	Deep CNN	MRS and CT medical imaging	[84]
Diabetes	Diabetes group	Functional Link CNN	Diabetes data	[85]
Parkinson	Healthy group Parkinson's disease patients	Deep CNN	Primary fibroblast from 91 Parkinson's disease patients and healthy controls	[86]
Ovarian tumors	Benign tumors Malignant tumors	Quantum CNN-Resnet	Benign and malignant tumor images	[87]
SARS-CoV-2	Non-COVID-19 Common Pneumonia COVID-19	Deep CNN	Normal pneumonia COVID-19 cases (Chest CT and X-ray)	[88]
Brain tumor	Brain tumor images	CNN-Resnet	Brain MRI images	[89]
Schizophrenia	Schizophrenia patients Healthy groups	CNN	Electroencephalographic data	[90]
Oxygenation in the brains of infants	Infant brain groups	Hybrid CNN	23,000 Near-Infrared Optical Tomography images	[91]
Abnormal heart sound	Heart sound signals	Parallel CNN	3240 heart sound signals	[92]
Benign vocal cord tumor	Cysts Granulomas Leukoplakia Nodules and polyps	CNN	2183 laryngoscopic images	[93]

3.2. New Drug Development

In the field of drug discovery, AI utilizes both supervised and unsupervised learning [94]. Supervised learning involves training data using labeled information, where there are explicitly defined answers. Conversely, unsupervised learning clusters unlabeled data based on similar features to predict outcomes for new data [95]. Supervised learning techniques can be further categorized into classification and regression algorithms, while unsupervised learning techniques are divided into clustering and dimensionality reduction algorithms. Each AI technique is detailed extensively in Chen et al. [1], and Table 3 categorizes the use of AI algorithms according to different types of drug development introduced later.

Table 3. Comparison of major AI techniques used in drug development.

Category	Algorithm	Application	Reference
Supervised learning	SVM	Drug screening	[96]
	RF	Drug target interaction Drug–drug interaction	[97]
	Decision tree	Drug–drug interaction Adverse drug reactions	[98]
Unsupervised learning	K-means clustering	Drug toxicity	[99]
	PCR	QSAR	[100]
Deep learning	CNN	Physiochemical property Drug target interaction Drug–drug interaction	[101,102]
	DNN	Drug screening	[103]
	RNN	De novo drug design Drug target interaction	[104,105]
	GAN	Molecule discovery	[106]
Reinforcement learning	Q-learning	De novo drug design Virtual screening	[107,108]
	Deep Q-network		
	DAN		

SVM: Support vector machine; RF: Random forest; PCR: principal-component analysis; CNN: Convolutional neural network; RNN: recurrent neural network; GAN: generative adversarial network; DAN: Deep Adversarial Networks.

3.2.1. Drug Screening

The traditional drug development process involves synthesizing and testing a large number of compounds to distinguish potential drug candidates, often taking over 10 years and costing an average of around $28 billion. Moreover, despite substantial investments, nine out of ten drug candidates fail in Phase II clinical trials and the regulatory approval process [109]. In response, powerful AI-based tools have emerged that can analyze large compound datasets to predict which treatments will work best for specific diseases. Databases such as ChEMBL, ChemDB, the Collection of Open Natural Products (COCONUT), the Drug–Gene Interaction Database (DGIdb), DrugBank, the Drug Target Commons (DTC), and the Intelligent Network Pharmacology Platform Unique for Traditional Chinese Medicine (INPUT) freely provide diverse information, including molecule names, molecular structures, structural characteristics, bioactive molecules, chemical compounds, bioactivity, and genetic data [110–116].

In drug screening, AI is employed to predict the toxicity, biological activity, and physicochemical properties of prospective novel drugs (Figure 5). The utilized algorithms encompass nearest-neighbor classifiers, random forest (RF), extreme learning machines, support vector machines (SVMs), and deep neural networks (DNNs). These computational methods are applied in VS with a focus on synthetic feasibility, offering predictions of in vivo activity and toxicity [117,118]. While these predictions demonstrate high accuracy, the cost of screening potential drug candidates from a vast array of natural compounds remains expensive. Several pharmaceutical companies, including Bayer, Roche, and Pfizer collaborate with IT companies to advance the development of diverse therapeutics solutions. The following subsections delve into the facets of integrating AI into VS [119].

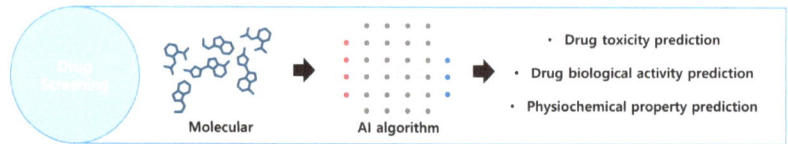

Figure 5. Overview of the drug screening process. By using AI technology to analyze the molecular structure of specific drugs, it is possible to predict drug usage. Created with BioRender.com.

Drug Toxicity Prediction

It is essential to predict the toxicity of drug molecules. Various advanced artificial intelligence techniques are used to identify substances that may have harmful effects on humans. Multiple computational methods were utilized to assess the toxicity of 12,707 environmental compounds and drugs in the Tox21 Challenge [120]. An ML algorithm named "DeepTox" is an ensemble model designed for predicting compound toxicity that combines multiple DNNs [121]. It encodes molecular shapes using 0D to 3D molecular structure models as the inputs for the DNNs. Comparative results based on the Tox21 dataset indicate that DeepTox, with 2500 toxicophore features, outperforms its competitors in toxicity prediction [121,122].

Drug Biological Activity Prediction

Predictions of the biological activity of prospective drugs are widely utilized in areas such as anticancer, antiviral, and antimicrobial drug development, playing a significant role in pharmaceutical discovery [123,124]. The efficacy of drug molecules is determined by their affinity for a target protein or receptor. However, novel drug molecules can also exhibit toxicity due to unintended interactions with target and nontarget proteins or receptors. Therefore, predicting drug–target binding affinity (DTBA) is crucial. Initially, artificial intelligence can be used to measure the binding affinity of a drug by considering the features or similarities between the drug and its target. Subsequently, recognizing the chemical components of the drug and its target is essential for determining feature vectors [125]. Various strategies, including ML and DL approaches such as KronRLS, SimBoost, DeepDTA, and PADME, have been employed to determine DTBA.

For example, Shen et al. [126] employed AutoMolDesigner to automatically design new antibiotics by considering compounds' structures and properties, addressing challenges such as antibiotic scarcity, inhibitory effects, and antibiotic resistance. This open-source tool has proven valuable for researchers seeking to develop novel antibiotics [126].

Physiochemical Property Prediction

The diverse nature of drug physicochemical properties, which includes solubility, partition coefficient (logP), degree of ionization, and intrinsic permeability, necessitates an indirect yet essential understanding of drug action [127]. In particular, the solubility of a drug has a significant impact on its pharmaceutical efficacy, influencing both the pharmacokinetic properties and the formulation of the drug [128]. Considerable investment has been made in developing AI-based solubility prediction models that leverage large datasets of physicochemical properties generated during compound optimization during training with DL models such as DNN or CNN [120].

Panapitiya et al. [129] compared various deep learning architectures, including fully connected neural networks, recurrent neural networks (RNNs), graph neural networks, and SchNet, presenting the strengths and weaknesses of each model. Fully connected neural networks, leveraging molecular descriptors, demonstrated the best performance in solubility prediction.

3.2.2. Key Areas for Drug Discovery

The pursuit of drug discovery involves the identification of active compounds that exhibit therapeutic effects for specific diseases. This subsection aims to elucidate the fundamental elements necessary for drug discovery, highlighting the utilization of AI in key areas such as de novo drug design, target structure prediction, and drug–target interaction (DTI) prediction. Figure 6 illustrates the application of AI technology in drug discovery, as previously discussed.

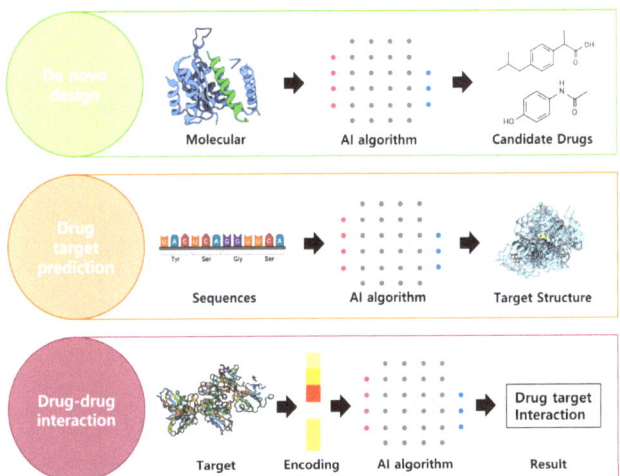

Figure 6. AI technologies to discover drugs. There are de novo design, drug target prediction, and drug–drug interaction elements to drug discovery. By utilizing AI technologies, it is possible to predict the structure of the desired drug target and determine the potential interactions. Created with BioRender.com.

De Novo Drug Design

Artificial intelligence can aid in designing new molecules optimized for specific therapeutic applications, significantly improving the drug discovery process. Computer-aided drug design methods (CADD) that utilize computers to develop drugs have recently become prominent [130]. De novo drug design, a technique for generating novel molecular structures, has garnered significant attention, and various artificial neural network models, such as the reinforcement-learning-based ReLeaSE, the encoder–decoder-based ChemVAE, the GAN-based GraphINVENT, and the RNN-based MolRNN, have been applied in de novo drug design [131–134]. Molecular representation plays a pivotal role in de novo drug design, with inputs for deep learning algorithms derived from simplified molecular-input line-entry system (SMILES), fingerprint, molecular-graph, and 3D geometry data [107,135]. Furthermore, when the structure of a receptor is known, molecular docking information is used, and in cases where the receptor's structure is unknown, quantitative structure–activity relationship (QSAR) and pharmacophore modeling can be employed to predict the 3D structure of the receptor. More recently, deep learning technology has also been applied to various aspects of drug discovery and development [136,137].

Drug Target Prediction

Drug target prediction has, to date, deciphered the structures of approximately 100,000 proteins, but this accounts for only a small fraction of the known protein universe [138]. However, succeeding in drug development with AI still requires addressing challenges such as difficulties in understanding protein tertiary structures. Predicting the 3D structure of proteins, which may consist of thousands of amino acids, demands significant time and resources [95]. To more efficiently predict protein structures, DeepMind has developed the neural-network-based tool AlphaFold, which can predict the 3D structure of proteins from amino acid sequences [138,139].

Predicting Drug–Drug Interactions

Drug target interaction (DTI) prediction, assessing the interactions between compounds and protein targets in an organism, is widely utilized through deep learning and is an essential process in drug development [140]. Prediction methods for DTI using biological data can be categorized into five approaches: ligand-based methods, docking simulations, genetic-algorithm-based methods, text-mining-based methods, and network-based methods [141].

Initially, the encoding of compounds and proteins is carried out using their respective features. Subsequently, the input for the deep learning methods involves the use of the feature embeddings of both compounds and proteins. Models based on deep confidence neural networks, CNNs, and multi-layer perceptrons are commonly employed for DTI prediction [125,142,143].

In in silico drug development, accurately predicting drug–protein interactions is a crucial step. This is essential for understanding the success of treatments and the efficacy and effects of drugs [128]. However, large-scale predictions for countless unknown interactions can involve complex processes. Therefore, semi-supervised learning techniques, primarily utilizing technologies that integrate compound structures, drug–protein interaction network data, and genome sequence data are commonly used [144].

Dhakal et al. [145] discussed that protein–ligand binding sites, ligand binding affinity, and binding structures can be predicted using various machine learning and deep learning techniques. Particularly noteworthy is the discussion on addressing data imbalance issues, where methods like multiple random undersampling and classifier ensembles are introduced to balance sample distributions and reduce information loss. Furthermore, the prediction accuracy is improved by leveraging successful prior research that used convolutional and recurrent neural network architectures to predict and interpret protein structures. The introduction of techniques such as the RF method for enhancing the prediction of ligand binding affinities and the utilization of various algorithms (RF, SVM, neural networks) is also highlighted.

Yaseen et al. [146] focused on predicting drug–target interactions based on text, utilizing data from drug databases and the drug–drug interaction corpus (DDI corpus). Both the CNN model and SVM used in this study demonstrated high performance, achieving excellent accuracy even when using an ensemble model. The study employed CNN models and machine-learning-based classifier SVMs. Both the single CNN and SVM models demonstrated high performances, achieving excellent accuracy even when using an ensemble model. Specifically, the single CNN model showed an F1-score of 0.82, and the ensemble model achieved an impressive 96.72% approved accuracy. The SVM model, in the machine-learning-based implementation, faced challenges due to the availability of negative DTI data but yielded good results in terms of area under the ROC curve (AUC) values. The paper also discusses techniques for addressing class imbalance issues and introduces threshold moving in ensemble models. Suggested future directions include integrating pre-trained embedding layers and position embeddings for improved performance.

4. Discussion

Over the past few years, the integration of AI technology into stem cell therapy, regenerative medicine, and drug development has made significant progress. Artificial intelligence has played a crucial role in recognizing and analyzing vast amounts of data that would be extremely challenging for humans, contributing greatly to the advancement of these fields [147]. However, there are many technological challenges to address before fully realizing the potential of AI in this field.

One of the limitations of AI technology in the fields of stem cell therapy, regenerative medicine, and drug development is the need for large-scale and high-quality data [124]. However, obtaining vast amounts of data is difficult for small patient populations, such as those with rare diseases [148].

In the field of stem cell research, experiments involving the stable culture of stem cells and their differentiation into desired cell types are time-consuming and costly, with inconsistent outcomes. These limitations can be addressed by incorporating AI technology. AI can analyze cell images to predict cell states and improve cell quality. These advantages can overcome the existing challenges of stem cell therapies, making them more effective [8]. In addition, the use of AI technology could ultimately bring about a paradigm shift in tests for development of the therapeutic stem cells, which require accurate and systematic technology in both preclinical and clinical trials.

The stem cells developed for patient treatment currently require substantial time and support. Continuous collaboration among researchers, healthcare providers, and AI developers is essential to continually advance AI technology in the field and achieve more effective personalized treatment solutions [149]. With the increasing availability of high-quality data, the opportunity to fine-tune and customize AI algorithms specifically for regenerative purposes will expand. The machine learning methods used on cells derived from patients with specific diseases have demonstrated high accuracy in predicting disease states, suggesting that the accurate classification capabilities of AI technology could revolutionize disease diagnosis, drug discovery, and the identification of pathological mechanisms in the future [73].

Furthermore, while many AI-based models are being developed, most of them lack freely available web servers or source codes. Even if some smart tools are developed, they are often only available commercially, restricting their application. Therefore, there is a need to develop open tools or packages that can serve as essential resources for applying these models in drug discovery and development [1].

In the field of drug development, it will be possible to predict drug side effects and develop personalized medications for individual diseases using vast amounts of data. Furthermore, the drug development process can be accelerated if collaboration among industry, academia, and regulatory agencies is established to collect, analyze, and validate large datasets.

However, AI algorithms must understand biological complexity to predict drug–target interactions, and the ability to interpret AI models is essential. In drug development, there are considerations for drug safety, efficacy, and ethical issues, requiring substantial investment in technology, infrastructure, and expertise. Despite these challenges, AI continues to advance and offers tremendous potential. As this field progresses and these obstacles are overcome, AI is expected to revolutionize the drug discovery process, leading to faster, more cost-effective, and personalized treatments, ultimately improving patient outcomes [148]. Therefore, the realization of AI's potential may pave the way for a new era in tackling intractable diseases, ultimately contributing to the advancement of medicine and societal well-being.

Author Contributions: Conceptualization, M.K. and S.H.; methodology, M.K.; writing—original draft preparation, M.K.; writing—review and editing, S.H.; supervision, S.H. All authors have read and agreed to the published version of the manuscript.

Funding: This research was funded by National Research Foundation of Korea, grant NRF-2020R1A2C1101294, and by Ministry of Health and Welfare of the government of the Republic of Korea, grant RS-2022-00060247.

Institutional Review Board Statement: Not applicable.

Informed Consent Statement: Not applicable.

Data Availability Statement: The authors declare that the data supporting the findings of this study are available within the article.

Conflicts of Interest: The authors declare no conflicts of interest.

References

1. Chen, W.; Liu, X.; Zhang, S.; Chen, S. Artificial intelligence for drug discovery: Resources, methods, and applications. *Mol. Ther.-Nucleic Acids* **2023**, *31*, 691–702. [CrossRef]
2. Hassanzadeh, P.; Atyabi, F.; Dinarvand, R. The significance of artificial intelligence in drug delivery system design. *Adv. Drug Deliv. Rev.* **2019**, *151*, 169–190. [CrossRef] [PubMed]
3. Gupta, R.; Srivastava, D.; Sahu, M.; Tiwari, S.; Ambasta, R.K.; Kumar, P. Artificial intelligence to deep learning: Machine intelligence approach for drug discovery. *Mol. Divers.* **2021**, *25*, 1315–1360. [CrossRef] [PubMed]
4. Duch, W.; Swaminathan, K.; Meller, J. Artificial intelligence approaches for rational drug design and discovery. *Curr. Pharm. Des.* **2007**, *13*, 1497–1508. [CrossRef] [PubMed]
5. You, Y.; Lai, X.; Pan, Y.; Zheng, H.; Vera, J.; Liu, S.; Deng, S.; Zhang, L. Artificial intelligence in cancer target identification and drug discovery. *Signal Transduct. Target. Ther.* **2022**, *7*, 156. [CrossRef]
6. Wei, M.; Zhang, X.; Pan, X.; Wang, B.; Ji, C.; Qi, Y.; Zhang, J.Z. HobPre: Accurate prediction of human oral bioavailability for small molecules. *J. Cheminform.* **2022**, *14*, 1. [CrossRef]
7. Agrawal, M.; Alexander, A.; Khan, J.; Giri, T.K.; Siddique, S.; Dubey, S.K.; Patel, R.J.; Gupta, U.; Saraf, S.; Saraf, S. Recent biomedical applications on stem cell therapy: A brief overview. *Curr. Stem Cell Res. Ther.* **2019**, *14*, 127–136. [CrossRef]
8. Mukherjee, S.; Yadav, G.; Kumar, R. Recent trends in stem cell-based therapies and applications of artificial intelligence in regenerative medicine. *World J. Stem Cells* **2021**, *13*, 521. [CrossRef] [PubMed]
9. Ntege, E.H.; Sunami, H.; Shimizu, Y. Advances in regenerative therapy: A review of the literature and future directions. *Regen. Ther.* **2020**, *14*, 136–153. [CrossRef]
10. Kwon, S.G.; Kwon, Y.W.; Lee, T.W.; Park, G.T.; Kim, J.H. Recent advances in stem cell therapeutics and tissue engineering strategies. *Biomater. Res.* **2018**, *22*, 36. [CrossRef]
11. Molofsky, A.V.; Pardal, R.; Morrison, S.J. Diverse mechanisms regulate stem cell self-renewal. *Curr. Opin. Cell Biol.* **2004**, *16*, 700–707. [CrossRef] [PubMed]
12. Lien, C.-Y.; Chen, T.-T.; Tsai, E.-T.; Hsiao, Y.-J.; Lee, N.; Gao, C.-E.; Yang, Y.-P.; Chen, S.-J.; Yarmishyn, A.A.; Hwang, D.-K. Recognizing the Differentiation Degree of Human Induced Pluripotent Stem Cell-Derived Retinal Pigment Epithelium Cells Using Machine Learning and Deep Learning-Based Approaches. *Cells* **2023**, *12*, 211. [CrossRef] [PubMed]
13. Nosrati, H.; Nosrati, M. Artificial intelligence in regenerative medicine: Applications and implications. *Biomimetics* **2023**, *8*, 442. [CrossRef] [PubMed]
14. Zakrzewski, W.; Dobrzyński, M.; Szymonowicz, M.; Rybak, Z. Stem cells: Past, present, and future. *Stem Cell Res. Ther.* **2019**, *10*, 68. [CrossRef] [PubMed]
15. Kolios, G.; Moodley, Y. Introduction to stem cells and regenerative medicine. *Respiration* **2012**, *85*, 3–10. [CrossRef] [PubMed]
16. Lo, B.; Parham, L. Ethical issues in stem cell research. *Endocr. Rev.* **2009**, *30*, 204–213. [CrossRef]
17. Romito, A.; Cobellis, G. Pluripotent stem cells: Current understanding and future directions. *Stem Cells Int.* **2016**, *2016*, 9451492. [CrossRef] [PubMed]
18. Takahashi, K.; Yamanaka, S. Induction of pluripotent stem cells from mouse embryonic and adult fibroblast cultures by defined factors. *Cell* **2006**, *126*, 663–676. [CrossRef]
19. Altyar, A.E.; El-Sayed, A.; Abdeen, A.; Piscopo, M.; Mousa, S.A.; Najda, A.; Abdel-Daim, M.M. Future regenerative medicine developments and their therapeutic applications. *Biomed. Pharmacother.* **2023**, *158*, 114131. [CrossRef]
20. Nosrati, H.; Aramideh Khouy, R.; Nosrati, A.; Khodaei, M.; Banitalebi-Dehkordi, M.; Ashrafi-Dehkordi, K.; Sanami, S.; Alizadeh, Z. Nanocomposite scaffolds for accelerating chronic wound healing by enhancing angiogenesis. *J. Nanobiotechnol.* **2021**, *19*, 1. [CrossRef]
21. Rajabzadeh, N.; Fathi, E.; Farahzadi, R. Stem cell-based regenerative medicine. *Stem Cell Investig.* **2019**, *6*, 19. [CrossRef] [PubMed]
22. Zhong, F.; Jiang, Y. Endogenous pancreatic β cell regeneration: A potential strategy for the recovery of β cell deficiency in diabetes. *Front. Endocrinol.* **2019**, *10*, 101. [CrossRef] [PubMed]
23. Gonçalves, A.I.; Costa-Almeida, R.; Gershovich, P.; Rodrigues, M.T.; Reis, R.L.; Gomes, M.E. Cell-based approaches for tendon regeneration. In *Tendon Regeneration*; Elsevier: Amsterdam, The Netherlands, 2015; pp. 187–203.
24. Farini, A.; Sitzia, C.; Erratico, S.; Meregalli, M.; Torrente, Y. Clinical applications of mesenchymal stem cells in chronic diseases. *Stem Cells Int.* **2014**, *2014*, 306573. [CrossRef] [PubMed]
25. Munir, H.; McGettrick, H.M. Mesenchymal stem cell therapy for autoimmune disease: Risks and rewards. *Stem Cells Dev.* **2015**, *24*, 2091–2100. [CrossRef] [PubMed]
26. Alzubaidi, L.; Zhang, J.; Humaidi, A.J.; Al-Dujaili, A.; Duan, Y.; Al-Shamma, O.; Santamaría, J.; Fadhel, M.A.; Al-Amidie, M.; Farhan, L. Review of deep learning: Concepts, CNN architectures, challenges, applications, future directions. *J. Big Data* **2021**, *8*, 53. [CrossRef] [PubMed]
27. Yamashita, R.; Nishio, M.; Do, R.K.G.; Togashi, K. Convolutional neural networks: An overview and application in radiology. *Insights Imaging* **2018**, *9*, 611–629. [CrossRef] [PubMed]
28. Ramakrishna, R.R.; Abd Hamid, Z.; Zaki, W.M.D.W.; Huddin, A.B.; Mathialagan, R. Stem cell imaging through convolutional neural networks: Current issues and future directions in artificial intelligence technology. *PeerJ* **2020**, *8*, e10346. [CrossRef] [PubMed]

29. Rawat, W.; Wang, Z. Deep convolutional neural networks for image classification: A comprehensive review. *Neural Comput.* **2017**, *29*, 2352–2449. [CrossRef] [PubMed]
30. Kusumoto, D.; Yuasa, S. The application of convolutional neural network to stem cell biology. *Inflamm. Regen.* **2019**, *39*, 14. [CrossRef]
31. Kim, M.; Namkung, Y.; Hyun, D.; Hong, S. Prediction of Stem Cell State Using Cell Image-Based Deep Learning. *Adv. Intell. Syst.* **2023**, *5*, 2300017. [CrossRef]
32. Waisman, A.; La Greca, A.; Möbbs, A.M.; Scarafía, M.A.; Velazque, N.L.S.; Neiman, G.; Moro, L.N.; Luzzani, C.; Sevlever, G.E.; Guberman, A.S. Deep learning neural networks highly predict very early onset of pluripotent stem cell differentiation. *Stem Cell Rep.* **2019**, *12*, 845–859. [CrossRef] [PubMed]
33. Zhang, J.; Chu, L.-F.; Hou, Z.; Schwartz, M.P.; Hacker, T.; Vickerman, V.; Swanson, S.; Leng, N.; Nguyen, B.K.; Elwell, A. Functional characterization of human pluripotent stem cell-derived arterial endothelial cells. *Proc. Natl. Acad. Sci. USA* **2017**, *114*, E6072–E6078. [CrossRef] [PubMed]
34. Kusumoto, D.; Lachmann, M.; Kunihiro, T.; Yuasa, S.; Kishino, Y.; Kimura, M.; Katsuki, T.; Itoh, S.; Seki, T.; Fukuda, K. Automated deep learning-based system to identify endothelial cells derived from induced pluripotent stem cells. *Stem Cell Rep.* **2018**, *10*, 1687–1695. [CrossRef] [PubMed]
35. Theagarajan, R.; Guan, B.X.; Bhanu, B. DeephESC: An automated system for generating and classification of human embryonic stem cells. In Proceedings of the 2018 24th International Conference on Pattern Recognition (ICPR), Beijing, China, 20–24 August 2018; pp. 3826–3831.
36. Orita, K.; Sawada, K.; Koyama, R.; Ikegaya, Y. Deep learning-based quality control of cultured human-induced pluripotent stem cell-derived cardiomyocytes. *J. Pharmacol. Sci.* **2019**, *140*, 313–316. [CrossRef] [PubMed]
37. Chang, Y.-H.; Abe, K.; Yokota, H.; Sudo, K.; Nakamura, Y.; Tsai, M.-D. Human induced pluripotent stem cell region detection in bright-field microscopy images using convolutional neural networks. *Biomed. Eng. Appl. Basis Commun.* **2019**, *31*, 1950009. [CrossRef]
38. Chang, Y.-H.; Abe, K.; Yokota, H.; Sudo, K.; Nakamura, Y.; Chu, S.-L.; Hsu, C.-Y.; Tsai, M.-D. Human induced pluripotent stem cell reprogramming prediction in microscopy images using LSTM based RNN. In Proceedings of the 2019 41st Annual International Conference of the IEEE Engineering in Medicine and Biology Society (EMBC), Berlin, Germany, 23–27 July 2019; pp. 2416–2419.
39. Chen, Y.-C.; Zhang, Z.; Yoon, E. Early prediction of single-cell derived sphere formation rate using convolutional neural network image analysis. *Anal. Chem.* **2020**, *92*, 7717–7724. [CrossRef]
40. Kegeles, E.; Naumov, A.; Karpulevich, E.A.; Volchkov, P.; Baranov, P. Convolutional neural networks can predict retinal differentiation in retinal organoids. *Front. Cell. Neurosci.* **2020**, *14*, 171. [CrossRef] [PubMed]
41. Liu, Y.Y.F.; Lu, Y.; Oh, S.; Conduit, G.J. Machine learning to predict mesenchymal stem cell efficacy for cartilage repair. *PLoS Comput. Biol.* **2020**, *16*, e1008275. [CrossRef]
42. Ahmadzadeh, E.; Jaferzadeh, K.; Shin, S.; Moon, I. Automated single cardiomyocyte characterization by nucleus extraction from dynamic holographic images using a fully convolutional neural network. *Biomed. Opt. Express* **2020**, *11*, 1501–1516. [CrossRef]
43. Zhang, Q.; Wang, S.; Chen, Z.; He, Y.; Liu, Q.; Huang, D.-S. Locating transcription factor binding sites by fully convolutional neural network. *Brief. Bioinform.* **2021**, *22*, bbaa435.
44. Hirose, T.; Kotoku, J.i.; Toki, F.; Nishimura, E.K.; Nanba, D. Label-free quality control and identification of human keratinocyte stem cells by deep learning-based automated cell tracking. *Stem Cells* **2021**, *39*, 1091–1100. [CrossRef]
45. Dursun, G.; Tandale, S.B.; Gulakala, R.; Eschweiler, J.; Tohidnezhad, M.; Markert, B.; Stoffel, M. Development of convolutional neural networks for recognition of tenogenic differentiation based on cellular morphology. *Comput. Methods Programs Biomed.* **2021**, *208*, 106279. [CrossRef]
46. Yan, R.; Fan, C.; Yin, Z.; Wang, T.; Chen, X. Potential applications of deep learning in single-cell RNA sequencing analysis for cell therapy and regenerative medicine. *Stem Cells* **2021**, *39*, 511–521. [CrossRef] [PubMed]
47. Au Yeung, W.K.; Maruyama, O.; Sasaki, H. A convolutional neural network-based regression model to infer the epigenetic crosstalk responsible for CG methylation patterns. *BMC Bioinform.* **2021**, *22*, 341. [CrossRef]
48. Matsuoka, Y.; Nakatsuka, R.; Fujioka, T. Automatic discrimination of human hematopoietic tumor cell lines using a combination of imaging flow cytometry and convolutional neural network. *Hum. Cell* **2021**, *34*, 1021–1024. [CrossRef] [PubMed]
49. Zhu, Y.; Huang, R.; Wu, Z.; Song, S.; Cheng, L.; Zhu, R. Deep learning-based predictive identification of neural stem cell differentiation. *Nat. Commun.* **2021**, *12*, 2614. [CrossRef] [PubMed]
50. Kim, G.; Jeon, J.H.; Park, K.; Kim, S.W.; Kim, D.H.; Lee, S. High throughput screening of mesenchymal stem cell lines using deep learning. *Sci. Rep.* **2022**, *12*, 17507. [CrossRef]
51. Mai, M.; Luo, S.; Fasciano, S.; Oluwole, T.E.; Ortiz, J.; Pang, Y.; Wang, S. Morphology-based deep learning approach for predicting adipogenic and osteogenic differentiation of human mesenchymal stem cells (hMSCs). *Front. Cell Dev. Biol.* **2023**, *11*, 1329840. [CrossRef]
52. Chu, S.-L.; Sudo, K.; Yokota, H.; Abe, K.; Nakamura, Y.; Tsai, M.-D. Human induced pluripotent stem cell formation and morphology prediction during reprogramming with time-lapse bright-field microscopy images using deep learning methods. *Comput. Methods Programs Biomed.* **2023**, *229*, 107264. [CrossRef]
53. Lan, Y.; Huang, N.; Fu, Y.; Liu, K.; Zhang, H.; Li, Y.; Yang, S. Morphology-based deep learning approach for predicting osteogenic differentiation. *Front. Bioeng. Biotechnol.* **2022**, *9*, 802794. [CrossRef]

54. Kim, H.; Park, K.; Yon, J.-M.; Kim, S.W.; Lee, S.Y.; Jeong, I.; Jang, J.; Lee, S.; Cho, D.-W. Predicting multipotency of human adult stem cells derived from various donors through deep learning. *Sci. Rep.* **2022**, *12*, 21614. [CrossRef] [PubMed]
55. Hanai, Y.; Ishihata, H.; Zhang, Z.; Maruyama, R.; Kasai, T.; Kameda, H.; Sugiyama, T. Temporal and Locational Values of Images Affecting the Deep Learning of Cancer Stem Cell Morphology. *Biomedicines* **2022**, *10*, 941. [CrossRef] [PubMed]
56. Marzec-Schmidt, K.; Ghosheh, N.; Stahlschmidt, S.R.; Küppers-Munther, B.; Synnergren, J.; Ulfenborg, B. Artificial intelligence supports automated characterization of differentiated human pluripotent stem cells. *Stem Cells* **2023**, *41*, 850–861. [CrossRef] [PubMed]
57. Mamaeva, A.; Krasnova, O.; Khvorova, I.; Kozlov, K.; Gursky, V.; Samsonova, M.; Tikhonova, O.; Neganova, I. Quality Control of Human Pluripotent Stem Cell Colonies by Computational Image Analysis Using Convolutional Neural Networks. *Int. J. Mol. Sci.* **2022**, *24*, 140. [CrossRef] [PubMed]
58. Jo, T.; Arai, Y.; Kanda, J.; Kondo, T.; Ikegame, K.; Uchida, N.; Doki, N.; Fukuda, T.; Ozawa, Y.; Tanaka, M. A convolutional neural network-based model that predicts acute graft-versus-host disease after allogeneic hematopoietic stem cell transplantation. *Commun. Med.* **2023**, *3*, 67. [CrossRef] [PubMed]
59. Witmer, A.; Theagarajan, R.; Bhanu, B. Triplet-net Classification of Contiguous Stem Cell Microscopy Images. *IEEE/ACM Trans. Comput. Biol. Bioinform.* **2023**, *20*, 2314–2327. [CrossRef]
60. He, L.; Li, M.; Wang, X.; Wu, X.; Yue, G.; Wang, T.; Zhou, Y.; Lei, B.; Zhou, G. Morphology-based deep learning enables accurate detection of senescence in mesenchymal stem cell cultures. *BMC Biol.* **2024**, *22*, 1. [CrossRef] [PubMed]
61. Christiansen, E.M.; Yang, S.J.; Ando, D.M.; Javaherian, A.; Skibinski, G.; Lipnick, S.; Mount, E.; O'neil, A.; Shah, K.; Lee, A.K. In silico labeling: Predicting fluorescent labels in unlabeled images. *Cell* **2018**, *173*, 792–803.e719. [CrossRef]
62. Edlund, C.; Jackson, T.R.; Khalid, N.; Bevan, N.; Dale, T.; Dengel, A.; Ahmed, S.; Trygg, J.; Sjögren, R. LIVECell—A large-scale dataset for label-free live cell segmentation. *Nat. Methods* **2021**, *18*, 1038–1045. [CrossRef]
63. Guo, Y.; Shen, D.; Zhou, Y.; Yang, Y.; Liang, J.; Zhou, Y.; Li, N.; Liu, Y.; Yang, G.; Li, W. Deep learning-based morphological classification of endoplasmic reticulum under stress. *Front. Cell Dev. Biol.* **2022**, *9*, 767866. [CrossRef]
64. Sarti, M.; Parlani, M.; Diaz-Gomez, L.; Mikos, A.G.; Cerveri, P.; Casarin, S.; Dondossola, E. Deep Learning for Automated Analysis of Cellular and Extracellular Components of the Foreign Body Response in Multiphoton Microscopy Images. *Front. Bioeng. Biotechnol.* **2022**, *9*, 797555. [CrossRef]
65. Niioka, H.; Asatani, S.; Yoshimura, A.; Ohigashi, H.; Tagawa, S.; Miyake, J. Classification of C2C12 cells at differentiation by convolutional neural network of deep learning using phase contrast images. *Hum. Cell* **2018**, *31*, 87–93. [CrossRef]
66. Buggenthin, F.; Buettner, F.; Hoppe, P.S.; Endele, M.; Kroiss, M.; Strasser, M.; Schwarzfischer, M.; Loeffler, D.; Kokkaliaris, K.D.; Hilsenbeck, O. Prospective identification of hematopoietic lineage choice by deep learning. *Nat. Methods* **2017**, *14*, 403–406. [CrossRef]
67. Ugawa, M.; Kawamura, Y.; Toda, K.; Teranishi, K.; Morita, H.; Adachi, H.; Tamoto, R.; Nomaru, H.; Nakagawa, K.; Sugimoto, K. In silico-labeled ghost cytometry. *eLife* **2021**, *10*, e67660. [CrossRef]
68. Juhola, M.; Joutsijoki, H.; Varpa, K.; Saarikoski, J.; Rasku, J.; Iltanen, K.; Laurikkala, J.; Hyyrö, H.; Ávalos-Salguero, J.; Siirtola, H. On computation of calcium cycling anomalies in cardiomyocytes data. In Proceedings of the 2014 36th Annual International Conference of the IEEE Engineering in Medicine and Biology Society, Chicago, IL, USA, 26–30 August 2014; pp. 1444–1447.
69. Fan, K.; Zhang, S.; Zhang, Y.; Lu, J.; Holcombe, M.; Zhang, X. A machine learning assisted, label-free, non-invasive approach for somatic reprogramming in induced pluripotent stem cell colony formation detection and prediction. *Sci. Rep.* **2017**, *7*, 13496. [CrossRef] [PubMed]
70. Sommer, C.; Gerlich, D.W. Machine learning in cell biology–teaching computers to recognize phenotypes. *J. Cell Sci.* **2013**, *126*, 5529–5539. [CrossRef] [PubMed]
71. Schaub, N.J.; Hotaling, N.A.; Manescu, P.; Padi, S.; Wan, Q.; Sharma, R.; George, A.; Chalfoun, J.; Simon, M.; Ouladi, M. Deep learning predicts function of live retinal pigment epithelium from quantitative microscopy. *J. Clin. Investig.* **2020**, *130*, 1010–1023. [CrossRef] [PubMed]
72. Kavitha, M.S.; Kurita, T.; Park, S.-Y.; Chien, S.-I.; Bae, J.-S.; Ahn, B.-C. Deep vector-based convolutional neural network approach for automatic recognition of colonies of induced pluripotent stem cells. *PLoS ONE* **2017**, *12*, e0189974. [CrossRef]
73. D'Sa, K.; Evans, J.R.; Virdi, G.S.; Vecchi, G.; Adam, A.; Bertolli, O.; Fleming, J.; Chang, H.; Leighton, C.; Horrocks, M.H. Prediction of mechanistic subtypes of Parkinson's using patient-derived stem cell models. *Nat. Mach. Intell.* **2023**, *5*, 933–946. [CrossRef]
74. Datta, S.; Barua, R.; Das, J. Application of artificial intelligence in modern healthcare system. In *Alginates-Recent Uses of This Natural Polymer*; InTechOpen: London, UK, 2019.
75. Sarraf, S.; Tofighi, G. Classification of alzheimer's disease using fmri data and deep learning convolutional neural networks. *arXiv* **2016**, arXiv:1603.08631.
76. Ma, H.; Liu, Z.-X.; Zhang, J.-J.; Wu, F.-T.; Xu, C.-F.; Shen, Z.; Yu, C.-H.; Li, Y.-M. Construction of a convolutional neural network classifier developed by computed tomography images for pancreatic cancer diagnosis. *World J. Gastroenterol.* **2020**, *26*, 5156. [CrossRef] [PubMed]
77. Rakhlin, A.; Shvets, A.; Iglovikov, V.; Kalinin, A.A. Deep convolutional neural networks for breast cancer histology image analysis. In Proceedings of the Image Analysis and Recognition: 15th International Conference, ICIAR 2018, Proceedings 15, Póvoa de Varzim, Portugal, 27–29 June 2018; pp. 737–744.

78. Hosseini-Asl, E.; Gimel'farb, G.; El-Baz, A. Alzheimer's disease diagnostics by a deeply supervised adaptable 3D convolutional network. *arXiv* **2016**, arXiv:1607.00556.
79. Zhang, Y.-D.; Pan, C.; Chen, X.; Wang, F. Abnormal breast identification by nine-layer convolutional neural network with parametric rectified linear unit and rank-based stochastic pooling. *J. Comput. Sci.* **2018**, *27*, 57–68. [CrossRef]
80. Yang, S.J.; Lipnick, S.L.; Makhortova, N.R.; Venugopalan, S.; Fan, M.; Armstrong, Z.; Schlaeger, T.M.; Deng, L.; Chung, W.K.; O'Callaghan, L. Applying deep neural network analysis to high-content image-based assays. *SLAS Discov. Adv. Life Sci. R D* **2019**, *24*, 829–841. [CrossRef] [PubMed]
81. Sivaranjini, S.; Sujatha, C. Deep learning based diagnosis of Parkinson's disease using convolutional neural network. *Multimed. Tools Appl.* **2020**, *79*, 15467–15479. [CrossRef]
82. Xie, F.; Yang, J.; Liu, J.; Jiang, Z.; Zheng, Y.; Wang, Y. Skin lesion segmentation using high-resolution convolutional neural network. *Comput. Methods Programs Biomed.* **2020**, *186*, 105241. [CrossRef]
83. Imamura, K.; Yada, Y.; Izumi, Y.; Morita, M.; Kawata, A.; Arisato, T.; Nagahashi, A.; Enami, T.; Tsukita, K.; Kawakami, H. Prediction model of amyotrophic lateral sclerosis by deep learning with patient induced pluripotent stem cells. *Ann. Neurol.* **2021**, *89*, 1226–1233. [CrossRef] [PubMed]
84. Mu, S.; Wang, J.; Gong, S. Application of medical imaging based on deep learning in the treatment of lumbar degenerative diseases and osteoporosis with bone cement screws. *Comput. Math. Methods Med.* **2021**, *2021*, 2638495. [CrossRef]
85. Jangir, S.K.; Joshi, N.; Kumar, M.; Choubey, D.K.; Singh, S.; Verma, M. Functional link convolutional neural network for the classification of diabetes mellitus. *Int. J. Numer. Methods Biomed. Eng.* **2021**, *37*, e3496. [CrossRef] [PubMed]
86. Schiff, L.; Migliori, B.; Chen, Y.; Carter, D.; Bonilla, C.; Hall, J.; Fan, M.; Tam, E.; Ahadi, S.; Fischbacher, B. Integrating deep learning and unbiased automated high-content screening to identify complex disease signatures in human fibroblasts. *Nat. Commun.* **2022**, *13*, 1590. [CrossRef]
87. Kodipalli, A.; Guha, S.; Dasar, S.; Ismail, T. An inception-ResNet deep learning approach to classify tumours in the ovary as benign and malignant. *Expert Syst.* **2022**, e13215. [CrossRef]
88. Reis, H.C.; Turk, V. COVID-DSNet: A novel deep convolutional neural network for detection of coronavirus (SARS-CoV-2) cases from CT and Chest X-Ray images. *Artif. Intell. Med.* **2022**, *134*, 102427. [CrossRef]
89. Xu, Y.; He, X.; Xu, G.; Qi, G.; Yu, K.; Yin, L.; Yang, P.; Yin, Y.; Chen, H. A medical image segmentation method based on multi-dimensional statistical features. *Front. Neurosci.* **2022**, *16*, 1009581. [CrossRef]
90. Korda, A.; Ventouras, E.; Asvestas, P.; Toumaian, M.; Matsopoulos, G.; Smyrnis, N. Convolutional neural network propagation on electroencephalographic scalograms for detection of schizophrenia. *Clin. Neurophysiol.* **2022**, *139*, 90–105. [CrossRef]
91. Ackermann, M.; Jiang, J.; Russomanno, E.; Wolf, M.; Kalyanov, A. Hybrid Convolutional Neural Network (hCNN) for Image Reconstruction in Near-Infrared Optical Tomography. In *Oxygen Transport to Tissue XLIII*; Springer: Berlin/Heidelberg, Germany, 2022; pp. 165–170.
92. Gharehbaghi, A.; Partovi, E.; Babic, A. Parralel Recurrent Convolutional Neural Network for Abnormal Heart Sound Classification. In *Caring Is Sharing–Exploiting the Value in Data for Health and Innovation*; IOS Press: Amsterdam, The Netherlands, 2023; p. 526.
93. Kim, G.H.; Hwang, Y.J.; Lee, H.; Sung, E.-S.; Nam, K.W. Convolutional neural network-based vocal cord tumor classification technique for home-based self-prescreening purpose. *BioMed. Eng. OnLine* **2023**, *22*, 81. [CrossRef]
94. Talevi, A.; Morales, J.F.; Hather, G.; Podichetty, J.T.; Kim, S.; Bloomingdale, P.C.; Kim, S.; Burton, J.; Brown, J.D.; Winterstein, A.G. Machine learning in drug discovery and development part 1: A primer. *CPT Pharmacomet. Syst. Pharmacol.* **2020**, *9*, 129–142. [CrossRef]
95. Dara, S.; Dhamercherla, S.; Jadav, S.S.; Babu, C.M.; Ahsan, M.J. Machine learning in drug discovery: A review. *Artif. Intell. Rev.* **2022**, *55*, 1947–1999. [CrossRef]
96. Heikamp, K.; Bajorath, J. Support vector machines for drug discovery. *Expert Opin. Drug Discov.* **2014**, *9*, 93–104. [CrossRef]
97. Shi, H.; Liu, S.; Chen, J.; Li, X.; Ma, Q.; Yu, B. Predicting drug-target interactions using Lasso with random forest based on evolutionary information and chemical structure. *Genomics* **2019**, *111*, 1839–1852. [CrossRef] [PubMed]
98. Hammann, F.; Gutmann, H.; Vogt, N.; Helma, C.; Drewe, J. Prediction of adverse drug reactions using decision tree modeling. *Clin. Pharmacol. Ther.* **2010**, *88*, 52–59. [CrossRef] [PubMed]
99. Wang, L.; Bai, M.; Zhao, H.; Qiu, S.; Wang, Z.; Zhao, H. Drug Toxicity Classification Based on ReliefF and K-means Algorithm. In Proceedings of the 2024 12th International Conference on Intelligent Control and Information Processing (ICICIP), Nanjing, China, 8–10 March 2024; pp. 95–99.
100. Yoo, C.; Shahlaei, M. The applications of PCA in QSAR studies: A case study on CCR5 antagonists. *Chem. Biol. Drug Des.* **2018**, *91*, 137–152. [CrossRef]
101. Zhang, C.; Lu, Y.; Zang, T. CNN-DDI: A learning-based method for predicting drug–drug interactions using convolution neural networks. *BMC Bioinform.* **2022**, *23*, 88. [CrossRef] [PubMed]
102. Wang, Y.; Zhang, Z.; Piao, C.; Huang, Y.; Zhang, Y.; Zhang, C.; Lu, Y.-J.; Liu, D. LDS-CNN: A deep learning framework for drug-target interactions prediction based on large-scale drug screening. *Health Inf. Sci. Syst.* **2023**, *11*, 42. [CrossRef] [PubMed]
103. Chen, S.; Yang, Y.; Zhou, H.; Sun, Q.; Su, R. DNN-PNN: A parallel deep neural network model to improve anticancer drug sensitivity. *Methods* **2023**, *209*, 1–9. [CrossRef] [PubMed]
104. Gupta, A.; Müller, A.T.; Huisman, B.J.; Fuchs, J.A.; Schneider, P.; Schneider, G. Generative recurrent networks for de novo drug design. *Mol. Inform.* **2018**, *37*, 1700111. [CrossRef]

105. Kavipriya, G.; Manjula, D. Drug–Target Interaction Prediction Model Using Optimal Recurrent Neural Network. *Intell. Autom. Soft Comput.* **2023**, *35*, 1676–1689. [CrossRef]
106. Blanchard, A.E.; Stanley, C.; Bhowmik, D. Using GANs with adaptive training data to search for new molecules. *J. Cheminform.* **2021**, *13*, 14. [CrossRef]
107. Mouchlis, V.D.; Afantitis, A.; Serra, A.; Fratello, M.; Papadiamantis, A.G.; Aidinis, V.; Lynch, I.; Greco, D.; Melagraki, G. Advances in de novo drug design: From conventional to machine learning methods. *Int. J. Mol. Sci.* **2021**, *22*, 1676. [CrossRef]
108. Edvinsson, F.; Jonsson, V. Autonomous Drug Design with Reinforcement Learning. Master's Thesis, Chalmers University of Technology, Göteborg, Sweden, 2023.
109. Kraljevic, S.; Stambrook, P.J.; Pavelic, K. Accelerating drug discovery: Although the evolution of '-omics' methodologies is still in its infancy, both the pharmaceutical industry and patients could benefit from their implementation in the drug development process. *EMBO Rep.* **2004**, *5*, 837–842. [CrossRef]
110. Mendez, D.; Gaulton, A.; Bento, A.P.; Chambers, J.; De Veij, M.; Félix, E.; Magariños, M.P.; Mosquera, J.F.; Mutowo, P.; Nowotka, M. ChEMBL: Towards direct deposition of bioassay data. *Nucleic Acids Res.* **2019**, *47*, D930–D940. [CrossRef]
111. Chen, J.; Swamidass, S.J.; Dou, Y.; Bruand, J.; Baldi, P. ChemDB: A public database of small molecules and related chemoinformatics resources. *Bioinformatics* **2005**, *21*, 4133–4139. [CrossRef] [PubMed]
112. Sorokina, M.; Merseburger, P.; Rajan, K.; Yirik, M.A.; Steinbeck, C. COCONUT online: Collection of open natural products database. *J. Cheminform.* **2021**, *13*, 2. [CrossRef]
113. Freshour, S.L.; Kiwala, S.; Cotto, K.C.; Coffman, A.C.; McMichael, J.F.; Song, J.J.; Griffith, M.; Griffith, O.L.; Wagner, A.H. Integration of the Drug–Gene Interaction Database (DGIdb 4.0) with open crowdsource efforts. *Nucleic Acids Res.* **2021**, *49*, D1144–D1151. [CrossRef] [PubMed]
114. Wishart, D.S.; Feunang, Y.D.; Guo, A.C.; Lo, E.J.; Marcu, A.; Grant, J.R.; Sajed, T.; Johnson, D.; Li, C.; Sayeeda, Z. DrugBank 5.0: A major update to the DrugBank database for 2018. *Nucleic Acids Res.* **2018**, *46*, D1074–D1082. [CrossRef] [PubMed]
115. Tang, J.; Ravikumar, B.; Alam, Z.; Rebane, A.; Vähä-Koskela, M.; Peddinti, G.; van Adrichem, A.J.; Wakkinen, J.; Jaiswal, A.; Karjalainen, E. Drug target commons: A community effort to build a consensus knowledge base for drug-target interactions. *Cell Chem. Biol.* **2018**, *25*, 224–229.e222. [CrossRef] [PubMed]
116. Li, X.; Tang, Q.; Meng, F.; Du, P.; Chen, W. INPUT: An intelligent network pharmacology platform unique for traditional Chinese medicine. *Comput. Struct. Biotechnol. J.* **2022**, *20*, 1345–1351. [CrossRef] [PubMed]
117. Álvarez-Machancoses, Ó.; Fernández-Martínez, J.L. Using artificial intelligence methods to speed up drug discovery. *Expert Opin. Drug Discov.* **2019**, *14*, 769–777. [CrossRef]
118. Dana, D.; Gadhiya, S.V.; St. Surin, L.G.; Li, D.; Naaz, F.; Ali, Q.; Paka, L.; Yamin, M.A.; Narayan, M.; Goldberg, I.D. Deep learning in drug discovery and medicine; scratching the surface. *Molecules* **2018**, *23*, 2384. [CrossRef] [PubMed]
119. Mak, K.-K.; Pichika, M.R. Artificial intelligence in drug development: Present status and future prospects. *Drug Discov. Today* **2019**, *24*, 773–780. [CrossRef]
120. Yang, X.; Wang, Y.; Byrne, R.; Schneider, G.; Yang, S. Concepts of artificial intelligence for computer-assisted drug discovery. *Chem. Rev.* **2019**, *119*, 10520–10594. [CrossRef]
121. Mayr, A.; Klambauer, G.; Unterthiner, T.; Hochreiter, S. DeepTox: Toxicity prediction using deep learning. *Front. Environ. Sci.* **2016**, *3*, 80. [CrossRef]
122. Pu, L.; Naderi, M.; Liu, T.; Wu, H.-C.; Mukhopadhyay, S.; Brylinski, M. etoxpred: A machine learning-based approach to estimate the toxicity of drug candidates. *BMC Pharmacol. Toxicol.* **2019**, *20*, 2. [CrossRef]
123. Basile, A.O.; Yahi, A.; Tatonetti, N.P. Artificial intelligence for drug toxicity and safety. *Trends Pharmacol. Sci.* **2019**, *40*, 624–635. [CrossRef]
124. Lysenko, A.; Sharma, A.; Boroevich, K.A.; Tsunoda, T. An integrative machine learning approach for prediction of toxicity-related drug safety. *Life Sci. Alliance* **2018**, *1*. [CrossRef]
125. Öztürk, H.; Özgür, A.; Ozkirimli, E. DeepDTA: Deep drug–target binding affinity prediction. *Bioinformatics* **2018**, *34*, i821–i829. [CrossRef]
126. Shen, T.; Guo, J.; Han, Z.; Zhang, G.; Liu, Q.; Si, X.; Wang, D.; Wu, S.; Xia, J. AutoMolDesigner for Antibiotic Discovery: An AI-based Open-source Software for Automated Design of Small-molecule Antibiotics. *J. Chem. Inf. Model.* **2023**, *64*, 575–583. [CrossRef]
127. Zang, Q.; Mansouri, K.; Williams, A.J.; Judson, R.S.; Allen, D.G.; Casey, W.M.; Kleinstreuer, N.C. In silico prediction of physicochemical properties of environmental chemicals using molecular fingerprints and machine learning. *J. Chem. Inf. Model.* **2017**, *57*, 36–49. [CrossRef]
128. Wan, F.; Zeng, J. Deep learning with feature embedding for compound-protein interaction prediction. *bioRxiv* **2016**, 086033. [CrossRef]
129. Panapitiya, G.; Girard, M.; Hollas, A.; Murugesan, V.; Wang, W.; Saldanha, E. Predicting aqueous solubility of organic molecules using deep learning models with varied molecular representations. *arXiv* **2021**, arXiv:2105.12638.
130. Lee, J.W.; Maria-Solano, M.A.; Vu, T.N.L.; Yoon, S.; Choi, S. Big data and artificial intelligence (AI) methodologies for computer-aided drug design (CADD). *Biochem. Soc. Trans.* **2022**, *50*, 241–252. [CrossRef]
131. Popova, M.; Isayev, O.; Tropsha, A. Deep reinforcement learning for de novo drug design. *Sci. Adv.* **2018**, *4*, eaap7885. [CrossRef]

132. Gómez-Bombarelli, R.; Wei, J.N.; Duvenaud, D.; Hernández-Lobato, J.M.; Sánchez-Lengeling, B.; Sheberla, D.; Aguilera-Iparraguirre, J.; Hirzel, T.D.; Adams, R.P.; Aspuru-Guzik, A. Automatic chemical design using a data-driven continuous representation of molecules. *ACS Cent. Sci.* **2018**, *4*, 268–276. [CrossRef]
133. Mercado, R.; Rastemo, T.; Lindelöf, E.; Klambauer, G.; Engkvist, O.; Chen, H.; Bjerrum, E.J. Graph networks for molecular design. *Mach. Learn. Sci. Technol.* **2021**, *2*, 025023. [CrossRef]
134. Li, Y.; Zhang, L.; Liu, Z. Multi-objective de novo drug design with conditional graph generative model. *J. Cheminform.* **2018**, *10*, 33. [CrossRef]
135. Wang, M.; Wang, Z.; Sun, H.; Wang, J.; Shen, C.; Weng, G.; Chai, X.; Li, H.; Cao, D.; Hou, T. Deep learning approaches for de novo drug design: An overview. *Curr. Opin. Struct. Biol.* **2022**, *72*, 135–144. [CrossRef]
136. Guo, J.; Janet, J.P.; Bauer, M.R.; Nittinger, E.; Giblin, K.A.; Papadopoulos, K.; Voronov, A.; Patronov, A.; Engkvist, O.; Margreitter, C. DockStream: A docking wrapper to enhance de novo molecular design. *J. Cheminform.* **2021**, *13*, 89. [CrossRef]
137. Wang, M.; Hsieh, C.-Y.; Wang, J.; Wang, D.; Weng, G.; Shen, C.; Yao, X.; Bing, Z.; Li, H.; Cao, D. Relation: A deep generative model for structure-based de novo drug design. *J. Med. Chem.* **2022**, *65*, 9478–9492. [CrossRef]
138. Jumper, J.; Evans, R.; Pritzel, A.; Green, T.; Figurnov, M.; Ronneberger, O.; Tunyasuvunakool, K.; Bates, R.; Žídek, A.; Potapenko, A. Highly accurate protein structure prediction with AlphaFold. *Nature* **2021**, *596*, 583–589. [CrossRef]
139. Aderinwale, T.; Bharadwaj, V.; Christoffer, C.; Terashi, G.; Zhang, Z.; Jahandideh, R.; Kagaya, Y.; Kihara, D. Real-time structure search and structure classification for AlphaFold protein models. *Commun. Biol.* **2022**, *5*, 316. [CrossRef]
140. Nag, S.; Baidya, A.T.; Mandal, A.; Mathew, A.T.; Das, B.; Devi, B.; Kumar, R. Deep learning tools for advancing drug discovery and development. *3 Biotech* **2022**, *12*, 110. [CrossRef]
141. Bagherian, M.; Sabeti, E.; Wang, K.; Sartor, M.A.; Nikolovska-Coleska, Z.; Najarian, K. Machine learning approaches and databases for prediction of drug–target interaction: A survey paper. *Brief. Bioinform.* **2021**, *22*, 247–269. [CrossRef]
142. Wen, M.; Zhang, Z.; Niu, S.; Sha, H.; Yang, R.; Yun, Y.; Lu, H. Deep-learning-based drug–target interaction prediction. *J. Proteome Res.* **2017**, *16*, 1401–1409. [CrossRef]
143. Lee, I.; Keum, J.; Nam, H. DeepConv-DTI: Prediction of drug-target interactions via deep learning with convolution on protein sequences. *PLoS Comput. Biol.* **2019**, *15*, e1007129. [CrossRef]
144. Xia, Z.; Wu, L.-Y.; Zhou, X.; Wong, S.T. Semi-supervised drug-protein interaction prediction from heterogeneous biological spaces. *BMC Syst. Biol.* **2010**, *4*, S6. [CrossRef]
145. Dhakal, A.; McKay, C.; Tanner, J.J.; Cheng, J. Artificial intelligence in the prediction of protein–ligand interactions: Recent advances and future directions. *Brief. Bioinform.* **2022**, *23*, bbab476. [CrossRef]
146. Yaseen, B.T. Drug Target Interaction Prediction Using Convolutional Neural Network (CNN). In Proceedings of the 2023 5th International Congress on Human-Computer Interaction, Optimization and Robotic Applications (HORA), Istanbul, Turkey, 8–10 June 2023; pp. 1–5.
147. Paul, D.; Sanap, G.; Shenoy, S.; Kalyane, D.; Kalia, K.; Tekade, R.K. Artificial intelligence in drug discovery and development. *Drug Discov. Today* **2021**, *26*, 80. [CrossRef]
148. Tiwari, P.C.; Pal, R.; Chaudhary, M.J.; Nath, R. Artificial intelligence revolutionizing drug development: Exploring opportunities and challenges. *Drug Dev. Res.* **2023**, *84*, 1652–1663. [CrossRef]
149. Alowais, S.A.; Alghamdi, S.S.; Alsuhebany, N.; Alqahtani, T.; Alshaya, A.I.; Almohareb, S.N.; Aldairem, A.; Alrashed, M.; Bin Saleh, K.; Badreldin, H.A. Revolutionizing healthcare: The role of artificial intelligence in clinical practice. *BMC Med. Educ.* **2023**, *23*, 689. [CrossRef]

Disclaimer/Publisher's Note: The statements, opinions and data contained in all publications are solely those of the individual author(s) and contributor(s) and not of MDPI and/or the editor(s). MDPI and/or the editor(s) disclaim responsibility for any injury to people or property resulting from any ideas, methods, instructions or products referred to in the content.

Article

Applications of Brain Wave Classification for Controlling an Intelligent Wheelchair

Maria Carolina Avelar [1], Patricia Almeida [1], Brigida Monica Faria [2,3,*] and Luis Paulo Reis [1,3]

1. Faculty of Engineering, University of Porto (FEUP), Rua Dr. Roberto Frias, s/n, 4200-465 Porto, Portugal; lpreis@fe.up.pt (L.P.R.)
2. ESS, Polytechnic of Porto (ESS-P.PORTO), Rua Dr. António Bernardino de Almeida, 400, 4200-072 Porto, Portugal
3. Artificial Intelligence and Computer Science Laboratory (LIACC—Member of LASI LA), Rua Dr. Roberto Frias, s/n, 4200-465 Porto, Portugal
* Correspondence: monica.faria@ess.ipp.pt

Abstract: The independence and autonomy of both elderly and disabled people have been a growing concern in today's society. Therefore, wheelchairs have proven to be fundamental for the movement of these people with physical disabilities in the lower limbs, paralysis, or other type of restrictive diseases. Various adapted sensors can be employed in order to facilitate the wheelchair's driving experience. This work develops the proof concept of a brain–computer interface (BCI), whose ultimate final goal will be to control an intelligent wheelchair. An event-related (de)synchronization neuro-mechanism will be used, since it corresponds to a synchronization, or desynchronization, in the mu and beta brain rhythms, during the execution, preparation, or imagination of motor actions. Two datasets were used for algorithm development: one from the IV competition of BCIs (A), acquired through twenty-two Ag/AgCl electrodes and encompassing motor imagery of the right and left hands, and feet; and the other (B) was obtained in the laboratory using an Emotiv EPOC headset, also with the same motor imaginary. Regarding feature extraction, several approaches were tested: namely, two versions of the signal's power spectral density, followed by a filter bank version; the use of respective frequency coefficients; and, finally, two versions of the known method filter bank common spatial pattern (FBCSP). Concerning the results from the second version of FBCSP, dataset A presented an F1-score of 0.797 and a rather low false positive rate of 0.150. Moreover, the correspondent average kappa score reached the value of 0.693, which is in the same order of magnitude as 0.57, obtained by the competition. Regarding dataset B, the average value of the F1-score was 0.651, followed by a kappa score of 0.447, and a false positive rate of 0.471. However, it should be noted that some subjects from this dataset presented F1-scores of 0.747 and 0.911, suggesting that the movement imagery (MI) aptness of different users may influence their performance. In conclusion, it is possible to obtain promising results, using an architecture for a real-time application.

Keywords: brain–computer interface; intelligent wheelchair; Emotiv EPOC headset

Citation: Avelar, M.C.; Almeida, P.; Faria, B.M.; Reis, L.P. Applications of Brain Wave Classification for Controlling an Intelligent Wheelchair. *Technologies* **2024**, *12*, 80. https://doi.org/10.3390/technologies12060080

Academic Editors: Juvenal Rodriguez-Resendiz, Gerardo I. Pérez-Soto, Karla Anhel Camarillo-Gómez, Saul Tovar-Arriaga and Luc de Witte

Received: 9 April 2024
Revised: 15 May 2024
Accepted: 20 May 2024
Published: 3 June 2024

Copyright: © 2024 by the authors. Licensee MDPI, Basel, Switzerland. This article is an open access article distributed under the terms and conditions of the Creative Commons Attribution (CC BY) license (https://creativecommons.org/licenses/by/4.0/).

1. Introduction

Independence and autonomy in mobility are two of the most important conditions for determining the quality of life of people with disabilities or with low mobility capacities [1]. Limited mobility could have origin in a broad range of situations, from accidents to disease to the ageing process. Currently, several mobility-related technologies are designed to achieve independent mobility, in particular powered orthosis, prosthetic devices, and exoskeletons. Notwithstanding these devices, wheeled mobility devices remain among the most used assistive devices [2]. According to the World Health Organization (WHO) [3], approximately 10% of the world's population, or around 740 million people, suffer from disabilities, and, among those people, almost 10% require a wheelchair. Therefore, it is estimated that about 1% of the total population needs wheelchairs, which translates into 74 million people worldwide [4].

The importance of providing multifaceted wheelchairs that can be adapted to the most diverse conditions of their users is thus emphasised. Different interfaces are being developed, enabling us to overcome existing barriers of use. In particular, special attention has been dedicated to voice control techniques, joysticks, and tongue or head movements. However, hand gesture recognition and brain–computer interface (BCI) systems are proving to be interesting methods of wheelchair control due to their accessible price and non-invasiveness. Therefore, BCIs seem the best option to bridge the users' will and the wheelchair, as they provide a direct pathway between the "mind" and the external world, just by interpreting the user's brain activity patterns into corresponding commands [5], and, thus, not requiring neuro-muscular control capabilities whatsoever. Furthermore, people desire to be in charge of their motion as much as possible, even if they have lost most of their voluntary muscle control; therefore, BCIs are an exceptional option [6]. Brain–computer interfaces provide control and communication between human intention and physical devices by translating the pattern of brain activity into commands [6]. The flow of a BCI consists of the acquisition of the information from the brain, followed by the data processing, and ending in the output of a control command [7]. Thus, usually, a BCI can be conceptually divided into signal acquisition, pre-processing, and feature extraction and classification; the last three are the interpretation of the first one.

This paper is structured into six sections, beginning with this introduction. The second section addresses the background and state of the art concerning brain–computer interfaces (BCIs) for acquiring and classifying brain activity. Section 3 details the methods and materials employed in the experimental work. In Section 4, the results obtained from various approaches, including a real-time application, are presented. A discussion of the findings is presented in Section 5, followed by conclusions and suggestions for future work.

2. Background and State of the Art of BCI Brain Activity Acquisition Methods

Understanding the acquisition methods for brain activity is crucial for the development of effective brain–computer interfaces (BCIs). There are several methods available, but the most commonly used and well-established method is electroencephalography (EEG). EEG is favoured for its low cost, convenience, standardized electrode placement, and well-documented acquisition techniques. Additionally, EEG offers known filtering methods to address noise and ocular artefacts, making it an attractive option for BCI applications.

2.1. Signal Acquisition

There are several methods to acquire brain activity that can be fed into a BCI; however, the most used acquisition method is EEG, as it is low-cost and convenient to use. Other factors that make it such an attractive tool are the standardisation of electrode placement, plentiful and well-documented information on acquisition techniques, and being a well-established method with known filtering [8]. Table 1, adapted from [7], compares the different types of methods used to acquire signals for BCI use.

Table 1. Properties of brain activity acquisition methods.

	EEG	MEG	NIRS	fMRI	ECoG	MEA	fTCD
Deployment	Non-invasive	Non-invasive	Non-invasive	Non-invasive	Invasive	Invasive	Non-invasive
Measured Activity	Electrical	Magnetic	Hemodynamic	Hemodynamic	Electrical	Electrical	Hemodynamic
Temporal Resolution	Good	Good	Low	Low	High	High	High
Spatial Resolution	Low	Low	Low	Good	Good	High	Low
Portability	High	Low	High	Low	High	High	High
Cost	Low	High	Low	High	High	High	

I. Magnetoencephalography (MEG) is a neuro-imaging technique, which uses the magnetic fields created by the natural currents that flow in the brain to map the brain activity. To do that, it uses magnetometers. The cerebral cortex's sites, which are activated by a stimulus, can be found from the detected magnetic field distribution [9].

II. Near-infrared spectroscopy (NIRS) is a spectroscopic method that uses the near-infrared (NIR) region of the electromagnetic spectrum (from 780 nm to 2500 nm). NIR light can penetrate human tissues; however, it suffers a relatively high attenuation due to the main chromophore haemoglobin (the oxygen transport red blood cell protein), which is presented in the blood. Therefore, when a specific area of the brain is activated, the localised blood volume in that area changes quickly and, if optical imaging is used, it is possible to measure the location and activity of specific regions of the brain. This is due to the continuous tracking of the haemoglobin levels through the determination of optical absorption coefficients [10].

III. Functional magnetic resonance imaging (fMRI), through variations associated with blood flow, can measure brain activity. This technique relies on the fact that cerebral blood flow and neuronal activation are coupled; thus, when an area of the brain is in use, the blood flow to that region also increases [11].

IV. Electrocorticography (ECoG) is a type of electrophysiological monitoring that records activity mainly from the cortical pyramidal cells (neurons). For that, it requires the electrodes to be placed directly on the exposed surface of the brain so that the recorded activity comes directly from the cerebral cortex [12].

V. Micro-electrode arrays (MEAs) are devices that contain multiple microelectrodes; the number can vary from ten to thousands, through which the neural signals are obtained. These arrays function as neural interfaces that connect neurons to electronic circuitry [13].

VI. Functional transcranial Doppler (fTCD) is a technique that uses ultrasound Doppler to measure the velocity of blood flow in the main cerebral arteries during local neural activity [14]. Changes in the velocity of the blood flow are correlated to changes in cerebral oxygen uptake, enabling fTCD to measure brain activity [15].

However, the robustness of all existing BCI systems is not satisfactory due to the non-stationary nature of non-invasive EEG signals. If a BCI system is unstable, other techniques should be further developed to improve the overall driving performance [6]. Usually, these concerns improve feature extraction and classification as the other option would fall on trading to an invasive approach. Although the range of existent commercial headsets is quite good, most of them lack in the number of available electrodes as they are more turned to improve the user's focus and to help to relax, or be used for gaming. Furthermore, the ones that present better characteristics are the Emotiv EPOC, Emotiv Flex, and the Open BCI [16]. Although the last two do not restrict the electrodes' configuration as Emotiv EPOC does, they are more expensive and complex. As for the open BCI one, it does not offer the same freedom of measurement and comfort as Emotiv ones, as these are wireless with a 12 h, for EPOC, and 9 h, for Flex, lasting battery [17]. The cost of an Emotiv is approximately USD 1000, and Open BCI can cost more than USD 2000 [16]. Hereupon, authors nowadays do not use commercial EEG headsets to obtain the signals that will feed the BCI; they prefer assembling their own EEG set through an amplifier and electrodes, as seen in Table 2.

Table 2. EEG headsets used in the literature.

Article	EEG Headset	Principle	Article	EEG Headset	Principle
[18]	12 Ag/Cl electrodes	ERP—P300	[19]	NuAmps and 12 electrodes	ERP—P300
[20]	NuAmps and 15 electrodes	ERP—P300	[21]	gTec EEG (16 electrodes and g.USBamp amplifier)	ERP—P300
[22]	16-channel electrode cap	ERP—P300	[23]	Biopac MP150 EEG system	ERP—P300
[24]	gTec EEG (12 electrodes and g.USBamp amplifier)	ERP—P300	[25]	Neuroscan (15 electrodes' cap)	ERP—P300

Table 2. *Cont.*

Article	EEG Headset	Principle	Article	EEG Headset	Principle
[26]	BioSemi ActiveTwo system 32 channels	SSVEP	[27]	g.USBamp amplifier with g.Butterfly active electrodes	SSVEP
[28]	8 gold electrodes connected to the g.USBamp amplifier	SSVEP	[29]	gTec EEG with g.USBamp amplifier	SSVEP
[30]	EEG Cap and g.USBamp amplifier	SSVEP	[31]	BrainNet-36 with 12 channels	SSVEP
[32]	BrainNet BNT-36 with 3 channels	SSVEP	[33]	6 channels EEG cap	SSVEP
[34]	NeuroSky mindset	ERD/ERS	[35]	Grass Telefactor EEG Twin3 Machine	ERD/ERS
[36]	G-TEC system with 5 Ag/AgCl electrodes	ERD/ERS	[37]	8 channels EEG cap	ERD/ERS
[38]	5 bipolar EEG channels and a g.tec amplifier	ERD/ERS	[39]	ActiveTwo 64-channel EEG system	ERD/ERS
[40]	Emotiv EPOC	ERD/ERS	[41]	Emotiv EPOC	ERD/ERS
[42]	Emotiv EPOC	ERD/ERS	[43]	Emotiv EPOC	ERD/ERS
[44]	EEG Cap—15 electrodes	ERD/ERS and SSVEP	[45]	Gtec Amplifier (15 channels)	ERD/ERS and SSVEP
[46]	g.BSamp amplifier (5 channels)	ERD/ERS and SSVEP	[47]	NuAmps device (15 channels)	ERD/ERS and ERP—P300
[48,49]	NeuroSky	ERP—P300; Eye Blinking (EMG)	[5]	SYMPTOM amplifier with 10 electrodes	ERP—P300 and SSVEP

ERP—event-related potential; ERS—event-related synchronization; SSVEP—steady-state visual evoked potential.

However, these are usually not wireless options. Nevertheless, there is still a significant group who use Emotiv EPOC, as this one offers a wider range of electrodes when compared with other commercial options, allowing obtaining of the signals from different brain lobes. It is possible to find several public EEG datasets related to motor imagery [50]. These datasets involve recordings of brain activity while subjects imagine performing specific motor tasks, such as moving a limb or making a particular gesture. These datasets are essential for studying motor control, brain–computer interfaces (BCIs), and rehabilitation. These datasets cover a wide range of tasks and experimental paradigms. With varying numbers of subjects, electrode configurations, and recording parameters, each dataset offers insights into different aspects of brain function and behaviour. From collections like the largest SCP data of motor imagery, with extensive EEG recordings spanning multiple sessions and participants, to focused datasets like the imagination of right-hand thumb movement, capturing specific motor imagery tasks, these datasets [50] serve as valuable resources for exploring the neural correlates of motor control, emotion processing, error monitoring, and other cognitive processes.

2.2. Signal Processing

The signal-processing module is divided into different parts [51]. The steps vary depending on whether the stage is training or testing; however, the training steps are broader than the testing ones, and, hence, these will be the ones to be discussed. The first step is to pre-process the signal, and it is further subdivided into band-pass and spatial filtering; afterwards, the features are extracted and selected. Finally, the classification is carried out, and the performance is evaluated. To perform this, techniques of machine learning must be applied, and thus the brief explanation of this concept.

Machine learning (ML) is based on data analytics that automates analytical model building. By using algorithms that iteratively learn from data, the computer can find hidden insights without being explicitly programmed where to look. This approach is used when the problem is complex and can be described by many variables. It creates an unknown target function that models the input into the desired output [52]. The learning algorithm receives a set of labelled examples (inputs with corresponding outputs) and learns by comparing its predicted output with correct outputs to find errors, modifying the model accordingly. The resulting model can predict future events. When exposed to new data, the model can adapt itself. In theory, if the algorithm works properly, the larger the amount of data there are, the better are the predictions. However, they are limited by bias in the algorithm and in the data, which can produce systematically skewed predictions. Therefore, the complexity of the learning algorithm is critical and should be balanced with the complexity of the data [52].

2.2.1. Pre-Processing

The EEG signal, per se, is very noisy, which is due to several aspects such as the low signal-to-noise ratio—as it is collected from the individual's scalp surface, the low spatial resolution, and other sources such as artefacts or interfering frequencies [53]. Artefact removal involves cancelling or correcting the artefacts without distorting the signal of interest and can be implemented in both the temporal and spatial domains [54]. Usually, the pre-processing concerns two types of filtering, in the frequency and the spatial domain [51]: band-pass filtering consists of removing some frequencies, or frequency bands, from the signal [53], outputting the frequency range of interest; and spatial filtering, which consists of combining the original sensor signals, usually linearly, which can result in a signal with a higher signal-to-noise ratio than that of individual sensors [51]. It combines the electrodes, which leads to more discriminating signals [54]. According to Pejas [55], approaches that rely on spatial filtering not only provide more true positives but also allow more flexibility when choosing the electrode placement. Spatial filters that linearly combine signals acquired from different EEG channels can extract and enhance the desired brain activity; thus, usually, it is enough to place the electrodes somewhere in the desired area and not in the exact location.

2.2.2. Feature Extraction and Classification

There are different types of features according to the domain from where they are extracted: time, frequency, or spatial. Different methods are used to extract the features from the EEG signal and further classify them so that the control commands can be obtained. Table 3, partly adapted from [56], presents a group of techniques used by different authors. It comprises several examples referring to the different principles: ERPs, SSVEP, ERD/ERS, and Hybrid.

Table 3. Summary of different authors' BCIs regarding the used EEG headset, the neuro-mechanism, the extracted features, the classification methods, the outputted commands, and accuracy.

Article	EEG Headset	Principle	Features	Classifier	Control	Accuracy
[18]	12 Ag/Cl electrodes	ERP—P300	Signal averaging and standard deviation	2 class Bayesian	L/R/F/B (45° or 90°)/S	95%
[19]	NuAmps and 12 electrodes	ERP—P300	Data vectors of concatenated epochs	BLDA (Bayesian)	9 destinations	89.6%
[20]	NuAmps and 15 electrodes	ERP—P300	Raw signal	SVM	7 locations, an 'application button' and lock	90%
[21]	gTec EEG (16 electrodes and g.USBamp amplifier)	ERP—P300	Moving average technique	LDA	15 locations, L/R and validate selection	94%
[22]	16-channel electrode cap	ERP—P300	Signal averaging	Linear classifier	6 for the IW (not specified)	92%
[23]	Biopac MP150 EEG system	ERP—P300	Signal averaging	Linear classifier	F/B/L/R	—

Table 3. *Cont.*

Article	EEG Headset	Principle	Features	Classifier	Control	Accuracy
[24]	gTec EEG (12 electrodes and g.USBamp amplifier)	ERP—P300	Optimal statistical spatial filter	Binary Bayesian	F/B/L/R (45° or 90°/S	88%
[25]	Neuroscan (15 electrodes' cap)	ERP—P300	Signal averaging	SVM	37 locations, validate or delete selection, stop and show extra locations	—
[26]	BioSemi ActiveTwo system 32 channels	SSVEP	Peaks in the frequency magnitude	—	L/R	>95%
[27]	g.USBamp amplifier with g.Butterfly active electrodes	SSVEP	Frequency band power (PSD)	SVM	L/R/F/S	95%
[28]	8 gold electrodes connected to the g.USBamp amplifier	SSVEP	—	LDA	L/R/B/F/S	90%
[29]	gTec EEG with g.USBamp amplifier	SSVEP	Frequency band power (PSD)	Threshold method not specified	L/R/B/F	93.6%
[30]	EEG Cap and g.USBamp amplifier	SSVEP	CCA	Bayesian	F/L/R/turn on/off	87%
[31]	BrainNet-36 with 12 channels	SSVEP	Frequency band power (PSD)	Decision trees	L/R/F/S	Qualitative evaluation
[32]	BrainNet BNT-36 with 3 channels	SSVEP	Frequency band power (PSD)	Statistical maximum	L/R/F/B	95%
[33]	6 channels EEG cap	SSVEP	FFT and CCA	CCA coefficient	L/R/F/B/S	>90%
[34]	NeuroSky mindset	ERD/ERS	Frequency band power (PSD)	NN	Game	91%
[35]	Grass Telefactor EEG Twin3 Machine	ERD/ERS	Coefficients from the wavelets	Radial basis function NN	L/R/F/B/rest	100%
[36]	G-TEC system with 5 Ag/AgCl electrodes	ERD/ERS	Common spatial frequency subspace decomposition (CSFSD)	SVM	L/R/F	91–95%
[37]	8 channels EEG cap	ERD/ERS	Mean, zero-crossing and energy from different levels of the DWT	ANN	L/R/F/S	91%
[38]	5 bipolar EEG channels and a g.tec amplifier	ERD/ERS	Logarithmic frequency band power	LDA	L/R	75%
[39]	ActiveTwo 64-channel EEG system	ERD/ERS	Frequency band power (PSD) and CSP	SVM	Exoskeleton control LH/LF/RH/RF	84%
[40]	Emotiv EPOC	ERD/ERS	PCA and average power of the wavelets' sub-bands	NN w/BP	L/R/F/B	91%
[42]	Emotiv EPOC	ERD/ERS	—	Emotiv program	L/R/F/S	70%
[56]	Emotiv EPOC	ERD/ERS	Frequency components	SVM	L/R/F/B/S	100%
[56]	Emotiv EPOC	ERD/ERS	Frequency components	NN	L/R/F/B/S	100%
[56]	Emotiv EPOC	ERD/ERS	Frequency components	Bayesian	L/R/F/B/S	94%
[56]	Emotiv EPOC	ERD/ERS	Frequency components	Decision trees	L/R/F/B/S	74%
[42]	Emotiv EPOC	ERD/ERS	Frequency band power (PSD)	LDA	L/R	70%
[43]	Emotiv EPOC	ERD/ERS	Metrics from the EEG signal	Decision trees	L/R	82%

Table 3. Cont.

Article	EEG Headset	Principle	Features	Classifier	Control	Accuracy
[57]	Emotiv EPOC	ERD/ERS	CSP	SVM	L/R	60%
[58]	Emotiv EPOC	ERD/ERS	—	LDA	L/R	60%
[59]	Emotiv EPOC	ERD/ERS	PSD, Hjort parameters, CWT and DWT—PCA for feature reduction	K-NN	L/R	86–92%
[60]	Emotiv EPOC	ERD/ERS	Energy distribution from the DWT	SVM	L/R/T/N	97%
[61]	Emotiv EPOC	ERD/ERS	CSP	LDA	L/R	68%
[61]	Emotiv EPOC	ERD/ERS	CSP	SVM	L/R	68%
[61]	Emotiv EPOC	ERD/ERS	CSP	Nu-SVC RBF Kernel	L/R	68%
[44]	EEG Cap—15 electrodes	ERD/ERS—(L/R) and SSVEP	CSP (ERD/ERS); CCA (SSVEP)	SVM (ERD/ERS); Canonical correlation coefficient (SSVEP)	L/R/A/DA, maintain an uniform velocity and turn on/off	—
[45]	Gtec Amplifier (15 channels)	ERD/ERS—(L/R) SSVEP-(Des)accelerate	CSP (ERD/ERS); CCA (SSVEP)	SVM	L/R/A/DA	—
[46]	g.BSamp amplifier (5 channels)	ERD/ERS and SSVEP	Frequency band power (PSD)	LDA	L/R	81%
[47,48]	NuAmps device (15 channels)	ERD/ERS and ERP—P300	CSP	LDA	L/R/A/DA	100%
[48,49]	NeuroSky	ERP—P300 and Eye Blinking (EMG)	Changes in the level	Threshold	L/R/F/B/S	—
[5]	SYMPTOM amplifier with 10 electrodes	ERP—P300 and SSVEP	PCA (ERP); PSD (SSVEP)	LDA	ERP—9 destinations SSVEP—confirm	99%

L—left; R—right; F—forward; B—backward; S—stop; A—accelerate; DA—decelerate; H—hand; F—foot; T—tongue; N—no imaging.

The ERD/ERS neuro-mechanism is a widely used one and has been producing noticeable results. This corresponds to a change in the power of specific frequency bands since the user is imagining or visualising a certain motor movement. The best combination is obtained with SVMs or NN as classifiers. Authors such as Abiyev et al. [57] and Khare et al. [35] achieved an impressive accuracy of 100%. The extracted features were all in the frequency domain, mostly from the frequency coefficients, band power, or spatial filtering. SSVEP BCIs can originate ace outcomes regardless of the classifier. This is probably due to the neuro-mechanism itself, as it is linked to a specific frequency, facilitating the extraction of the feature vector. However, contrarily to the ERD/ERS BCIs, these require some sort of hardware, usually flashing buttons (each one at a unique frequency rate), which will act as the stimulus for the user. The latter will focus on the button, which represents the desired direction; hence, proportionally amplifying the EEG signal band corresponding to the button frequency. The extracted features fall in the frequency domain and regard the power in specific frequency bands (corresponding to the respective button). ERPs are short amplitude deflections in the brain signal that are timestamped to an event. They are identified by the triggering event, direction of deflection, observed location, and latency [7]. That is why these BCIs usually use temporal features, whereas ERD and SSVEP BCIs employ frequency features [6]. Concerning the used classifier, the BCI performance does not seem to depend upon this choice. Regarding hybrid BCIs, it can be deduced that methods that aim to decompose the signal are preferentially used to extract the features. Concerning the classification, the used classifiers are mainly SVMs and LDA. It is possible to conclude that, depending on the chosen neuro-mechanism, the type of extracted features will differ.

However, for the classifiers, the same cannot be applied, although it is possible to infer that some classifiers have a better performance than others have, namely, SVMs, NN, and LDA.

A BCI provides control and communication between human intention and physical devices by translating the pattern of the brain activity into commands. The goal is to use this as a way of controlling an IW, which will eventually lead to an increase in the quality of life of people with disorders and limitations. A BCI has different main blocks: signal acquisition, signal processing, and the application of the output commands. The first one aims to collect the brain signals to feed them to the signal processing unit. There are several ways of achieving this, with EEG being the most common, affordable, and well-documented way. To make it even more accessible and portable for the patient, the EEG headset should be wireless; hence, the Emotiv EPOC is the chosen one.

Moreover, the aim is to use the Emotiv EPOC headset as a way to record the user's brain activity, as it is rapidly installed and portable. Although many authors have already proposed several solutions, none of them meet the required criteria to be commercialised, either by the lack of portability or the lack of accuracy. Therefore, the final goal would be to have a portable, comfortable, affordable, and reliable solution for an end-user consumer, so that the system would ideally be prepared for an out-of-the-laboratory application. Thus, this work contributes to the conceptualisation of the BCI system, regarding its architecture and algorithms.

3. Materials and Methods

A motor imagery (MI) neuro-mechanism is proposed, as it allows the user to focus on the path instead of focusing on the user interface, as the last two are stimulus-dependent neuro-mechanisms. Figure 1 presents the overall scheme of the BCI architecture along with its constituent parts.

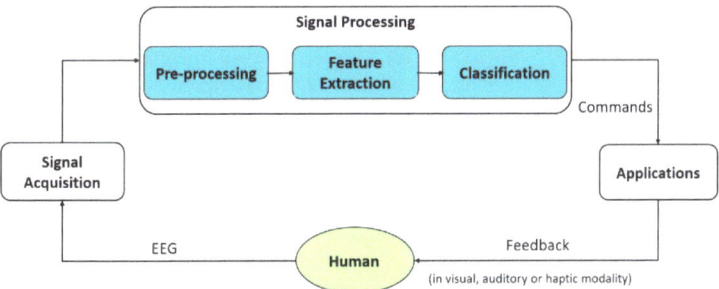

Figure 1. Scheme of the BCI architecture and its parts.

Three classes for the commands are used, namely left (0), right (1), and neutral (2). The first two correspond to changes in the direction, whereas the last one implies that the subject wishes to maintain the same direction. This choice relies on the fact that the left and right are the basic commands to control a moving device and, since the system is working in a continuum, the necessity of a neutral class to maintain the direction of movement arises. According to Tang et al. [39], some subjects present a better ability to distinguish between the feet and hands, rather than the left hand from the right one. Consequently, three different runs are tested, where the subject can substitute one of the hands for the thought of feet. More specifically, the subject may have a better performance while differentiating the left hand from the feet, and it may be advantageous to use the thought of the feet to turn to the right.

Moreover, the experiments are divided into two main parts: the validation of the concept and the corresponding execution or testing. Regarding the first part, two datasets are used, dataset 2a from the BCI competition IV (dataset A) [62] and another one acquired in our laboratory using the Emotiv EPOC headset (dataset B). Concerning the execution of the algorithm, a real-time acquisition from the headset is attempted and evaluated.

3.1. Datasets

Dataset A contains a four-class MI for different body parts: the left and right hand (LH/RH), feet (F), and tongue. This dataset corresponds to dataset 2a of the BCI competition IV and comprises 2 sessions of 288 trials from 9 different subjects. In each session, there were 6 smaller sessions of 48 trials, each separated by breaks. It also encompasses an evaluation dataset with the same characteristics as the previously described one. For this work, the tongue MI was discarded, as it was not of interest.

The acquisition protocol for each trial can be seen in Figure 2 and it is a sequence composed of a fixation cross (2 s), followed by an arrow representing the desired MI (1.5 s), a period of blank screen for the subject to imagine the asked cue (2.75 s), and it finishes with a break (~2 s). Furthermore, there is a sound alerting for the beginning and ending of the MI period (4 s).

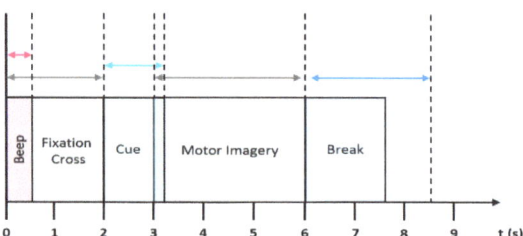

Figure 2. Acquisition protocol for dataset A.

The signals were obtained using 22 Ag/AgCl electrodes, which were positioned following the 10/20 system shown in Figure 3a. These were placed mostly at the central part of the cortex, where the sensorimotor part is located.

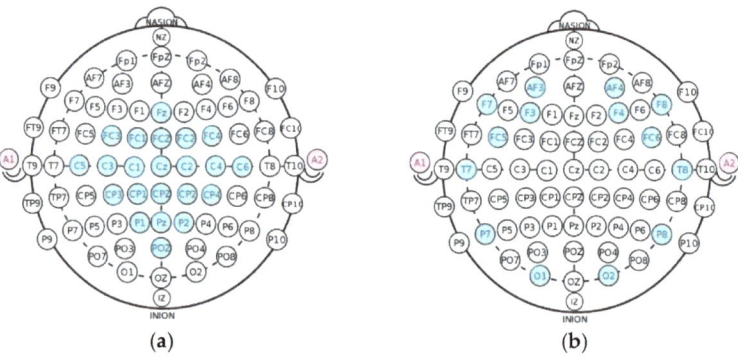

Figure 3. Electrodes' placement, according to the 10/20 system, for both datasets: (**a**) dataset A; (**b**) dataset B.

The acquisition protocol for dataset B was approximately the same as for dataset A, with two differences inspired by Tang et al. [39], Dharmasena et al. [42], and Stock and Balbinot [63]. More specifically, in the MI cue, the arrows were displayed on the screen for the whole period, as shown in the diagram presented in Figure 4. Furthermore, the indication of the start of a cue was not used to simplify the process. There were three different cues: right hand (right arrow), left hand (left arrow), and foot (down arrow). The set of sessions comprised 360 trials, 120 for each MI.

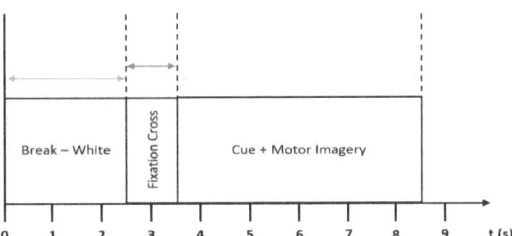

Figure 4. Acquisition protocol for dataset B.

In total, signals from nine different healthy subjects were acquired, where subjects 4 and 6 are left-handed, while the others are right-handed. All subjects are below 25 years old, except subject 5, who is 51. During acquisition, the subjects were seated comfortably in a chair, in a quiet room, with their hands on the top of the table while looking at the screen; they were also asked to keep their movements, such as eye gazing, sniffing, or coughing, to a minimum. All the procedures were performed under the ethical standards of the 1964 Helsinki Declaration.

The electrode placement can be seen in Figure 3b. Although none of these match the placement for dataset 2a, they still cover part of the central, parietal, and frontal locations of the cortex, which are known for contributing to the MI [14]. However, it is expected that the results will not be as satisfactory, as the centre of the cortex is not covered [61]. It is also important to state that the recorded points may also depend on the format of the subject's head, as electrode placement on narrower heads will not be the same as for wider ones, because the electrodes in the headset are fixed.

3.2. Data Processing

The two datasets were divided into training and test as follows:

- Dataset A: the training data supplied by the BCI Competition IV were used as a train and the evaluation one as a test. The duration of the epochs was two seconds, as explained in [62].
- Dataset B: 100 trials of each MI were used as training data, and the remaining 20 were used as tests. Usually, each subject would do a 20-trial session, which results in 5 sessions for training and 1 session for testing. For each visual cue and motor imagery moment, as these had a duration of 5 s, two epochs of two seconds each were extracted, allowing to double the data, ending up with 240 epochs, in total, for each class.
- A subject-oriented approach was followed, requiring the model to undergo training specific to each subject before being tested. However, it should be noted that the sessions utilised for testing differed from those used for training purposes.

3.2.1. Pre-Processing and Feature Extraction

Filtering the EEG signal is already enough to remove noise and ocular artefacts, which are the most common. The first comprises high frequencies, which are discarded, as these are not included in the bands of interest. Moreover, ocular artefacts mainly appear in the theta band, which, once again, is not a band of interest for the MI paradigm. Thus, for every feature extraction approach, presented in the next section, a filtering step is always applied to eliminate these artefacts. The main methods for feature extraction regarding the MI paradigm are spatial filtering using the common spatial pattern (CSP) approach and the use of the signal's frequency band or the frequency coefficients as features. The different approaches were tested, but with some variations. The next steps, feature selection and classification, were the same for all the approaches.

1. Filter Bank Common Spatial Pattern I

As dataset A is from a competition, the first approach was to develop an algorithm based on the winning method, denoted the filter bank common spatial pattern (FBCSP), as described in [62]. The goal is to maximise the best band for each user, which results in dividing the alpha and beta bands into nine sub-bands, from 4 Hz to 40 Hz. Although Ang et al. [62] use a Chebyshev II filter, in this work a Butterworth filter of order five and zero phase was applied. This choice lies in the facts that this filter is known for being the flattest in the passing band, the zero phase provides zero group distortion, and the order five is a nice compromise with respect to speed. The FBCSP algorithm applies the CSP procedure to each sub-band of the signal. The algorithm generates a linear filter, which is used to extract features that best discriminate between classes, by maximising the ratio between their covariance matrices [64].

2. Filter Bank Common Spatial Pattern II

This approach follows the same principles as the first approach but, after obtaining a spatial filter, the average power of Z is computed and used as features.

3. Power Spectral Density

The signal is filtered using a Butterworth filter, for the reasons previously enunciated, from 4 to 35 Hz to comprise the alpha and beta bands. Afterwards, epochs of two seconds are obtained and normalised. The latter consists of centring each channel to have zero mean. For that, the mean of each epoch for each channel is calculated and then subtracted [57]. Afterwards, the Welch method, with a Hanning window, is applied to obtain the power spectral density for each epoch, which is used as the features vector. The Welch method consists of dividing the signal into overlapping segments, which are further windowed. Then, the signal periodogram, which is an estimate of the signal spectral density, is calculated, resorting to the discrete Fourier transform. Windowing the segments, for example with the Hanning window, allows for mitigating spectral leakage. This is because the Fourier transform assumes that the signal is periodic, and non-periodic signals lead to sudden transitions that have a broad frequency response [65]. Different methods for choosing the most significant features were tested, namely a method based on a mutual information criterion, the ANOVA F test, and the extra trees classifier, to compute the features' importance. The first measures the dependency between two random variables and relies on non-parametric methods based on entropy estimation, such as from K-nearest neighbours, to improve the selection. The second assesses the amount of variability between each class mean, in the context of the variation within the different classes, to determine whether the mean differences are statistically significant or not. Finally, the extra trees classifier is used to compute the importance of the features, allowing the irrelevant ones to be discarded. For either of the methods, only the K best features are selected. This is performed by a 10-fold cross-validation, using 5–70% of the features. The 70% limit is imposed to prevent overfitting.

3.2.2. Classification

The classifiers were trained to differentiate between three different classes. Due to slower computational time and the fact that they might generate overfitting, non-linear classifiers were not used as a first approach [56]. Thus, four classifiers were trained: Gaussian Naive Bayes (GNB), linear discriminant analysis (LDA), linear support vector machines (LSVMs), and logistic regression (LR). Using these four classifiers, different combinations were tested, as represented in Table 4.

Table 4. Testing a combination of classifiers.

Number of Classifiers	Type	Number of Classifiers	
1	Non-probabilistic	2—Ensemble	Voting Hard
	Probabilistic—F1		Voting Soft
2	Non-probabilistic		AdaBoost
	Probabilistic—F1		

When using a single classifier to predict the result, there are two main approaches: predict a class or predict the probability of belonging to each class. In the latter approach, the ideal value for the probability threshold can be obtained through different metrics. The F1-score was the chosen one, as it considers both the precision and the recall of the classifier. When using two classifiers to predict the final command, different approaches were applied, which are further explained:

- Two classifiers: if both classifiers predict that the class is 0, then the class is 0; the same is applied for classes 1 and 2. However, if they do not agree with the classification, then the trial is classified as 2 in order to decrease the number of false positives, which in this case are the trials miss-classified as MI to the left or right;
- Two classifiers with variable probability: the idea behind this approach is the same as before; however, the output of each classifier is a probability and not a class label. Thus, a threshold is estimated for each one of the classifiers to output a label, and then the same method is applied, as explained for the two classifiers;
- Ensemble methods: these methods are already developed and are widely used to combine the different predictions so that a more generalised and robust model can be obtained. These methods can be divided into two main groups: averaging and boosting. Regarding the first one, the different classifiers are built independently and only after that are their outputs combined to reduce the variance. Concerning the boosting methods, the classifiers are built sequentially so that the next classifier can try to decrease the bias of the combined. Voting classifier: this combines the predictions of the different classifiers and outputs a final prediction as the result of a majority vote. This majority vote can be hard or soft. Hard: each classifier predicts the class, and the final prediction is the one that most of them predicted. The final prediction can be obtained using a weighted averaging procedure if the classifiers have different weights. Soft: each classifier has a weight and predicts the probability of each class, and then the final prediction is obtained using a weighted averaging procedure. AdaBoost: considers several initial classifiers, called weak learners, and combines their predictions through a weighted majority vote. This process is repeated, and at each iteration/boost the data are modified. Each sample starts with a weight, and if it is incorrectly classified its weight increases for the classifier to notice it more; on the other hand, correctly classified samples have their weights reduced. After several iterations, the overall classifier, or strong learner, is expected to be better than the individual ones.

To further improve results, several non-linear classifiers were also tested. These include the K-nearest neighbours (K-NNs), kernel support vector machines (KSVMs), decision trees (DTs), neural networks (NNs), and, finally, random forest (RF). Similarly to linear classifiers, the same combinations of classifiers were tested as well. As these classifiers require more data to obviate overfitting, for each trial of MI, which had a duration of five seconds, more epochs were extracted. For each trial, two epochs of two seconds were extracted. The first second of the signal was not used, as a preventive way, since the image of the arrow could act as a stimulus for the pretended direction. Hereupon, this time is sufficient for the person to assimilate which MI must do. Nevertheless, the approach of doubling the number of epochs ended up also being used for the linear and statistical classifiers. This was to ensure that both methods used the same amount of data. Table 5 summarises all the optimised hyper-parameters for the respective classifiers, as well as a brief description of their function. The optimisation process was carried out using a grid search with a 5-fold cross-validation decided by the F1-score.

Table 5. Optimised hyper-parameters for the different classifiers.

Supervised Classifier	Hyper-Parameter	Grid Search Space	Description
LR	C	logspace −4 to 6, step size 1	Regularisation parameter, which has a significant effect on the generalisation performance of the classifier
K-NN	n_neighbors	1 to 50, step size 10	Number of neighbours to use
SVM	Kernel	rbf—Gaussian kernel function	Function used to compute the kernel matrix for classification
	gamma	logspace −3 to 6, step size 1	Regularisation parameter used in RBF kernel, which has significant impact on the performance of the kernel
	C	logspace −3 to 7, step size 1	Regularisation parameter, which has a significant effect on the generalisation performance of the classifier
DT	max_depth	1 to 20, step size 2	The maximum depth of the tree. If none, then nodes are expanded until all leaves are pure or until all leaves contain less than min_samples_split samples.
	min_samples_split	10 to 500, step size 20	Minimum number of samples required to split a node
	min_samples_leaf	1 to 10, step size 2	Minimum number of samples required in a newly created leaf after the split
NN	hidden_layers	5 to 55, step 10	The i element represents the number of neurons in the i hidden layer
	activation	relu—rectified linear unit function	Activation function for the hidden layer
	solver	adam—stochastic gradient-based optimiser	The solver for weight optimisation
	learning_rate	constant	Learning rate schedule for weight updates. If 'constant', the learning rate is given by learning_rate_init
	learning_rate_init	logspace −4 to 4, step 1	The initial learning rate used. It controls the step size in updating the weights.
	alpha	logspace −4 to 4, step 1	L2 penalty (regularisation term) parameter.
RF	n_estimators	10 to 100, step 20	Number of trees in the forest
	max_depth	None or 2 to 10, step size 2	The maximum depth of the tree. If none, then nodes are expanded until all leaves are pure or until all leaves contain less than min_samples_split samples.
	min_samples_split	10 to 500, step size 20	The minimum number of samples required to split a node
	min_samples_leaf	1 to 10, step size 2	The minimum number of samples required in a newly created leaf after the split

LR—logistic regression; K-NN—K-nearest neighbour; SVM—support vector machine; DT—decision tree; NN—neural networks; RF—random forest.

3.2.3. Evaluation

To evaluate the results from the different approaches on the two datasets, the F1-score, the kappa score, and the false positive (FP) rate were used. The F1-score is the average of the precision and recall, and it reaches its best value at 1 (perfect precision and recall) and worst at 0. The kappa score expresses the level of agreement between two annotators. Although it is not usually used to compare a prediction with a ground truth, it was the only metric provided by the IV BCI Competition. A kappa value between −1 and 0 denotes a random classifier, while a value near 1 means a perfect one. Concerning the FP rate, a new metric was developed, since it is more important, for the final application, to penalise the FPs from classes 0 and 1, than from class 2. Nevertheless, a high rate of true positives is still desirable, independently of the class. Thus, the false positive rate was used for this evaluation. For an FP rate higher than 1, the classifier produces more false positives than true positives; hence, a rate smaller than 1 is desirable. For each subject, the best run out of the three was obtained based on the F1-score. Then, for that run, the respective kappa score and FP rate are also presented. Furthermore, for all the approaches, the feature selector was the extra trees. Table 6 contains the used linear and statistical classifiers (0–3), and the non-linear classifiers (4–8).

Table 6. Labels of the classifiers.

Number	Name
0	Gaussian Naive Bayes (GNB)
1	Linear discriminant analysis (LDA)
2	Linear support vector machines (LSVMs)
3	Logistic regression (LR)
4	K-nearest neighbours (K-NNs)
5	Kernel support vector machines (KSVMs)
6	Decision trees (DTs)
7	Neural networks (NNs)
8	Random forest (RF)

The first step consisted of only applying the linear and statistical classifiers. Afterwards, with the intent of improving even more the performance of the approach that presented the best results, the non-linear classifiers, along with the first set of classifiers, were only applied to the correspondent approach. This is because these classifiers take longer to run and optimise.

3.3. Hardware and Software

Python version 3 was the programming language used for the experimental work related to signal processing and classification, along with Numpy, Pandas, Seaborn, and Scikit-Learn libraries. For real-time testing, an interface was required to deliver the raw data acquired by the headset to Python. Given that the headset is an Emotiv EPOC, the pyemotiv Python library was applied. This library interfaces with the Emotiv EPOC research SDK, provided by the distributor, enabling the output of raw EEG data for the experimental setup [17].

4. Results

This section presents the results obtained from the different approaches previously explained, as well as the outcomes of a real-time application.

4.1. Filter Bank Common Spatial Pattern I—FBCSP I

In this approach, only the linear and statistical classifiers are used to build the different classifier combinations, since the obtained results were not the best.

4.1.1. FBCSP I Approach Using Dataset A

Table 7 presents the obtained F1-score for the different combinations, using only linear or statistical classifiers. The row "Best" corresponds to the best score for each subject. Most subjects presented a preferable run, regardless of the combinations, except for subject 1, who chose, at least once, one of the three runs. The highest F1-score, on average, was obtained by the ensemble voting hard, which can be explained as it corresponds to the major vote between the two best classifiers; thus, by combining their predictions, it comes out as more accurate.

Table 7. F1-score, kappa, and FP rate for dataset A and FBCSP I, using the first set of classifiers.

		Average	1	2	3	4	5	6	7	8	9
F1-score	1 Classifier	0.729	0.763	0.560	0.906	0.708	0.615	0.601	0.903	0.774	0.733
	Prob. F1	0.729	0.797	0.663	0.883	0.714	0.603	0.575	0.896	0.662	0.768
	2 Classifiers	0.712	0.782	0.496	0.920	0.679	0.610	0.546	0.848	0.776	0.751
	Prob. F1	0.738	0.792	0.623	0.932	0.689	0.634	0.567	0.874	0.768	0.760
	Soft	0.730	0.843	0.597	0.950	0.692	0.563	0.592	0.848	0.749	0.735
	Hard	0.748	0.801	0.632	0.928	0.719	0.647	0.615	0.869	0.759	0.758
	Ada	0.682	0.694	0.588	0.856	0.616	0.551	0.509	0.894	0.676	0.750
	Best	0.765	0.843	0.663	0.950	0.719	0.647	0.615	0.903	0.776	0.768
Kappa	1 Classifier	0.587	0.632	0.333	0.854	0.556	0.410	0.396	0.847	0.660	0.597
	Prob. F1	0.573	0.694	0.493	0.819	0.556	0.299	0.354	0.840	0.472	0.632
	2 Classifiers	0.583	0.653	0.236	0.875	0.500	0.403	0.546	0.764	0.660	0.611
	Prob. F1	0.590	0.674	0.438	0.896	0.535	0.382	0.347	0.792	0.646	0.604
	Soft	0.586	0.764	0.389	0.924	0.521	0.347	0.354	0.771	0.618	0.590
	Hard	0.607	0.701	0.361	0.889	0.569	0.472	0.417	0.785	0.639	0.632
	Ada	0.535	0.542	0.493	0.785	0.424	0.326	0.264	0.840	0.514	0.625
	Best	0.656	0.764	0.493	0.924	0.569	0.472	0.546	0.847	0.660	0.632
	Winner	0.570	0.680	0.420	0.750	0.480	0.400	0.270	0.770	0.750	0.610
FP rate	1 Classifier	0.295	0.206	0.475	0.092	0.355	0.565	0.527	0.031	0.228	0.177
	Prob. F1	0.271	0.134	0.294	0.111	0.178	0.878	0.423	0.123	0.200	0.098
	2 Classifiers	0.264	0.090	0.613	0.071	0.250	0.415	0.607	0.066	0.144	0.119
	Prob. F1	0.195	0.095	0.415	0.045	0.282	0.165	0.516	0.027	0.145	0.063
	Soft	0.334	0.126	0.578	0.044	0.361	0.582	0.699	0.098	0.280	0.242
	Hard	0.271	0.145	0.734	0.060	0.188	0.414	0.500	0.038	0.189	0.172
	Ada	0.386	0.307	0.685	0.157	0.474	0.571	0.755	0.109	0.260	0.154
	Best	0.159	0.090	0.294	0.044	0.178	0.165	0.423	0.027	0.144	0.063

Table 7 presents the respective kappa score, corroborating with the ensemble voting hard being the best combination. The kappa score from the winner of the IV BCI competition is also presented. However, the competition involved the classification of four classes: left hand, right hand, foot, and tongue; as for this work, there are only three classes: left, right, and neutral. Thereafter, the results from the competition are here exposed just as a qualitative comparison. Hereupon, the obtained kappa value of 0.604 is in the same order of magnitude as the result from the competition, 0.57, and thus higher than 0.5, which surely reflects the no randomness of the classifiers. Moreover, the FP rate had its lowest value, on average, for the one probabilistic classifier, whose threshold was decided based on the maximisation of the F1-score. This result is logical, as, by maximising the F1-score, there is an implicit maximisation of the precision and the recall, thus minimising the FP rate. However, the lowest FP rate was expected to belong to the ensemble voting hard because it was the combination with the highest F1-score. These different approaches correspond to different combinations of several classifiers. Table 8 presents the best ones for the different approaches and subjects. It can be concluded that the best algorithms correspond to the Gaussian Naive Bayesian classifier (0), linear discriminant analysis (1) and logistic regression (3), which was not presumed, as the LR was seldom mentioned during the literature review. Concerning the linear SVM (2), it was never picked, suggesting that it is not a good classifier for this dataset, using these features, as it is not capable of accurately distinguishing the three classes.

Table 8. Best classifiers, from the first set, for each combination for dataset A and FBCSP I.

	1	2	3	4	5	6	7	8	9
1 Classifier	0	0	3	3	3	3	3	3	3
Prob. F1	0	3	3	3	3	0	0	1	3
2 Classifiers	3 0	3 1	3 1	1 3	0 1	3 0	0 1	3 0	3 1
Prob. F1	0 1	3 1	0 3	3 1	0 3	0 3	3 0	1 3	1 0
Soft	0 3	3 1	3 1	3 0	3 1	0 3	0 3	3 0	1 3
Hard	3 0	0 1	0 3	3 0	3 0	3 0	3 0	1 3	3 1

0—Gaussian Naïve Bayesian; 1—linear discriminant analysis; 3—logistic regression.

4.1.2. FBCSP I Approach Using Dataset B

Similarly to what was presented for dataset A, Table 9 introduces the F1-score for the best run in each approach. Contrary to A, several subjects picked all three runs at least once as their best. Only subjects 5, 6, and 7 picked one or two. This already suggests that the extracted features were not very strongly indicative of the class. Once again, the best F1-score was obtained by the ensemble voting hard approach, followed by the two classifiers. However, since the F1-score varies from 0 to 1, the obtained result is not satisfactory as it stays in the bottom half of the spectrum. Similarly, for the kappa score, the value of 0.218 is closer to 0 than to 1, indicating that the classifier is closer to random than to perfect, as presumed. The FP rate is quite high, reaching almost 1, that is to say, the number of FP is almost the same as TP, thus manifesting that this approach is not adequate for the ultimate goal of controlling an IW. A very low FP rate is mandatory to maintain the safety of the IW driver. Nevertheless, subjects 3 and 1 presented a better performance than the others did, presenting scores equivalent to dataset A, which corroborates that people have different aptness regarding MI [40]. Nevertheless, it is also important to consider inter-individual differences, such as distinct brainwave patterns, cognitive abilities, and learning speeds, among others. This major difference between subjects may also be due to the positioning of the headset, as the electrodes are fixed, which may lead to more coverage of the motor cortex in some subjects than in others.

Table 10 contains the chosen classifiers for the different methods. Similarly to A, LR (3) and LDA (1) presented the best performance. However, the Gaussian Naive Bayes (0) did not perform well enough to be chosen. Once again, the linear SVM (2) was not picked.

Table 9. F1-score, kappa, and FP rate for dataset B and FBCSP I, using the first set of classifiers.

		Average	1	2	3	4	5	6	7	8	9
F1-score	1 Classifier	0.444	0.507	0.386	0.636	0.400	0.375	0.405	0.456	0.390	0.439
	Prob. F1	0.438	0.470	0.416	0.662	0.316	0.492	0.334	0.437	0.361	0.452
	2 Classifiers	0.448	0.442	0.456	0.622	0.391	0.419	0.361	0.442	0.404	0.496
	Prob. F1	0.427	0.427	0.353	0.640	0.444	0.403	0.333	0.442	0.303	0.496
	Soft	0.435	0.485	0.358	0.592	0.430	0.455	0.401	0.406	0.330	0.453
	Hard	0.484	0.597	0.401	0.626	0.410	0.547	0.386	0.470	0.483	0.440
	Ada	0.428	0.417	0.375	0.625	0.392	0.433	0.383	0.358	0.375	0.492
	Best	0.507	0.597	0.456	0.662	0.444	0.547	0.405	0.470	0.483	0.496
Kappa	1 Classifier	0.160	0.262	0.075	0.450	0.075	0.063	0.100	0.175	0.075	0.162
	Prob. F1	0.150	0.175	0.000	0.488	0.075	0.250	0.012	0.125	0.050	0.175
	2 Classifiers	0.133	0.162	0.188	0.438	0.037	0.125	0.050	0.000	0.137	0.063
	Prob. F1	0.121	0.137	−0.012	0.438	0.037	0.150	0.000	0.137	−0.038	0.238
	Soft	0.146	0.225	0.037	0.387	0.137	0.175	0.113	0.075	−0.012	0.175
	Hard	0.218	0.387	0.100	0.438	0.088	0.325	0.063	0.200	0.200	0.162
	Ada	0.142	0.125	0.063	0.438	0.088	0.150	0.075	0.037	0.063	0.238
	Best	0.253	0.387	0.188	0.488	0.137	0.325	0.113	0.200	0.200	0.238

Table 9. Cont.

		Average	1	2	3	4	5	6	7	8	9
FP rate	1 Classifier	0.908	0.672	1.239	0.355	1.370	0.867	0.646	0.852	1.435	0.736
	Prob. F1	1.202	1.056	2.000	0.380	1.587	0.767	1.610	1.240	1.273	0.907
	2 Classifiers	0.796	0.811	0.727	0.373	0.744	0.680	0.955	1.250	1.000	0.622
	Prob. F1	0.888	0.804	1.949	0.240	0.209	0.596	0.975	1.176	1.486	0.559
	Soft	0.918	0.793	1.116	0.451	1.059	0.611	0.857	1.043	1.590	0.741
	Hard	0.730	0.521	1.125	0.440	0.511	0.439	1.156	0.839	0.821	0.717
	Ada	0.916	0.720	1.222	0.373	1.426	0.692	1.065	1.302	1.000	0.441
	Best	0.543	0.521	0.727	0.240	0.209	0.439	0.646	0.839	0.821	0.441

Table 10. Best classifiers, from the first set, for each combination for dataset B and FBCSP I.

Classifiers	1		2		3		4		5		6		7		8		9	
1 Classifier	3		1		3		3		3		3		1		3		3	
Prob. F1	1		3		3		3		3		3		3		3		3	
2 Classifiers	1	3	3	1	3	1	3	1	3	1	3	1	3	1	3	1	3	1	3
Prob. F1	3	1	3	1	3	1	0	3	3	1	3	1	3	1	3	1	3	1
Soft	3	1	3	1	1	3	3	1	3	1	1	3	1	3	1	3	3	1
Hard	3	1	1	3	3	1	3	1	3	1	3	1	3	1	3	1	1	3

0—Gaussian Naïve Bayesian; 1—linear discriminant analysis; 3—logistic regression.

4.2. Filter Bank Common Spatial Pattern II—FBCSP II

Next, we describe the implementation of the FBCSP II approach using dataset A, providing detailed analysis and results of the classification performance. Following that, we examine the application of the same approach using dataset B, shedding light on its efficacy and comparative performance metrics.

4.2.1. FBCSP II Approach Using Dataset A

Table 11 presents the F1-score for dataset A, regarding the use of only linear or statistical classifiers. Most of the subjects present a preference concerning the run; for others, such as 3 and 5, it is not clear, as the three runs were chosen as the best one, at least once. Globally, the different classifier combinations presented resulted more or less in the same ranking and behaved as expected. The best F1-score was obtained, once again, for the ensemble voting hard approach, followed by the one classifier approach, which was not prospected, as it is intuitive that the output of two classifiers would be more accurate than just one. The worst score corresponds to the AdaBoost, thus clearly excluding it as a recommended approach.

The best kappa was from the ensemble voting hard approach, 0.693, which is higher than the one from the previous approach, 0.607, and in the same order of magnitude as 0.57, the kappa score of the competition's winner. It was already prospected that both the two classifier approaches would present a lower FP rate because they prevent the FP for classes 0 and 1. This is also one of the reasons why their F1-score is slightly lower than for the other approaches. AdaBoost presented the lower F1-score and kappa, and therefore presented the highest FP rate. It can be concluded (Table 12) that the best algorithms correspond once again to linear discriminant analysis (1) and logistic regression (3). The Gaussian Naive Bayesian classifier (0) was chosen a few times; however, it was almost always the second-best classifier, whereas the linear SVM (2) was never picked. Therefore, it is possible to conclude that the latter is not a good classifier for this dataset using these features, as it is not capable of accurately distinguishing the three classes.

Table 11. F1-score, kappa, and FP rate for dataset A and FBCSP II, using the first set of classifiers.

		Average	1	2	3	4	5	6	7	8	9
F1-score	1 Classifier	0.793	0.856	0.697	0.950	0.838	0.606	0.668	0.898	0.782	0.846
	Prob. F1	0.792	0.854	0.643	0.928	0.849	0.629	0.663	0.923	0.788	0.855
	2 Classifiers	0.785	0.840	0.643	0.933	0.840	0.604	0.686	0.859	0.842	0.817
	Prob. F1	0.783	0.840	0.643	0.930	0.840	0.644	0.686	0.859	0.793	0.817
	Soft	0.787	0.878	0.682	0.921	0.825	0.662	0.608	0.888	0.805	0.818
	Hard	0.797	0.829	0.698	0.949	0.833	0.618	0.679	0.924	0.774	0.871
	Ada	0.724	0.801	0.620	0.894	0.769	0.616	0.542	0.819	0.745	0.713
	Best	0.818	0.878	0.698	0.950	0.849	0.662	0.686	0.924	0.842	0.871
Kappa	1 Classifier	0.683	0.785	0.535	0.924	0.757	0.382	0.500	0.833	0.667	0.764
	Prob. F1	0.685	0.764	0.465	0.928	0.764	0.431	0.479	0.882	0.674	0.778
	2 Classifiers	0.596	0.681	0.361	0.896	0.611	0.292	0.382	0.639	0.764	0.743
	Prob. F1	0.658	0.757	0.458	0.889	0.750	0.396	0.521	0.757	0.681	0.715
	Soft	0.623	0.757	0.472	0.875	0.736	0.493	0.306	0.826	0.701	0.438
	Hard	0.693	0.743	0.542	0.924	0.750	0.424	0.514	0.882	0.653	0.806
	Ada	0.586	0.701	0.431	0.840	0.653	0.424	0.313	0.729	0.618	0.569
	Best	0.720	0.785	0.542	0.928	0.764	0.493	0.521	0.882	0.764	0.806
	Winner	0.570	0.680	0.420	0.750	0.480	0.400	0.270	0.770	0.750	0.610
FP rate	1 Classifier	0.218	0.108	0.383	0.029	0.127	0.646	0.347	0.016	0.238	0.071
	Prob. F1	0.161	0.038	0.317	0.040	0.066	0.403	0.255	0.025	0.237	0.065
	2 Classifiers	0.192	0.153	0.532	0.035	0.131	0.395	0.276	0.024	0.132	0.050
	Prob. F1	0.150	0.083	0.370	0.080	0.067	0.186	0.279	0.011	0.194	0.080
	Soft	0.238	0.127	0.529	0.081	0.152	0.336	0.581	0.042	0.197	0.097
	Hard	0.195	0.145	0.340	0.044	0.156	0.429	0.295	0.020	0.247	0.080
	Ada	0.260	0.116	0.463	0.109	0.229	0.474	0.624	0.068	0.118	0.136
	Best	0.121	0.038	0.317	0.029	0.066	0.186	0.255	0.011	0.132	0.050

Table 12. Best classifiers, from the first set, for each combination for dataset A and FBCSPII.

Classifiers	1		2		3		4		5		6		7		8		9	
1 Classifier	3		3		3		3		3		3		3		3		3	
Prob. F1	3		3		1		3		3		3		3		3		3	
2 Classifiers	3	1	3	1	3	1	3	1	3	1	3	1	3	1	3	1	3	1
Prob. F1	3	0	3	1	1	3	3	1	3	1	3	1	3	1	3	1	3	1
Soft	0	3	3	0	3	1	3	0	3	0	3	1	3	1	3	0	3	0
Hard	3	0	3	1	3	1	3	0	3	0	3	1	3	1	3	0	3	1

0—Gaussian Naïve Bayesian; 1—linear discriminant analysis; 3—logistic regression.

Due to their promising results, the use of non-linear classifiers was also tested. However, the results, contrary to what was expected, did not improve; instead, they stayed roughly the same. Furthermore, the best combination was not the ensemble voting hard but the two classifiers, which reflected in a lower FP rate. Regarding the kappa score, its average value is very close to the one previously obtained for the first group of classifiers.

4.2.2. FBCSP II Approach Using Dataset B

Similarly to what was presented for dataset A, Table 13 introduces the F1-score for the best run in each approach. Most of the subjects presented a preference regarding a run or two. The best classifier was the AdaBoost, with an average F1-score of 0.504, immediately followed by the ensemble voting soft, with a score of 0.497. Moreover, the highest F1-score value was lower than for A but higher than for the previous approach, 0.484. Once again, it is important to emphasise that subject 1 and subject 3 presented scores equivalent to dataset A.

Table 13. F1-score, kappa, and FP rate for dataset B and FBCSP II, using the first set of classifiers.

		Average	1	2	3	4	5	6	7	8	9
F1-score	1 Classifier	0.461	0.612	0.451	0.712	0.438	0.515	0.479	0.472	0.000	0.471
	Prob. F1	0.478	0.556	0.431	0.710	0.394	0.506	0.467	0.456	0.316	0.463
	2 Classifiers	0.454	0.606	0.435	0.697	0.431	0.527	0.471	0.468	0.000	0.455
	Prob. F1	0.457	0.605	0.433	0.700	0.437	0.506	0.490	0.456	0.000	0.489
	Soft	0.497	0.624	0.425	0.716	0.414	0.526	0.481	0.481	0.337	0.471
	Hard	0.486	0.612	0.433	0.705	0.408	0.515	0.422	0.472	0.332	0.471
	Ada	0.504	0.517	0.417	0.742	0.475	0.500	0.508	0.467	0.342	0.567
	Best	0.524	0.624	0.451	0.742	0.475	0.527	0.508	0.481	0.342	0.567
Kappa	1 Classifier	0.243	0.400	0.175	0.550	0.137	0.275	0.225	0.200	0.000	0.225
	Prob. F1	0.200	0.262	0.100	0.513	0.088	0.225	0.200	0.188	0.000	0.225
	2 Classifiers	0.232	0.387	0.150	0.513	0.125	0.288	0.213	0.200	0.012	0.200
	Prob. F1	0.233	0.375	0.150	0.513	0.137	0.225	0.238	0.188	0.025	0.250
	Soft	0.244	0.425	0.137	0.563	0.113	0.288	0.225	0.213	0.013	0.225
	Hard	0.224	0.400	0.150	0.538	0.100	0.275	0.125	0.200	0.000	0.225
	Ada	0.256	0.275	0.125	0.613	0.213	0.250	0.262	0.200	0.012	0.350
	Best	0.285	0.425	0.175	0.613	0.213	0.288	0.262	0.213	0.025	0.350
FP rate	1 Classifier	0.515	0.306	0.741	0.167	0.51	0.597	0.69	0.804	0.05	0.776
	Prob. F1	0.459	0.361	0.563	0.136	0.511	0.241	0.804	0.727	0.15	0.638
	2 Classifiers	0.49	0.296	0.731	0.148	0.5	0.524	0.667	0.714	0.024	0.804
	Prob. F1	0.444	0.373	0.563	0.173	0.549	0.241	0.695	0.618	0.071	0.717
	Soft	0.544	0.311	0.882	0.188	0.592	0.556	0.724	0.772	0.098	0.776
	Hard	0.561	0.306	0.827	0.169	0.646	0.597	0.8	0.804	0.125	0.776
	Ada	0.513	0.468	0.68	0.124	0.404	0.55	0.77	0.804	0.293	0.529
	Best	0.385	0.296	0.563	0.124	0.404	0.241	0.667	0.618	0.024	0.529

Furthermore, the best value for the average kappa score, 0.256, is both lower than the one for dataset A and also lower than the value of the competition; however, it is higher than the value of 0.218 from FBCSP I. Nevertheless, is still not a desirable value as it is too close to 0. As presumed from the previous scores, the FP rate is higher for dataset B than for dataset A. The lowest value, 0.444, was obtained with the one probabilistic classifier approach. The FP rate for the best approach, AdaBoost, was 0.513, which is high but still lower than the best for approach FBCSP I, 0.730. Concerning the selected classifiers, it is clear, from Table 14, that LDA and LR were preferred over the other two. Moreover, the Gaussian Naive Bayes classifier was not chosen as the best one for either of the combinations. Despite what was concluded from dataset A, the linear SVM was picked for this one, even if only once, which endorses that LDA and LR are the most suitable classifiers for this approach.

Table 14. Best classifiers, from the first set, for each combination for dataset B and FBCP II.

	S1	S2	S3	S4	S5	S6	S7	S8	S9
1 Classifier	3	1	2	3	1	3	1	1	3
Prob. F1	3	1	3	3	3	3	1	1	3
2 Classifiers	3 1	1 3	3 1	3 1	1 3	3 1	1 3	1 3	3 1
Prob. F1	3 1	1 3	3 1	3 1	1 3	3 1	1 3	1 3	3 1
Soft	3 1	1 3	3 1	3 1	1 3	3 1	1 3	1 3	3 1
Hard	3 1	1 3	3 1	3 1	1 3	1 3	1 3	1 3	3 1

1—Linear discriminant analysis; 2—linear support vector machines; 3—logistic regression.

Because the formerly exposed results were not satisfactory, in the sense that they are not acceptable for a real application, further approaches demanded to be tested. Moreover, as FBCSP II was better than FBCSP I, non-linear classifiers were added to the previous classifiers to be trained and tried out. Table 15 contains the F1-score from this method. As expected, the results improved compared with the linear and statistical classifiers approach.

Two combinations obtained the same average F1-score, the one classifier, and the ensemble voting hard, which was not foreseen as it was presumed that two classifiers would predict a more accurate result rather than just one.

Table 15. F1-score, kappa, and FP rate for dataset B and FBCSP II, using both sets of classifiers.

		Average	1	2	3	4	5	6	7	8	9
F1-score	1 Classifier	0.651	0.747	0.595	0.911	0.604	0.660	0.706	0.604	0.517	0.515
	Prob. F1	0.648	0.718	0.631	0.912	0.628	0.667	0.737	0.642	0.450	0.447
	2 Classifiers	0.588	0.712	0.586	0.773	0.582	0.601	0.613	0.545	0.460	0.418
	Prob. F1	0.599	0.712	0.560	0.771	0.619	0.650	0.658	0.623	0.374	0.418
	Soft	0.646	0.758	0.626	0.892	0.589	0.644	0.678	0.622	0.485	0.519
	Hard	0.651	0.747	0.595	0.911	0.604	0.660	0.706	0.604	0.517	0.515
	Ada	0.668	0.758	0.631	0.912	0.628	0.667	0.737	0.642	0.517	0.519
	Best	0.651	0.747	0.595	0.911	0.604	0.660	0.706	0.604	0.517	0.515
Kappa	1 Classifier	0.426	0.463	0.375	0.863	0.387	0.488	0.525	0.387	0.125	0.225
	Prob. F1	0.433	0.625	0.175	0.863	0.425	0.488	0.563	0.463	0.100	0.200
	2 Classifiers	0.359	0.550	0.313	0.625	0.400	0.390	0.413	0.325	0.050	0.162
	Prob. F1	0.413	0.575	0.325	0.638	0.413	0.475	0.488	0.438	0.175	0.188
	Soft	0.447	0.625	0.450	0.838	0.375	0.463	0.450	0.425	0.150	0.250
	Hard	0.443	0.600	0.375	0.863	0.387	0.488	0.538	0.387	0.125	0.225
	Ada	0.478	0.625	0.450	0.863	0.425	0.488	0.563	0.463	0.175	0.250
	Best	0.426	0.463	0.375	0.863	0.387	0.488	0.525	0.387	0.125	0.225
FP rate	1 Classifier	0.471	0.519	0.314	0.101	0.690	0.468	0.422	0.648	0.080	1.000
	Prob. F1	0.411	0.322	0.111	0.101	0.595	0.506	0.400	0.506	0.125	1.036
	2 Classifiers	0.314	0.190	0.200	0.078	0.486	0.178	0.288	0.561	0.068	0.774
	Prob. F1	0.399	0.370	0.364	0.110	0.534	0.397	0.466	0.440	0.071	0.836
	Soft	0.471	0.300	0.461	0.121	0.700	0.506	0.500	0.581	0.135	0.933
	Hard	0.471	0.519	0.314	0.101	0.690	0.468	0.422	0.648	0.080	1.000
	Ada	0.290	0.190	0.111	0.078	0.486	0.178	0.288	0.440	0.068	0.774
	Best	0.471	0.519	0.314	0.101	0.690	0.468	0.422	0.648	0.080	1.000

Regarding the kappa score, it was higher for the ensemble voting hard, which is better than the score for just the linear and statistical classifiers, but still lower than for dataset A, which corroborates with that previously stated about the headset used to acquire these signals. The FP rate was lower than previously, as was anticipated due to the rise of the F1-score. Despite these results, it is still important to mention that subjects 1, 3, and 6 produced results comparable to the ones from dataset A, even if with slightly higher FP rates than the ones from dataset A. Table 16 presents the chosen classifier(s) for each combination. These encompass mainly the kernel SVMs, followed by the neural networks. In some cases, the K-NN is chosen as the second-best classifier. None of the linear or statistical classifiers were chosen, which indicates that this dataset requires more complex models to predict better results.

Table 16. Best classifiers, from both sets, for each combination for dataset B and FBCSP II.

Individual (I)	I1		I2		I3		I4		I5		I6		I7		I8		I9	
1 Classifier	5		5		5		5		5		5		5		5		5	
Prob. F1	5		5		5		5		5		5		5		5		5	
2 Classifiers	5	7	5	7	5	4	5	7	5	7	5	7	5	7	5	7	5	7
Prob. F1	5	7	5	4	5	4	5	7	5	7	5	7	5	7	5	4	5	7
Soft	5	7	5	4	5	4	5	8	5	7	5	7	5	7	5	4	5	4
Hard	5	7	5	7	5	4	5	7	5	7	5	7	5	7	5	4	5	7

4—K-nearest neighbours; 5—kernel support vector machines; 7—neural networks; 8—random forest.

4.3. Power Spectral Density

The next subsections present the findings and insights from the analysis of results obtained using power spectral density.

4.3.1. Power Spectral Density Approach Using Dataset A

As mentioned before, since the best results with linear classifiers were not obtained with the PSD approach, non-linear strategies were not employed. Moreover, as no single linear classifier produced satisfactory results on its own, combinations thereof were not evaluated either. The obtained F1-score was 0.510, which is much lower than the F1-score of 0.793 for the combination of one classifier and FBCSP II. Moreover, the FP rate is also much higher than the one for FBCSP II, thus endorsing the idea that this approach is not the best for this dataset.

4.3.2. Power Spectral Density Approach Using Dataset B

Similarly to dataset A, only the one classifier approach was tested due to unsatisfactory results. However, the difference between this and the FBCSP II was not as large as the one for dataset A. The average F1-score, 0.443, was quite similar to the one for dataset A, 0.510, which did not happen for the other approaches. However, the kappa score and the FP rate were worse. The latter presents a high value of 0.844, which lies close to 1, that is to say, there is almost more FP than TP, which is not the goal.

4.4. Real-Time Application

The subject with the best performance for dataset B was chosen to perform the real-time testing. The subject was asked to sit still and maintain movements to a minimum, similar to the training phase. An external person experimented and asked the subject to imagine a certain MI. Every 2 s an epoch was sent to the system and a class was predicted. The person conducting the experiment waited for ten predictions to appear before asking for the next one, as a way of allowing the system to stabilise. Again, due to stabilisation, the first three outputs after a new MI were discarded. As subject 3 was the one who presented the best results, the steps described previously were applied in a real-time scenario, leading to the results in Table 17. The final system consisted of applying the FBCSP II approach, which produced the best score. Then, 70% of the features were extracted and fed to an ensemble voting hard classifier built with the major vote between the kernel SVM and the K-NN, where the vote percentage was 2 to 1, respectively.

Table 17. Cues and respective outputs from subject 3's real-time applications.

MI	Output										Majority	%	MI	Output										Majority	%
N	2	1	1	2	2	2	0	0	2	2	2	60%	R	1	2	1	1	1	1	1	1	1	1	1	90%
R	1	1	0	1	2	1	1	0	0	1	1	60%	N	2	2	2	2	1	2	0	0	2	2	2	70%
N	2	0	2	2	2	2	2	0	0	2	2	70%	L	2	1	1	1	1	1	0	0	1	0	1	30%
L	0	0	0	1	0	0	1	1	2	2	0	50%	N	2	2	2	0	2	2	2	2	2	2	2	90%
L	0	1	0	0	0	1	0	1	2	2	0	50%	R	1	0	1	1	1	0	1	1	1	2	1	70%
N	2	0	2	2	0	2	2	2	0	2	2	70%	N	2	2	2	2	2	2	2	2	2	2	2	100%
R	1	0	1	2	1	1	2	1	1	0	1	60%	L	0	0	2	0	2	0	2	0	2	2	0	50%
L	0	0	2	0	0	0	1	1	0	1	0	60%	N	2	1	2	2	2	1	2	2	2	2	2	80%
N	2	2	2	1	2	2	1	2	0	2	2	70%	R	1	2	1	2	1	2	0	2	0	2	2	30%
R	1	1	1	1	0	0	2	1	1	1	1	70%	N	2	2	2	2	2	2	2	2	2	2	2	100%
N	0	2	2	0	2	2	0	0	2	2	2	60%	L	0	0	0	1	0	0	0	0	2	2	0	70%

Overall, the results were satisfactory and as anticipated from the performance previously analysed. From twenty-two cues, there were only two cues incorrectly classified, represented in red, one for the left and another for the right. The left one was misclassified as right, which is a problem since, if it was an IW, it would go in the opposite direction. The right cue was classified as neutral. Despite being misclassified, it is not the worst since a hypothetical IW would have maintained the same direction. Nevertheless, none of the neutral cues were incorrectly classified. Although some limitations related to the accuracy, these results show that the system is evolving towards the right direction, suggesting that a new headset and a refinement of the algorithm would deliver promising results.

5. Discussion

Overall, dataset A, independently of the approach, produced better results than dataset B. The only exception might be for the PSD approaches, where the average F1-score value for both datasets was very similar. There are several possibilities for this discrepancy in the results, such as dataset A was obtained using a stable headset, with movable electrodes. This leads to the electrodes always assuming the same known position throughout the different sessions. Moreover, the results were obtained by professionals in a carefully controlled environment. Dataset B was collected using an Emotiv EPOC. There are several problems regarding the EPOC, such as it has fixed electrodes. This leads to the absolute position being different for all the subjects, and even varies from session to session. Moreover, it did not fit everyone's head, and four subjects did not make it into this dataset due to limitations in positioning the electrodes correctly. Some sensors were starting to be oxidised, which led to noisy acquisitions, which hampered the already difficult EEG processing, as the signal is very sensitive. The Emotiv EPOC does not cover the motor cortex, which is critical for the tasks in this study. While the literature suggests it can work for parietal and frontal areas, its performance is not optimal for motor cortex tasks. Nevertheless, the Emotiv EPOC was chosen for its balance of cost, ease of use, and functionality, offering a reasonable number of electrodes, wireless operation, extended battery life, and affordability. Due to the fact that the process of training is very time-consuming and this work was merely a proof of concept, only the subject with the best performance for dataset B was chosen for the real-time testing.

Comparing the different approaches, it was already expected that the best method would be related to the CSP, as it was the winning method of the competition. This suggests that spatial methods perform better than the others do, which may be related to the elimination of existent artefacts in the bands of interest. However, it was interesting that FBCSP II produced slightly better scores than FBCSP I, implying that feeding the whole spatial filtered signal to the feature selector works better than feeding a transformed version of the signal filtered by just the columns of the spatial filter. Although the results from the competition are merely qualitative, the results from FBCSP II also indicate that using the extra tree classifier to obtain the features' importance and the ensemble voting hard, employing the LR and the GNB, or LDA, to classify the epochs, represents a valuable update. This led to a final average kappa score of 0.69, which is 20% higher than the winner value of 0.57. Despite the 20% not being a real quantitative evaluation, the value of 0.69 already suggests that the algorithm is considerably better than a random classifier and can correctly classify the epochs, presenting an F1-score of 0.797 and the smallest FP rate of all the tested approaches, 0.150. Similarly, dataset B also presented better results for the FBCSP II than for the FBCSP I. Moreover, contrarily to dataset A, dataset B improved its results by allowing the use of non-linear classifiers. Fakhruzzaman et al. [66] and Muñoz et al. [61] used the Emotiv EPOC headset and the CSP method as a features extractor. Fakhruzzaman et al. [66] obtained an average accuracy of 60%, whereas Muñoz et al. [61] obtained an average accuracy of 67.5% using the LDA classifier, 68.3% using the SVM, and 96.7% using Nu-SVC RBF kernel. Overall, the result of 65% from dataset B regarding the FBCSP II approach with the ensemble voting hard classifier falls within the mean values presented by these authors, except for the last method of Muñoz et al. [61]. The latter is greatly higher than the others are, suggesting that this classifier is indicated for this type of feature, and should be considered for further implementation in future work. Furthermore, it is important to state that the signals in dataset B had constraints in AF3 and AF4 electrodes, which may be important electrodes according to Lin and Lo [60] and Muñoz et al. [61], thus decreasing the obtained accuracy and the F1-score. The average accuracy of other authors using the EPOC and the magnitude of frequency components or the power spectral density (square of the magnitude) as features was 74–100% for Abiyev et al. [57], 70% for Hurtado-Rincon et al. [59], and 86–92% for Lin and Lo [60], and Siribunyaphat and Punsawad [67]. More recent works [49,68,69] have also achieved important F1-scores, using different EEG headsets. This reflects a promising area to explore for controlling intelligent wheelchairs.

6. Conclusions and Future Work

The goal of this work was being able to decode MI intentions from the users, using an Emotiv EPOC as the headset to extract the EEG signals. The intentions were left, right, and neutral, which would be further translated into control commands for an intelligent wheelchair. This headset has higher constraints in terms of accessing data in a less controlled environment; however, overall, this work allowed the development of a proof of concept for future projects and a thorough study regarding the different algorithms. Although the real-time results are still not suitable for the actual application, they validate the concept and the developed architecture to connect the different parts of the system. For future work, utilising a broader and more diverse dataset may contribute to enhancing the model's generality. Another interesting future work could be applying different methods for noise removal, such as independent component analysis. Additionally, conducting long-term usage and testing across diverse environments will be essential for assessing system stability and applicability.

Author Contributions: Conceptualization, L.P.R. and B.M.F.; methodology, M.C.A., P.A., L.P.R. and B.M.F.; software, M.C.A. and P.A.; validation, M.C.A. and P.A.; formal analysis, M.C.A., P.A. and L.P.R.; investigation, M.C.A., P.A. and L.P.R.; resources, L.P.R. and B.M.F.; data curation, M.C.A. and P.A.; writing—original draft preparation, M.C.A. and P.A.; writing—review and editing, L.P.R. and B.M.F.; visualization, M.C.A., P.A., L.P.R. and B.M.F.; supervision, L.P.R. and B.M.F. project administration, L.P.R. and B.M.F.; funding acquisition, L.P.R. and B.M.F. All authors have read and agreed to the published version of the manuscript.

Funding: This work was financially supported by: Base Funding—UIDB/00027/2020 of the Artificial Intelligence and Computer Science Laboratory (LIACC) funded by national funds through the FCT/MCTES (PIDDAC) and IntellWheels2.0: Intelligent Wheelchair with Flexible Multimodal Interface and Realistic Simulator (POCI-01-0247-FEDER-39898), supported by NORTE 2020, under PT2020.

Institutional Review Board Statement: The study was conducted in accordance with the Declaration of Helsinki, and approved by the Institutional Review Board of Optimizer/LITEC (under project POCI-01-0247-FEDER-39898 approved on 28 May 2019).

Informed Consent Statement: Informed consent was obtained from all subjects involved in the study.

Data Availability Statement: Data from the BCI competitions are available at https://www.bbci.de/competition/iv/#datasets first accessed on 1 June 2019. Other data are available on request from the corresponding author.

Conflicts of Interest: The authors declare no conflicts of interest. The funders had no role in the design of the study; in the collection, analyses, or interpretation of data; in the writing of the manuscript; or in the decision to publish the results.

References

1. Davies, A.; Souza, L.H.; Frank, A.O. Changes in the quality of life in severely disabled people following provision of powered indoor/outdoor chairs. *Disabil. Rehabil.* **2003**, *25*, 286–290. [CrossRef] [PubMed]
2. Koontz, A.M.; Ding, D.; Jan, Y.; Groot, S.; Hansen, A. Wheeled mobility. *BioMed Res. Int.* **2015**, *2015*, 138176. [CrossRef] [PubMed]
3. Guidelines on the Provision of Manual Wheelchairs in Less Resourced Settings. Available online: https://www.who.int/publications/i/item/9789241547482 (accessed on 5 January 2024).
4. Worldwide Need—Wheelchair Foundation. Available online: https://www.wheelchairfoundation.org/fth/analysis-of-wheelchair-need (accessed on 5 January 2024).

5. Xin-an, F.; Luzheng, B.; Teng, T.; Ding, H.; Liu, Y. A brain–computer interface-based vehicle destination selection system using p300 and ssvep signals. *IEEE Trans. Intell. Transp. Syst.* **2015**, *16*, 274–283. [CrossRef]
6. Luzheng, B.; Xin-An, F.; Liu, Y. EEG-based brain-controlled mobile robots: A survey. *IEEE Trans. Hum.-Mach. Syst.* **2013**, *43*, 161–176. [CrossRef]
7. Gürkök, H.; Nijholt, A. Brain–computer interfaces for multimodal interaction: A survey and principles. *Int. J. Hum.-Comput. Interact.* **2012**, *28*, 292–307. [CrossRef]
8. Major, T.C.; Conrad, J.M. A survey of brain computer interfaces and their applications. In Proceedings of the SOUTHEASTCON 2014, Lexington, KY, USA, 13–16 March 2014; pp. 1–8. [CrossRef]
9. Hämäläinen, M.; Hari, R.; Ilmoniemi, R.J.; Knuutila, J.; Lounasmaa, O.V. Magnetoencephalography—Theory, instrumentation, and applications to noninvasive studies of the working human brain. *Rev. Mod. Phys.* **1993**, *65*, 413. [CrossRef]
10. Ferrari, M.; Quaresima, V. A brief review on the history of human functional near-infrared spectroscopy (fnirs) development and fields of application. *Neuroimage* **2012**, *63*, 921–935. [CrossRef] [PubMed]
11. Logothetis, N.K.; Pauls, J.; Augath, M.; Trinath, T.; Oeltermann, A. Neurophysiological investigation of the basis of the fmri signal. *Nature* **2001**, *412*, 150. [CrossRef] [PubMed]
12. Hashiguchi, K.; Morioka, T.; Yoshida, F.; Miyagi, Y.; Nagata, S.; Sakata, A.; Sasaki, T. Correlation between scalp-recorded electroencephalographic and electrocorticographic activities during ictal period. *Seizure* **2007**, *16*, 238–247. [CrossRef]
13. Buitenweg, J.R.; Rutten, W.L.C.; Marani, E. Geometry-based finiteelement modeling of the electrical contact between a cultured neuron and a microelectrode. *IEEE Trans. Biomed. Eng.* **2003**, *50*, 501–509. [CrossRef]
14. Trambaiolli, L.R.; Falk, T.H. Hybrid brain–computer interfaces for wheelchair control: A review of existing solutions, their advantages and open challenges. In *Smart Wheelchairs and Brain-Computer Interfaces*; Elsevier: Amsterdam, The Netherlands, 2018; pp. 229–256. [CrossRef]
15. Hage, B.; Alwatban, M.R.; Barney, E.; Mills, M.; Dodd, M.D.; Truemper, E.J.; Bashford, G.R. Functional transcranial doppler ultrasound for measurement of hemispheric lateralization during visual memory and visual search cognitive tasks. *IEEE Trans. Ultrason. Ferroelectr. Freq. Control.* **2016**, *63*, 2001–2007. [CrossRef] [PubMed]
16. Open Bci. Available online: https://openbci.com/ (accessed on 5 January 2024).
17. Emotiv Epoc. Available online: https://www.emotiv.com/ (accessed on 10 January 2024).
18. Pires, G.; Castelo-Branco, M.; Nunes, U. Visual p300-based bci to steer a wheelchair: A bayesian approach. In Proceedings of the Annual International Conference of the IEEE Engineering in Medicine and Biology Society, Vancouver, BC, Canada, 20–25 August 2008; pp. 658–661. [CrossRef]
19. Zhang, R.; Wang, Q.; Li, K.; He, S.; Qin, S.; Feng, Z.; Chen, Y.; Song, P.; Yang, T.; Zhang, Y. A bci-based environmental control system for patients with severe spinal cord injuries. *IEEE Trans. Biomed. Eng.* **2017**, *64*, 1959–1971. [CrossRef]
20. Rebsamen, B.; Burdet, E.; Guan, C.; Teo, C.L.; Zeng, Q.; Ang, M.; Laugier, C. Controlling a wheelchair using a bci with low information transfer rate. In Proceedings of the 2007 IEEE 10th International Conference on Rehabilitation Robotics, Noordwijk, The Netherlands, 13–15 June 2007; pp. 1003–1008. [CrossRef]
21. Iturrate, I.; Antelis, J.M.; Kubler, A.; Minguez, J. A noninvasive brainactuated wheelchair based on a p300 neurophysiological protocol and automated navigation. *IEEE Trans. Robot.* **2009**, *25*, 614–627. [CrossRef]
22. Alqasemi, R.; Dubey, R. A 9-dof wheelchair-mounted robotic arm system: Design, control, brain-computer interfacing, and testing. In *Advances in Robot Manipulators*; InTech: Rijeka, Croatia, 2010. [CrossRef]
23. Shin, B.; Kim, T.; Jo, S. Non-invasive brain signal interface for a wheelchair navigation. In Proceedings of the International Conference on Control Automation and Systems, Gyeonggi-do, Republic of Korea, 27–30 October 2010. [CrossRef]
24. Lopes, A.C.; Pires, G.; Nunes, U. Assisted navigation for a brain-actuated intelligent wheelchair. *Robot. Auton. Syst.* **2013**, *61*, 245–258. [CrossRef]
25. Zhang, R.; Li, Y.; Yan, Y.; Zhang, H.; Wu, S.; Yu, T.; Gu, Z. Control of a wheelchair in an indoor environment based on a brain–computer interface and automated navigation. *IEEE Trans. Neural Syst. Rehabil. Eng.* **2016**, *24*, 128–139. [CrossRef] [PubMed]
26. Wang, Y.; Wang, R.; Gao, X.; Hong, B.; Gao, S. A practical vepbased brain-computer interface. *IEEE Trans. Neural Syst. Rehabil. Eng.* **2006**, *14*, 234–240. [CrossRef] [PubMed]
27. Dasgupta, S.; Fanton, M.; Pham, J.; Willard, M.; Nezamfar, H.; Shafai, B.; Erdogmus, D. Brain controlled robotic platform using steady state visual evoked potentials acquired by eeg. In Proceedings of the 2010 Conference Record of the Forty Fourth Asilomar Conference on Signals, Systems and Computers (ASILOMAR), Pacific Grove, CA, USA, 7–10 November 2010; pp. 1371–1374. [CrossRef]
28. Prueckl, R.; Guger, C. Controlling a robot with a brain-computer interface based on steady state visual evoked potentials. In Proceedings of the 2010 International Joint Conference on Neural Networks (IJCNN), Barcelona, Spain, 18–23 July 2010; pp. 1–5. [CrossRef]
29. Mandel, C.; Lüth, T.; Laue, T.; Röfer, T.; Gräser, A.; KriegBrückner, B. Navigating a smart wheelchair with a brain-computer interface interpreting steady-state visual evoked potentials. In Proceedings of the IEEE/RSJ International Conference on Intelligent Robots and Systems, St. Louis, MO, USA, 10–15 October 2009; pp. 1118–1125. [CrossRef]

30. Xu, Z.; Li, J.; Gu, R.; Xia, B. Steady-state visually evoked potential (ssvep)-based brain-computer interface (bci): A low-delayed asynchronous wheelchair control system. In *Neural Information Processing*; Springer: Berlin/Heidelberg, Germany, 2012; pp. 305–314. [CrossRef]
31. Müller, S.M.; Bastos, T.F.; Filho, M.S. Proposal of a ssvep-bci to command a robotic wheelchair. *J. Control. Autom. Electr. Syst.* **2013**, *24*, 97–105. [CrossRef]
32. Diez, P.F.; Müller, S.M.; Mut, V.A.; Laciar, E.; Avila, E.; Bastos, T.F.; Filho, M.S. Commanding a robotic wheelchair with a high-frequency steady-state visual evoked potential based brain–computer interface. *Med. Eng. Phys.* **2013**, *35*, 1155–1164. [CrossRef]
33. Duan, J.; Li, Z.; Yang, C.; Xu, P. Shared control of a brain-actuated intelligent wheelchair. In Proceedings of the 11th World Congress on Intelligent Control and Automation (WCICA), Shenyang, China, 29 June–4 July 2014; pp. 341–346. [CrossRef]
34. Larsen, E.A. Classification of Eeg Signals in a Brain-Computer Interface System. Master's Thesis, Institutt for Datateknikk og Informasjonsvitenskap, Trondheim, Norway, 2011.
35. Khare, V.; Santhosh, J.; Anand, S.; Bhatia, M. Brain computer interface based real time control of wheelchair using electroencephalogram. *Int. J. Soft Comput. Eng. IJSCE* **2011**, *1*, 41–45.
36. Choi, K. Control of a vehicle with eeg signals in real-time and system evaluation. *Eur. J. Appl. Physiol.* **2012**, *112*, 755–766. [CrossRef]
37. Barbosa, A.O.; Achanccaray, D.R.; Meggiolaro, M.A. Activation of a mobile robot through a brain computer interface. In Proceedings of the 2010 IEEE International Conference on Robotics and Automation (ICRA), Anchorage, AK, USA, 3–7 May 2010; pp. 4815–4821. [CrossRef]
38. Tsui, C.S.; Gan, J.Q.; Roberts, S.J. A self-paced brain–computer interface for controlling a robot simulator: An online event labelling paradigm and an extended kalman filter based algorithm for online training. *Med. Biol. Eng. Comput.* **2009**, *47*, 257–265. [CrossRef]
39. Tang, Z.; Sun, S.; Zhang, S.; Chen, Y.; Li, C.; Chen, S. A brain-machine interface based on erd/ers for an upper-limb exoskeleton control. *Sensors* **2016**, *16*, 2050. [CrossRef]
40. Bahri, Z.; Abdulaal, S.; Buallay, M. Sub-band-power-based efficient brain computer interface for wheelchair control. In Proceedings of the 2014 World Symposium on Computer Applications & Research (WSCAR), Sousse, Tunisia, 18–20 January 2014; pp. 1–7. [CrossRef]
41. Li, M.; Zhang, Y.; Zhang, H.; Hu, S. An eeg based control system for intelligent wheelchair. In *Applied Mechanics and Materials*; Trans Tech Publications: Bäch, Switzerland, 2013; Volume 300, pp. 1540–1545. [CrossRef]
42. Dharmasena, S.; Lalitharathne, K.; Dissanayake, K.; Sampath, A.; Pasqual, A. Online classification of imagined hand movement using a consumer grade eeg device. In Proceedings of the 2013 8th IEEE International Conference on Industrial and Information Systems (ICIIS), Peradeniya, Sri Lanka, 17–20 December 2013; pp. 537–541. [CrossRef]
43. Batres-Mendoza, P.; Ibarra-Manzano, M.; Guerra-Hernandez, E.; Almanza-Ojeda, D.; Montoro-Sanjose, C.; Romero-Troncoso, R.; Rostro-Gonzalez, H. Improving eeg-based motor imagery classification for real-time applications using the qsa method. *Comput. Intell. Neurosci.* **2017**, *2017*, 9817305. [CrossRef]
44. Cao, L.; Li, J.; Ji, H.; Jiang, C. A hybrid brain computer interface system based on the neurophysiological protocol and brain-actuated switch for wheelchair control. *J. Neurosci. Methods* **2014**, *229*, 33–43. [CrossRef]
45. Li, J.; Ji, H.; Cao, L.; Zang, D.; Gu, R.; Xia, B.; Wu, Q. Evaluation and application of a hybrid brain computer interface for real wheelchair parallel control with multi-degree of freedom. *Int. J. Neural Syst.* **2014**, *24*, 1450014. [CrossRef]
46. Allison, B.Z.; Brunner, C.; Kaiser, V.; Müller-Putz, G.; Neuper, C.; Pfurtscheller, G. Toward a hybrid brain–computer interface based on imagined movement and visual attention. *J. Neural Eng.* **2010**, *7*, 026007. [CrossRef]
47. Long, J.; Li, Y.; Wang, H.; Yu, T.; Pan, J. Control of a simulated wheelchair based on a hybrid brain computer interface. In Proceedings of the 2012 Annual International Conference of the IEEE Engineering in Medicine and Biology Society (EMBC), San Diego, CA, USA, 28 August–1 September 2012; pp. 6727–6730. [CrossRef]
48. Rani, B.J.; Umamakeswari, A. Electroencephalogram-based brain controlled robotic wheelchair. *Indian J. Sci. Technol.* **2015**, *8* (Suppl. 9), 188–197. [CrossRef]
49. Almeida, P.; Faria, B.M.; Reis, L.P. Brain Waves Classification Using a Single-Channel Dry EEG Headset: An Application for Controlling an Intelligent Wheelchair. In *Advances in Practical Applications of Agents, Multi-Agent Systems, and Cognitive Mimetics*; The PAAMS Collection; Lecture Notes in Computer Science; Mathieu, P., Dignum, F., Novais, P., De la Prieta, F., Eds.; Springer: Cham, Switzerland, 2023; Volume 13955. [CrossRef]
50. Open Bci-Publicly Available EEG Datasets. Available online: https://openbci.com/community/publicly-available-eeg-datasets/ (accessed on 5 February 2024).
51. Lotte, F.; Bougrain, L.; Cichocki, A.; Clerc, M.; Congedo, M.; Rakotomamonjy, A.; Yger, F. A review of classification algorithms for eeg-based brain–computer interfaces: A 10 year update. *J. Neural Eng.* **2018**, *15*, 031005. [CrossRef]
52. Bzdok, D.; Krzywinski, M.; Altman, N. Points of significance: Machine learning: A primer. *Nat. Methods* **2017**, *14*, 1119–1120. [CrossRef]
53. Johansson, M. Novel cluster-based svm to reduce classification error in noisy eeg data: Towards real-time brain-robot interfaces. *Comput. Biol. Med.* **2018**, *148*, 105931. [CrossRef]

54. Singh, A.; Lal, S.; Guesgen, H.W. Architectural review of co-adaptive brain computer interface. In Proceedings of the 2017 4th Asia-Pacific World Congress on Computer Science and Engineering (APWC on CSE), Mana Island, Fiji, 11–13 December 2017; pp. 200–207. [CrossRef]
55. Pejas, J.; El Fray, I.; Hyla, T.; Kacprzyk, J. (Eds.) *Advances in Soft and Hard Computing*; Springer: Berlin/Heidelberg, Germany, 2018. [CrossRef]
56. Fernández-Rodríguez, A.; Álvarez, F.; Angevin, R. Review of real brain-controlled wheelchairs. *J. Neural Eng.* **2016**, *13*, 061001. [CrossRef] [PubMed]
57. Abiyev, A.H.; Akkaya, N.; Aytac, E.; Günsel, I.; Çagman, A. Brain based control of wheelchair. In Proceedings of the International Conference on Artificial Intelligence (ICAI), Las Vegas, NV, USA, 27–30 July 2015; p. 542. Available online: https://api.semanticscholar.org/CorpusID:13468342 (accessed on 9 April 2024).
58. Carrino, F.; Dumoulin, J.; Mugellini, E.; Khaled, O.A.; Ingold, R. A self-paced bci system to control an electric wheelchair: Evaluation of a commercial, lowcost eeg device. In Proceedings of the 2012 ISSNIP Biosignals and Biorobotics Conference: Biosignals and Robotics for Better and Safer Living (BRC), Manaus, Brazil, 9–11 January 2012; pp. 1–6. [CrossRef]
59. Rincon, J.; Jaramillo, S.; Cespedes, Y.; Meza, A.M.; Domínguez, G. Motor imagery classification using feature relevance analysis: An emotiv-based bci system. In Proceedings of the 2014 XIX Symposium on Image, Signal Processing and Artificial Vision, Armenia, Colombia, 17–19 September 2014; pp. 1–5. [CrossRef]
60. Lin, J.; Lo, C. Mental commands recognition on motor imagery-based brain computer interface. *Int. J. Comput. Consum. Control.* **2016**, *25*, 18–25.
61. Muñoz, J.; Ríos, L.H.; Henao, O. Low cost implementation of a motor imagery experiment with bci system and its use in neurorehabilitation. In Proceedings of the Annual International Conference of the IEEE Engineering in Medicine and Biology Society, Chicago, IL, USA; 2014. [CrossRef]
62. Ang, K.; Chin, Z.; Wang, C.; Guan, C.; Zhang, H. Filter bank common spatial pattern algorithm on bci competition iv datasets 2a and 2b. *Front. Neurosci.* **2012**, *6*, 39. [CrossRef] [PubMed]
63. Stock, V.; Balbinot, A. Movement imagery classification in emotiv cap based system by naïve bayes. In Proceedings of the 2016 38th Annual International Conference of the IEEE Engineering in Medicine and Biology Society (EMBC), Orlando, FL, USA, 16–20 August 2016; pp. 4435–4438. [CrossRef]
64. Kim, P.; Kim, K.; Kim, S. Using common spatial pattern algorithm for unsupervised real-time estimation of fingertip forces from semg signals. In Proceedings of the 2015 IEEE/RSJ International Conference on Intelligent Robots and Systems (IROS), Hamburg, Germany, 28 September–2 October 2015; pp. 5039–5045. [CrossRef]
65. Siemens PLM Community. Spectral Leakage. Available online: https://community.plm.automation.siemens.com/t5/Testing-Knowledge-Base/Windows-and-Spectral-Leakage/ta-p/432760 (accessed on 5 December 2023).
66. Fakhruzzaman, M.N.; Riksakomara, E.; Suryotrisongko, H. Eeg wave identification in human brain with Emotiv EPOC for motor imagery. *Procedia Comput. Sci.* **2016**, *72*, 269–276. [CrossRef]
67. Siribunyaphat, N.; Punsawad, Y. Brain–Computer Interface Based on Steady-State Visual Evoked Potential Using Quick-Response Code Pattern for Wheelchair Control. *Sensors* **2023**, *23*, 2069. [CrossRef] [PubMed]
68. Ramírez-Arias, F.J.; García-Guerrero, E.E.; Tlelo-Cuautle, E.; Colores-Vargas, J.M.; García-Canseco, E.; López-Bonilla, O.R.; Galindo-Aldana, G.M.; Inzunza-González, E. Evaluation of Machine Learning Algorithms for Classification of EEG Signals. *Technologies* **2022**, *10*, 79. [CrossRef]
69. Sabio, J.; Williams, N.S.; McArthur, G.M.; Badcock, N.A. A scoping review on the use of consumer-grade EEG devices for research. *PLoS ONE* **2024**, *19*, e0291186. [CrossRef] [PubMed] [PubMed Central]

Disclaimer/Publisher's Note: The statements, opinions and data contained in all publications are solely those of the individual author(s) and contributor(s) and not of MDPI and/or the editor(s). MDPI and/or the editor(s) disclaim responsibility for any injury to people or property resulting from any ideas, methods, instructions or products referred to in the content.

Article

An End-to-End Lightweight Multi-Scale CNN for the Classification of Lung and Colon Cancer with XAI Integration

Mohammad Asif Hasan [1], Fariha Haque [1], Saifur Rahman Sabuj [2,*], Hasan Sarker [1], Md. Omaer Faruq Goni [3], Fahmida Rahman [4] and Md Mamunur Rashid [5]

1. Department of Electronics & Telecommunication Engineering, Rajshahi University of Engineering & Technology, Rajshahi 6204, Bangladesh; asifhasan2189@gmail.com (M.A.H.); farihatanzim16@gmail.com (F.H.); hasan.ruet.ete@gmail.com (H.S.)
2. Department of Electrical and Electronic Engineering, Brac University, Dhaka 1212, Bangladesh
3. Department of Electrical & Computer Engineering, Rajshahi University of Engineering & Technology, Rajshahi 6204, Bangladesh; omaerfaruq@ece.ruet.ac.bd
4. Department of Computer Science & Engineering, International Islamic University Chittagong, Chittagong 4318, Bangladesh; fahmida.rahman103@gmail.com
5. School of Information Technology, Deakin University, Victoria 3125, Australia; mamunrashid.ete88@gmail.com
* Correspondence: s.r.sabuj@ieee.org

Citation: Hasan, M.A.; Haque, F.; Sabuj, S.R.; Sarker, H.; Goni, M.O.F.; Rahman, F.; Rashid, M.M. An End-to-End Lightweight Multi-Scale CNN for the Classification of Lung and Colon Cancer with XAI Integration. *Technologies* 2024, 12, 56. https://doi.org/10.3390/technologies12040056

Academic Editors: Juan Gabriel Avina-Cervantes, Juvenal Rodriguez-Resendiz, Gerardo I. Pérez-Soto, Karla Anhel Camarillo-Gómez and Saul Tovar-Arriaga

Received: 15 February 2024
Revised: 11 April 2024
Accepted: 18 April 2024
Published: 21 April 2024

Copyright: © 2024 by the authors. Licensee MDPI, Basel, Switzerland. This article is an open access article distributed under the terms and conditions of the Creative Commons Attribution (CC BY) license (https://creativecommons.org/licenses/by/4.0/).

Abstract: To effectively treat lung and colon cancer and save lives, early and accurate identification is essential. Conventional diagnosis takes a long time and requires the manual expertise of radiologists. The rising number of new cancer cases makes it challenging to process massive volumes of data quickly. Different machine learning approaches to the classification and detection of lung and colon cancer have been proposed by multiple research studies. However, when it comes to self-learning classification and detection tasks, deep learning (DL) excels. This paper suggests a novel DL convolutional neural network (CNN) model for detecting lung and colon cancer. The proposed model is lightweight and multi-scale since it uses only 1.1 million parameters, making it appropriate for real-time applications as it provides an end-to-end solution. By incorporating features extracted at multiple scales, the model can effectively capture both local and global patterns within the input data. The explainability tools such as gradient-weighted class activation mapping and Shapley additive explanation can identify potential problems by highlighting the specific input data areas that have an impact on the model's choice. The experimental findings demonstrate that for lung and colon cancer detection, the proposed model was outperformed by the competition and accuracy rates of 99.20% have been achieved for multi-class (containing five classes) predictions.

Keywords: convolutional neural network; cancer classification; deep learning; explainable artificial intelligence; Gradio; gradient-weighted class activation mapping; lightweight; Shapley additive explanation

1. Introduction

Cancer is a health condition that is characterized by the unregulated growth of abnormal cells that can develop in any tissue or organ inside the body. The World Health Organization says that it was the second most common cause of death in the world in 2020, with about 10 million deaths [1]. Compared to other cancer types, colorectal cancer accounts for 1.80 million new cases and 783 thousand fatalities, whereas lung cancer contributes to 1.76 million new cases and 1.76 million fatalities. The two varieties of lung cancer that spread and grow quickly are small-cell lung cancer (SCLC) and non-small-cell lung cancer (NSCLC) [2,3]. Cells with neuroendocrine characteristics cause SCLC, which accounts for 15% of all instances of lung cancer and remains a hazardous form of the disease. The three pathologic types of NSCLC, including immense cell carcinoma, adenocarcinoma, and squamous cell carcinoma, account for 85% of all cases [4]. The most common cause of death is colorectal cancer, which accounts for 10.7% of all instances [1].

To examine the therapy possibilities in the early stages of the disease, a more precise diagnosis of various cancer subtypes is required. For lung cancer, radiography, computed tomography (CT) imaging, flexible sigmoidoscopy, and CT colonoscopy are among the non-invasive diagnostic techniques [4]. Histopathology is one simple test that can be required to effectively diagnose the disease and improve the quality of treatment. However, non-invasive techniques may not always produce effective classifications of these cancers. Additionally, pathologists may grow exhausted from manually grading histological images. Additionally, expert pathologists are required for the precise classification of lung and colon cancer (LCC) subtypes; manual grading may be prone to mistakes. To lessen the workload on pathologists, automated image processing techniques for LCC subtype screening are necessary [5].

There are a lot of different methods for diagnosing cancer symptoms. The amount of data kept in archives is increasing daily because of technical improvements [5]. The rising accessibility of healthcare data offers researchers opportunities to enhance current methods for more in-depth clinical analysis [6]. Artificial intelligence (AI) techniques like machine learning (ML) and DL are the foundation of automatic diagnosis approaches. Researchers have solved numerous health challenges and applications using a variety of traditional machine learning techniques [7,8]. The traditional method for utilizing ML to retrieve and categorize photos in the medical area is solely dependent on manually constructed features created through the feature engineering process. All kinds of characteristics must be used to automatically classify LCC. Filtering and segmentation algorithms can retrieve intensity values and texture descriptors, which are examples of low-level features that are important aspects of an image. Additionally, low-level characteristics can be extracted automatically from LCC images using feature extraction methods, including Haralick characteristics and local binary patterns (LBPs). They function as a base for representations of higher-level features [9].

The characteristics of the surrounding tissue, as well as the tumor's location, size, and shape, are important classification criteria for LCC. Both low-level and high-level features must be considered for automatic tumor categorization to be accurate and reliable. The basic characteristics of images are captured by low-level features, while high-level features offer general and meaningful data [10]. Thus, due to its capacity to deal with these drawbacks and its powerful discrimination capabilities, DL has gained popularity for use in medical testing [11]. These features can be automatically extracted using DL techniques, and they are necessary to organize treatments and make correct diagnoses. Combining these features is required to obtain high accuracy. Convolutional neural network (CNN) is a well-known DL architecture that is frequently employed in this context [12]. Through their numerous deep layers, CNN models may identify high-level features in raw data. In this manner, CNNs can successfully analyze complicated and challenging data. These models have an increasing number of parameters, along with substantial complexity [13]. The complexity and depth of the CNN architecture are what make the models so successful.

Based on the CNN model, explainable artificial intelligence (XAI) is a useful tool in the medical industry that increases the transparency of automatically generated prediction models. It speeds up the creation of predictive models, utilizing expertise in the field and helping to produce results that are understandable to humans [14,15]. There are several ways to show the most active areas and to make a model more explicable. A few examples of these techniques include utilizing XAI algorithms, Shapley Additive Explanation (SHAP), and gradient-weighted class activation mapping (Grad-CAM) [16] for the model's explanatory categorization [17].

It is not easy to process LCC datasets using conventional methods as there are various challenges, such as the following:

- Most of these techniques have substantial computing costs and require a lot of labeled training data.

- Overfitting can happen when the model works well with training data but poorly with new, untested data.
- Risk of poor performance brought on by inaccurate or biased training data.
- DL models' decision-making process is not explainable.

To avoid overfitting or inaccurate diagnosis, it is essential to use DL models that have been thoroughly tested and proven on large, diverse datasets. It is also critical to use techniques like cross-validation (CV) to account for any possible biases in the data used for training to guarantee the model's wider applicability. The overall objective of the proposed method is to be improved by offering precise and early diagnoses that allow for quick and efficient treatment. This paper focuses on categorizing lung and colon cancer subtypes using histopathological images from a dataset named LC25000, which is publicly available on Kaggle. There are a total of five classes in the dataset, which are benign lung tissue, lung adenocarcinomas, lung squamous cell carcinoma, benign colon tissue, and colon adenocarcinomas.

Compared to DL models, the suggested model is less complex and very lightweight, with only eight layers, which makes it appropriate for real-time applications and mobile applications. The originality of the proposed work can be summed up as follows:

- A novel lightweight multi-scale (LW-MS) end-to-end CNN model for the identification of LCC is introduced. The proposed model has 1.1 million trainable parameters and is superior to other models in this field, which need deeper layers to achieve acceptable detection accuracy. This reduces processing time and model complexity, making the system suitable for real-time applications.
- To increase the accuracy and efficiency of multi-class predictions, predictions from multiple layers are concatenated to produce a range of feature maps that function at different resolutions.
- XAI techniques have been integrated into the proposed LW-MS CNN model with its performance metrics analysis. This aspect has frequently been neglected in prior studies.
- A web application system has been developed with the purpose of aiding pathologists and doctors in the diagnosis of histological pictures and offering substantiation for their scientific findings.

The remainder of the article is organized as follows. Section 2 includes a review of the literature on the most current DL advancements related to LCC detection. Section 3 discusses the proposed method in detail. Section 4 presents and thoroughly discusses the experimental setup and achieved results. Section 5 summarizes the proposed model's findings and discusses the model. Finally, Section 6 concludes the research and suggests potential future research directions.

2. Related Works

To classify images of lung tissues from histopathology, a computer-aided system (CAD) method was developed by Nishio et al. [18]. They extracted visual characteristics from two datasets using homology-based processing of images (HI) and traditional texture analysis (TA), and then they assessed the effectiveness of eight ML algorithms. In both datasets, the HI-equipped CAD system outperformed the TA system. They concluded that for CAD systems, HI was significantly more advantageous than TA and that this could lead to the development of an accurate CAD system. Similarly, Mangal et al. [19] developed a CAD system by looking at digital pathology images and using CNN to identify lung and colon cancer. In comparison to deep CNN models that employ TL trained on a similar collection and classical ML models, their experimental results on the LC25000 showed a decent accuracy of 96.61% for the colon and 97.89% for the lung, which were acquired by the CNN using the most recent feature descriptions. Shandilya et al. [20] have created a CAD technique to categorize lung tissue histology pictures. They employed a dataset of histological pictures of lung tissue that was made publicly available for the development and validation of CAD. Multi-scale processing was used to extract image features. Seven

CNN models that had been hyper-tuned before were used in a comparative analysis to predict lung cancer, with ResNet101 achieving the greatest overall accuracy at 98.67%. Masud et al. [21] used DL on histopathology pictures to present a categorization system for five different types of lung and colon tissues. First, image sharpening was applied to pathological example images. A CNN model that was manually tweaked was trained using these features. This model's accuracy performance was reported to be 96.33%.

Similarly, Hatuwal et al. [22] stated a CNN-based technique for classifying histological images to diagnose cancer. They built and trained a neural network with a specific shape. The accuracy in training and validation were reported to be 96.11% and 97.20%, respectively. Similar to this, three CNN models were introduced by Tasnim et al. [23] to assess colon cell imaging data. To calculate the learning rate, the models were developed and put to the test at various epochs. It was demonstrated that the maximum pooling layer has an accuracy of 97.49%, while the average pooling layer has an accuracy of 95.48%. MobileNetV2 outperforms the previous two versions, with a 99.67% accuracy rate and a 1.24% loss rate. However, Sikder et al. [24] have suggested a novel technique for separating, recognizing, classifying, and spotting various malignant cell types in RGB and MRI images. They merged a CNN model with a SegNet method that employs anatomical changes that were better than the regular SegNet model to shorten training times and enhance segmentation results. The proposed method identified cancer cells from several cancer datasets with an average accuracy rate of 93%. They were able to overcome the drawbacks of using different cancer detection methods for MRI and histopathology data.

A CNN model for predicting colon cancer developed by Qasim et al. [25] is notable for its speed and accuracy, with few parameters. They used two separate strategies in their model and then 256 feature maps were created by each. By increasing the number of features at different levels, they were able to increase the accuracy and sensitivity. The same dataset was used to develop and train the VGG16, which was used to evaluate the effectiveness of the suggested strategy. The proposed model's achieved accuracy is 99.6%, while the VGG16's is 96.2%. The results suggest that it was effective in detecting colon cancer. To classify different forms of lung and colon cancer, Talukder et al. [1] have introduced a combination of ensemble attribute-obtaining techniques. Ensemble learning for image filtering and the deep feature extraction method were combined. The proposed hybrid model reportedly had a 99.05% accuracy rate in identifying the possibility of cancer. Hanan et al. [26] have presented the Marine Predator Algorithm with DL (MPADL-LC3) method for classifying lung and colon cancer. This method leveraged MobileNet to generate feature vectors and used CLAHE-based contrast enhancement as a preprocessing step. They introduced MPA as a hyper-parameter optimizer, and a deep belief network was applied for classification. With a maximum accuracy of 99.27%, the comparison research emphasized the improved results of the MPADL-LC3 approach.

Attallah et al. [27] have created a lightweight DL method. To achieve feature reduction and provide a more comprehensive representation of the data, the architecture uses various transformation techniques. In that sense, the SqueezeNet, ShuffleNet, and MobileNet algorithms are fed with HSI. Thus, the features extracted from the model are decreased by using PCA models and the fast Walsh–Hadamard transform (FHWT). It obtained 99.6% accuracy. Al-Jabbar et al. [28] have suggested a method that combines ANN with fusion features and CNN models. The ANN achieved an accuracy of 99.64% with VGG-19 fusion features and handcrafted features. By analyzing the LC2500 dataset, Sameh et al. [29] have built a unique deep network for LCC fine-tuning using pre-trained ResNet101. Hyper-parameter optimizations were used to make these improvements. They obtained 99.84%, 99.85%, 99.84%, 99.96%, and 99.94% scores for their model's precision, recall, F1-score, specificity, and accuracy, respectively. Imran et al. [30] have proposed a deep CNN model for the automated detection and characterization of colon cancer, in which textured images are trained in high resolution without being converted into low-resolution images by changing the classification of binary data in the resultant activation layer to the sigmoid function. They achieved 99.80% recall, 99.87% F1-score, 99.80% accuracy, and 100% precision. Two methods were presented by Kumar et al. [3]. Six approaches for extracting handcrafted aspects based on color, texture, shape, and structure are provided in one method. They also employed seven frameworks for DL that extract features from deep data from histopathology pictures, with the idea of transfer learning. However, compared to manually created features, deep CNN network features show a considerable boost in classifier performance. The LCC tissue was recognized by the Random Forest classifier, with DenseNet-121 retrieving deep features with an accuracy and recall of 98.60%, precision of 98.63%, F1 score of 0.985, and receiver operating characteristic curve (ROC)—area under the ROC Curve (AUC) of 1.

Even though numerous research works show the outstanding accuracy in limited-class and binary classification scenarios, their performance steadily deteriorates as the number of classes rises. This phenomenon results from the growing difficulty of differentiating between many diseases with precisely various characteristics. This restriction makes the models less useful in actual clinical settings where patients may present with a range of lung diseases. Consequently, to perform a multi-class classification of lung and colon diseases with high accuracy and confidence for real-life scenarios, a customized and reliable deep learning framework is needed. In this study, a LW-MS CNN with 1.1 million parameters has been proposed to produce a more promising outcome than the state-of-the-art (SOTA) models. Nevertheless, Grad-CAM and SHAP have been used for showing the effectiveness of the model by detecting ROI despite all the challenges. Also, to the best of the authors' knowledge, only a small number of studies have so far demonstrated these explainable AI methods to show interpretability.

3. Proposed Method

This section discusses the proposed method in depth. This section also covers the datasets used in this research. Furthermore, a detailed discussion of each step of the suggested process, including how the images have been pre-processed, and clear insight into the lightweight multi-scale convolutional cancer network (LW-MS-CCN) is included. Moreover, this section discusses the explainable AI methods used in this paper. Figure 1 outlines the suggested framework for detecting LCC from histopathological images.

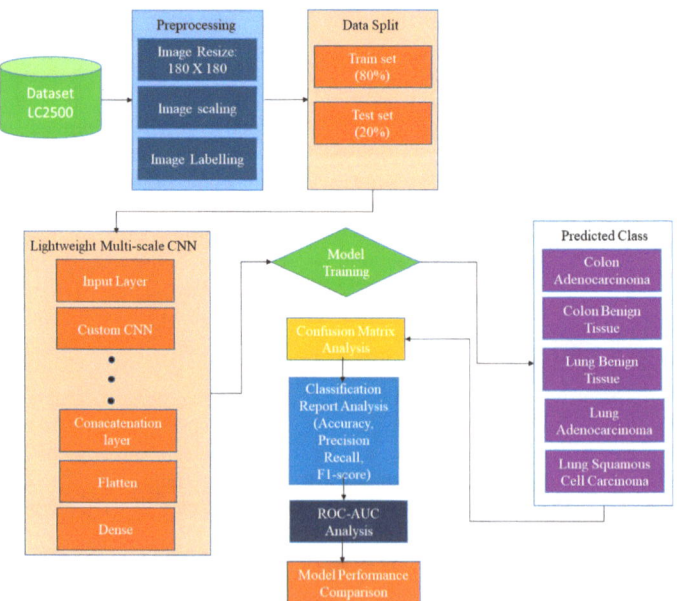

Figure 1. Proposed workflow for visualization of multi-class.

3.1. Dataset Description

The publicly accessible dataset LC25000 [31] was used for this research. For this dataset, a Leica microscope LM190 HD camera coupled to an Olympus BX41 microscope was used to collect each subtype of LCC, taking 250 color photos, 1250 total photos in total, and without any data augmentation. Next, the 250 samples for every subtype of cancer were multiplied via augmentation techniques, like left and right rotations and horizontal and vertical flips, to create 5000 images. Consequently, upon data augmentation, there were already 25,000 typical histopathology photos in the dataset, with 10,000 images showing colon cancer and 15,000 showing lung cancer. Three cell labels, including adenocarcinomas, squamous cell carcinoma, and benign tissue, were present in the lung cancer dataset. Adenocarcinomas and benign tissue are two examples of the two cell labels found in the colon cancer dataset. Prior to applying data augmentation, the photos were resized from their original 1024 × 768 resolution to 768 × 768.

Images of lung, colon, and both types of cancer were used in this work. Hence, there were 25,000 photos in total—10,000 for colon tissue and 15,000 for lung tissue for both the colon and the lung. The distribution of the cancerous dataset is presented in Table 1.

Table 1. Number of images in each class.

Classes	No. of Images
Benign lung tissue	5000
Lung adenocarcinomas	5000
Lung squamous cell carcinoma	5000
Benign colon tissue	5000
Colon adenocarcinomas	5000

3.2. Data Pre-Processing

For medical image analysis, image preprocessing is essential, since the classification performance varies depending on how well the image has been preprocessed [32]. To train the model, the input image was reduced to 180 × 180. To work with image intensity values, the resized image was converted into bgr2rgb, and then the images were converted into a NumPy array. After that, a technique called scaling was used to normalize image intensity values between 0 and 1. By dividing the image array by 255, the image was scaled to reduce computing complexity. Finally, the image label has been added, which is a crucial step because it enables us to recognize cancerous images.

3.3. Lightweight Multi-Scale Convolution Cancer Network

A compact, straightforward, and yet efficient model has been created. The LW-MS-CCN has a single input layer that is 180 × 180 × 3 in size. Globalmaxpooling2D was utilized after each convolutional layer to pick out the most crucial details from each feature map, make them smaller, combine them, and then use this combined information to understand the image better. Globalmaxpooling2D obtains critical values using the max operation. One method to address overfitting difficulties is the dropout layer [33]. Figure 2 shows the design of the LW-MS-CCN model proposed in this method. The dataset was divided according to an 80/20 rule: 80% of the data were used for training and 20% were utilized for testing. There are 12 convolution layers in the backbone CNN. To avoid high dimensionality, using more filters in higher layers was avoided when designing the custom CNN. It can automatically extract characteristics from input images without requiring human interaction [34]. The model's capacity to extract discriminative features from data by utilizing a lightweight CNN as its foundation, convolutional layers as the head for feature extraction across many scales, and filter size optimizations histopathological pictures was improved.

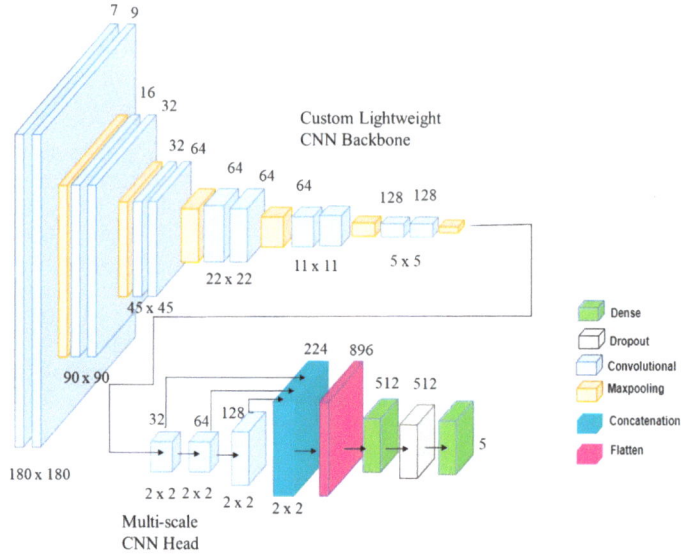

Figure 2. The proposed model visualization for multi-class.

Several convolutional layers are stacked and added to the top of the backbone model to form the CNN head. These convolutional layers are added to the model head, allowing the architecture to be tailored for the extraction of features at various scales. Deeper layers in the CNN head learn more complicated and abstract characteristics, whereas the CNN layers closer to the input learn low-level features like edges and textures. This is essential for classification, since unpredictability in the picture can have unusual appearances. Through multi-scale feature mapping, the model's overall accuracy in classification is improved, making it more resilient and able to recognize unpredictability of different sizes.

The network starts with "Layer 1", which consists of two convolutional layers, both using a 3×3 kernel size and a 180×180 input size. The first layer has 7 filters, and the second layer has 9 filters. After each convolutional layer, a max pool layer with a 2×2 kernel size is applied. In this research, filters of 2×2 size were used to apply max pooling. The maximum value is chosen for each window as the pooling window advances over the feature maps. The spatial dimensions of the feature map are cut in half by utilizing a pooling window of 2×2 and a stride of 2 [35]. In order to achieve translation invariance and resilience against minute spatial alterations, max pooling is successful in capturing the most important characteristics within local areas. Next, "Layer 2" has two convolutional layers, where the input size is reduced to 90×90. The first layer in this layer has 16 filters, and the second layer has 32 filters. Again, after each convolutional layer, a max pool layer is used. "Layer 3" further reduces the input size to 45×45 and contains two convolutional layers with 256 filters each. After each convolution, max pool is applied. "Layer 4" consists of three convolutional layers, which results in a reduction in the input size to 22×22. The initial two layers are equipped with 32 filters each, while the third layer is equipped with 64 filters. "Layer 5" consists of two convolutional layers with an input size of 11×11. The starting layer is equipped with 64 filters, while the subsequent layer also employs 64 filters. Finally, "Layer 6" includes two convolutional layers with an input size of 5×5. Both layers have 128 filters. Max pool is used after each convolution. In summary, the custom CNN architecture is built with a consistent pattern of 3×3 convolutional layers, followed by 2×2 max pool layers, progressively reducing the input size and increasing the number of filters in deeper layers.

This design aims to capture and learn hierarchical features from the input image data, ultimately leading to more accurate predictions for the given classification task. From

Table 2 and Figure 2, it can be seen that the max_pooling2d_5 (last max pooling layer of CNN block) is connected to conv2d_11 (head of multi-scale CNN). conv2d_11, conv2d_12, and conv2d_13 are all the layers connected together in a concatenate layer, which enables the network to simultaneously use data from several scales. After that, flatten layer is used to convert all the features into 1D vector. Then, as the flattened characteristics move through dense layers, high-level abstraction and pattern detection are made easier. After using dropout regularization to reduce overfitting, a final dense layer with softmax activation is used to produce class probabilities. The model can learn rich representations of the input data across multiple levels of abstraction according to the multi-scale CNN design, which efficiently combines features from different scales.

Table 2. An overview of the proposed model that includes the number of parameters and information about each layer.

Layer (Type)	Output Shape	Params	Connected to
Input 1	(None, 180, 180, 3)	0	
conv2d	(None, 180, 180, 7)	196	Input 1
conv2d_1	(None, 180, 180, 9)	576	conv2d
max_pooling2d	(None, 90, 90, 9)	0	conv2d_1
conv2d_2	(None, 90, 90, 16)	1312	max_pooling2d
conv2d_3	(None, 90, 90, 32)	4640	conv2d_2
max_pooling2d_1	(None, 45, 45, 32)	0	conv2d_3
conv2d_4	(None, 45, 45, 32)	9248	max_pooling2d_1
conv2d_5	(None, 45, 45, 64)	18,496	conv2d_4
max_pooling2d_2	(None, 22, 22, 64)	0	conv2d_5
conv2d_6	(None, 22, 22, 64)	36,928	max_pooling2d_2
conv2d_7	(None, 22, 22, 64)	36,928	conv2d_6
max_pooling2d_3	(None, 11, 11, 64)	0	conv2d_7
conv2d_8	(None, 11, 11, 64)	36,928	max_pooling2d_3
conv2d_9	(None, 11, 11, 128)	73,856	conv2d_8
max_pooling2d_4	(None, 5, 5, 128)	0	conv2d_9
conv2d_10	(None, 5, 5, 128)	147,584	max_pooling2d_4
conv2d_11	(None, 5, 5, 128)	147,584	conv2d_10
max_pooling2d_5	(None, 2, 2, 128)	.	conv2d_11
conv2d_11	(None, 2, 2, 32)	32,896	max_pooling2d_5
conv2d_12	(None, 2, 2, 64)	18,496	conv2d_11
conv2d_13	(None, 2, 2, 128)	73,856	conv2d_12
concatenate	(None, 2, 2, 224)	0	conv2d_13, conv2d_12, conv2d_11
flatten	(None, 896)	0	concatenate
dense	(None, 512)	459,264	flatten
dropout	(None, 512)	0	dense
dense_1	(None, 5)	2565	dropout
Total params	11,05,353		
Trainable params	11,05,353		
Non-trainable params	0		

Table 2 provides an overview of the proposed LW-MS-CCN model architecture, detailing the number of parameters and information about each layer. The model comprises a series of convolutional and pooling layers, culminating in fully connected layers. The total number of parameters is 1,105,353, all of which are trainable. Table 3 outlines the hyper-parameters utilized in training the LW-MS-CCN model. Selecting hyper-parameters for a CNN involves a combination of domain knowledge, empirical experimentation, and sometimes trial and error. For classification problems with multiple classes, a categorical cross-entropy technique is commonly used. The 'sparse' variant is used when the labels are integers rather than one-hot encoded. A small learning rate, like 0.0001, is often chosen to ensure stable convergence. Learning rates are often tuned through experimentation, and techniques like learning rate schedules or adaptive learning rate methods may be employed. Too few epochs may result in under fitting, while too many may lead to overfitting. The optimal number of epochs is determined through training on a validation set, and techniques like early stopping may be used to prevent overfitting. Smaller batch sizes often lead to faster convergence, and larger batch sizes can provide a regularizing effect and speed up training. Batch sizes are selected based on computational constraints and experimentation to find a balance between speed and model performance. Shuffling the training data every epoch helps prevent the model from memorizing the order of examples and improves generalization. These hyper-parameters contribute to the effective learning and convergence of the LW-MS-CCN model, ensuring its successful application to the given task.

Table 3. Hyper-parameters of the proposed LW-MS-CCN model.

Parameters	Value
Loss function	Sparse-categorical-cross-entropy
Initial learning rate	0.0001
No. of epochs	100
Batch size	16
Shuffle	Every epoch

3.4. XAI

XAI describes the ability of an AI system to provide understandable and interpretable explanations for its decisions and actions, enhancing transparency and trustworthiness [36]. XAI in medical imaging helps bridge the gap between AI technology and medical practitioners, making AI-assisted diagnosis and treatment more trustworthy, understandable, and reliable. In XAI, the "black-box" concept refers to AI models that make decisions without providing clear or understandable reasons, while the "white-box" concept pertains to AI models that are transparent and provide interpretable explanations for their decisions, making their internal workings accessible and comprehensible to humans [37]. The XAI methods that have been used in this paper are explained below.

3.4.1. Grad-CAM

Grad-CAM is a visualization method used in DL for understanding model decisions, especially in computer vision tasks. The approach utilizes the gradients of the target class with respect to the final convolutional layer to generate a heatmap. The final convolutional layers are chosen for their balance between spatial information and high-level semantics, allowing for the visualization of class-specific details in the input image [17,38].

By emphasizing regions where the model focuses its attention to create distinctive patterns, Grad-CAM leverages the rich information in the final layer. The algorithm computes gradients of the class score with respect to feature maps, performs weighted combinations, and generates a heatmap, effectively highlighting key areas in the input image that contribute to the target class prediction [39].

$$L_c^{CAM} = \sum_i \sum_j w_c^k A_k^{ij}$$ (1)

where L_c^{CAM} is the localization map for class c in GradCAM. w_c^k is the weight associated with the k-th feature map for class c. A_k^{ij} is the activation of the k-th feature map at a spatial location (i, j). $\sum_i \sum_j$ is double summation over spatial dimensions. This equation represents the GradCAM formulation for obtaining a class-specific localization map by combining the feature weights $\left(w_c^k\right)$ with the activations $\left(A_k^{ij}\right)$ from different spatial locations.

3.4.2. SHAP Visualization

SHAP aims to explain predictions in ML models by calculating the contribution of each feature to a given prediction instance. It utilizes coalitional game theory to derive Shapley values, representing the fair contributions of individual features to the prediction. In this technique, the feature values of a data instance act as players in a coalition, and Shapley values help distribute the prediction fairly among these features. Players can be single feature values or collections of feature values, such as super-pixels in images. SHAP introduces a novel approach by presenting Shapley values as a linear model and linking them with the values of local interpretable model-agnostic explanations. This additive feature attribution model provides a comprehensive explanation of the prediction [16].

$$g(z') = \phi_0 + \sum_{j=1}^{M} \phi_j z'_j$$ (2)

where g is the explanation model, $z' \in \{0,1\}^M$ is the coalition vector, M is the maximum coalition size, and $\phi_j \in \mathbb{R}$ is the feature attribution for a feature j; the Shapley values. The expression is a linear combination of input features z'_j weighted by coefficients ϕ_j, and the result is adjusted by an intercept term ϕ_0.

4. Result Analysis

In this section, all the experimental setups and results of this research will be described in detail.

4.1. Experimental Setup

The experimental setup of the proposed system is described in this subsection. Table 4 accommodates the system specifications upon which the proposed work has been based. All coding operations have been performed in Google Colab, which has a backend of Keras with TensorFlow, and the disk space for it is 78.2 GB. The GPU used was a Nvidia Tesla T4 with a RAM size of 15 GB. In this study, the operating system was Windows 11, and for visualization in web environments, Gradio Library was used.

4.2. Performance Metrics of the Proposed Framework

The confusion matrix is a technique for assessing how well ML categorization works. The terms TP (true positive) and TN (true negative) accurately reflect expected positive values. TP represents a correctly predicted positive value, FP (false positive) represents a false positive value, and FN (false negative) represents a false negative value. They are highly helpful in determining the ROC curve, F1-score, accuracy, recall, and precision.

The most obvious performance statistic is accuracy, which is directly proportional to the number of properly predicted observations over the total number of observations [40,41].

$$\text{Accuracy} = \frac{TP + TN}{TP + FP + TN + FN}$$ (3)

Table 4. System specifications of the proposed framework.

Features	Specifications
Programming Language	Python (version-3.10.12)
Environment	Google Colab
Backend	Keras with TensorFlow
Disk Space	78.2 GB
GPU RAM	15 GB
GPU	Nvidia Tesla T4
System RAM	12.72 GB
Operating System	windows 11
Input	LCC Images
Input Size	180 × 180
Web Development Tool	Gradio Library

Precision is defined as the proportion of accurately anticipated positive values to all positively predicted values. It is shown as follows:

$$\text{Precision} = \frac{\text{TP}}{\text{TP} + \text{FP}} \quad (4)$$

Recall [42] is defined as the ratio of all the actual values to the values that were positively predicted and successfully made. It is demonstrated as follows:

$$\text{Recall} = \frac{\text{TP}}{\text{TP} + \text{FN}} \quad (5)$$

The harmonic mean of a classification problem's precision and recall scores is known as the F1-score [43]. The F1-score is shown as follows:

$$\text{F1} - \text{score} = \frac{2 \times \text{precision} \times \text{recall}}{\text{precision} \times \text{recall}} \quad (6)$$

ROC curves are two-dimensional graphs that are used for evaluating and understanding classifier performance [44]. Classifiers are graded and chosen according to particular user requirements, which are often associated with changeable error costs and accuracy expectations [45,46]. The sensitivity or specificity interchanges in a classifier for all possible classification thresholds are displayed in detail on the ROC graphs. The AUC measures the degree of distinction, whereas the ROC is a likelihood curve. It demonstrates a model's ability to discriminate across various groups. Plotting the false positive rate on the x-axis corresponds to the genuine positive rate on the y-axis. An AUC near 1 suggests that the expected model performs well in terms of class label separability, whereas an AUC near 0 denotes a poorly anticipated model. Actually, the word "lousy" means that the effect is being reflected [47]. It is a method for demonstrating the effectiveness of a classification [48]. The best classifiers are those with greater ROC curves [49].

$$\text{AUC} = \frac{1}{2}\left(\frac{\text{TP}}{\text{TP} + \text{FN}} + \frac{\text{TN}}{\text{TN} + \text{FP}}\right) \quad (7)$$

Specificity is a metric that evaluates the ability of a model to correctly detect true negatives within each available class. The mathematical expression can be expressed as follows [48].

$$\text{Specificity} = \frac{\text{TN}}{\text{TN} + \text{FP}} \quad (8)$$

The XAI performance metrics include normalized root mean square error (nRMSE), which is a standardized form of the root mean square error (RMSE). The metric calculates the mean size of the discrepancies between projected and actual values, which is then adjusted based on the data's range. It offers a standardized way to quantify errors, allowing for meaningful comparisons across diverse datasets. The structural similarity index (SSIM) is a perceptual model that takes into account brightness, contrast, and structure. The normalized index quantifies the degree of structural similarity between two images. The values go from -1 to 1, with a value of 1 denoting photo that are identical. The multi-scale structural similarity index (MS-SSIM) is an extension of SSIM that takes into account changes in image resolution by using multiple scales. It offers a more adaptable assessment of structural similarity by taking into account variations in image viewing conditions. Using the k-fold CV technique, k, smaller sets are created from a training set. The plan is to train a model on each of the k "folds" and then validate it using the remaining data. Using k-fold CV, the average of the values computed in the loop is then included as an evaluation metric. For LCC detection experiments, k-fold CV with a value of $k = 5$ has been used. Five distinct folds are created from the dataset, and each is used as a testing component while the dataset is being folded. The dataset is divided into 80% for training and the remaining 20% for testing in a k-fold.

4.3. Performance Evaluations

In this section, the performance of the proposed model on the LC25000 dataset is demonstrated. The performance is evaluated with different performance metrics as well as by using XAI like Grad-CAM and SHAP to evaluate the proposed model based on which portion of the image the decision is made on and what predicting the class is.

Performance Evaluation of Lung and Colon Cancer

Table 5 shows the fold-wise outcomes for each class using the LW-MS-CCN network. The results consistently demonstrate the validity and robustness of the model at all folds. Notably, Fold 4 comes out with superior accuracy and specificity when looking at the average findings over all five folds. Thus, for emphasis, the improved performance measures in Fold 4 have been bolded. This highlights how important Fold 4 is for demonstrating the potential of the model. This emphasizes the significance of Fold 4 in showcasing the model's capabilities.

In Figure 3, the confusion matrix of the LW-MS-CCN model on (a) Fold 1, (b) Fold 2, (c) Fold 3, (d) Fold 4, and (e) Fold 5 for lung and colon cancer classifications is shown, and the insightful analysis of the confusion matrices reveals noteworthy findings regarding the performance of different folds. More precisely, the analysis of the confusion matrix for Fold 4, as shown in Figure 3d, reveals that this fold has performed exceptionally well in terms of its ability to make accurate predictions. The focus of the observation is on the model's capacity to reduce false positive values, indicating a high level of precision in its predictions. When the true label is Col_Ade, the LW-CNN model correctly predicted Col_Ade instances 1021 times. Furthermore, within the same category, it exhibited 1000 accurate predictions for Col_Ben. More precisely, the LW-CNN model correctly identified 985 instances of Lun_Ben where the real label was Lun_Ben. Furthermore, within the same category, it accurately predicted 985 instances of Lun_Ade when the true label was also Lun_Ade. Once again, when the actual label was "Lun_Squ", and the model correctly predicted the "Lun_Squ" class 1001 times. The model produced inaccurate predictions on four occasions where the actual label was "Lun_Ade", and it incorrectly predicted "Lun_Squ". The emphasis on Fold 4 as the top performer was based on the comparison of false positive rates across different folds. The lower incidence of false positives in Fold 4, as evidenced in the confusion matrix, signifies a superior ability of the model to avoid incorrect positive predictions. This characteristic is particularly crucial in medical applications, where minimizing false positives is essential for ensuring the accuracy and reliability of diagnostic outcomes.

Overall, the findings from the confusion matrix in Figure 3d highlight the commendable performance of Fold 4, making it a standout in terms of predictive accuracy and reliability.

Table 5. Performance metrics analysis of LW-MS-CCN model for each class of the LCC dataset.

Fold Number	Class	Accuracy (%)	Precision (%)	Recall (%)	F1-Score (%)	Specificity (%)	AUC
Fold = 1	Col_Ade	99.92	99.71	99.90	99.80	99.37	100
	Col_Ben	99.92	99.90	99.70	99.80	99.68	100
	Lun_Ben	99.98	100	99.90	99.95	99.57	100
	Lun_Ade	99.24	97.69	98.48	98.09	100	100
	Lun_Squ	99.26	98.50	97.81	98.15	100	100
	Average	99.66	99.16	99.16	98.16	99.72	100
Fold = 2	Col_Ade	99.80	99.42	99.61	99.51	99.40	100
	Col_Ben	99.86	99.80	99.50	99.65	99.81	100
	Lun_Ben	99.94	99.80	99.90	99.85	99.71	100
	Lun_Ade	99.08	97.52	97.72	97.62	97.89	100
	Lun_Squ	99.16	98.01	97.82	97.91	99.61	100
	Average	99.57	98.91	98.91	98.91	99.68	100
Fold = 3	Col_Ade	99.84	99.58	99.58	99.58	99.68	100
	Col_Ben	99.90	99.80	99.71	99.75	99.57	100
	Lun_Ben	99.94	99.80	99.90	99.85	99.79	100
	Lun_Ade	99.10	97.89	97.59	97.74	100	100
	Lun_Squ	99.14	97.78	98.06	97.92	100	100
	Average	99.58	98.97	98.97	98.97	99.81	100
Fold = 4	Col_Ade	99.82	100	99.80	99.80	100	100
	Col_Ben	99.94	99.70	99.60	99.80	100	100
	Lun_Ben	99.98	99.90	99.90	99.95	100	100
	Lun_Ade	99.37	98.69	99.50	98.09	99.60	100
	Lun_Squ	99.45	98.89	98.89	98.15	100	100
	Average	99.71	99.39	99.54	99.16	99.92	100
Fold = 5	Col_Ade	99.75	99.42	99.58	99.40	99.80	100
	Col_Ben	99.80	99.75	99.71	99.30	100	100
	Lun_Ben	99.90	99.66	99.90	99.40	100	100
	Lun_Ade	99.50	97.77	97.59	98.55	98.98	100
	Lun_Squ	99.30	97.90	98.06	98.92	98.67	100
	Average	99.65	98.90	97.77	99.11	99.49	100

Figure 4 shows the ROC curve that was achieved for each fold, and this shows that Fold 4 has achieved AUC 1 for each class, showing the best result. The other folds have also achieved great results in the ROC curves. Figure 5 shows the training and testing accuracy curve and Figure 6 shows the training and testing loss for all the folds. Figure 5d shows the accuracy curve for Fold 4, and it shows least fluctuations in the curve, making it also best training and test accuracy result. Seeing the curves, the performance of the proposed model on the dataset can be visualized. The consistent and stable natures of the Fold 4 accuracy curve and loss curve suggest a robust and reliable performance, signifying its strong generalization capability to classify correctly LCC even further. In Figure 6d, the loss curve is depicted for Fold 4. This time, the loss is reduced, with the least sudden fluctuations. The curve proves the proposed model's capability to reduce loss with time and increase accuracy with time.

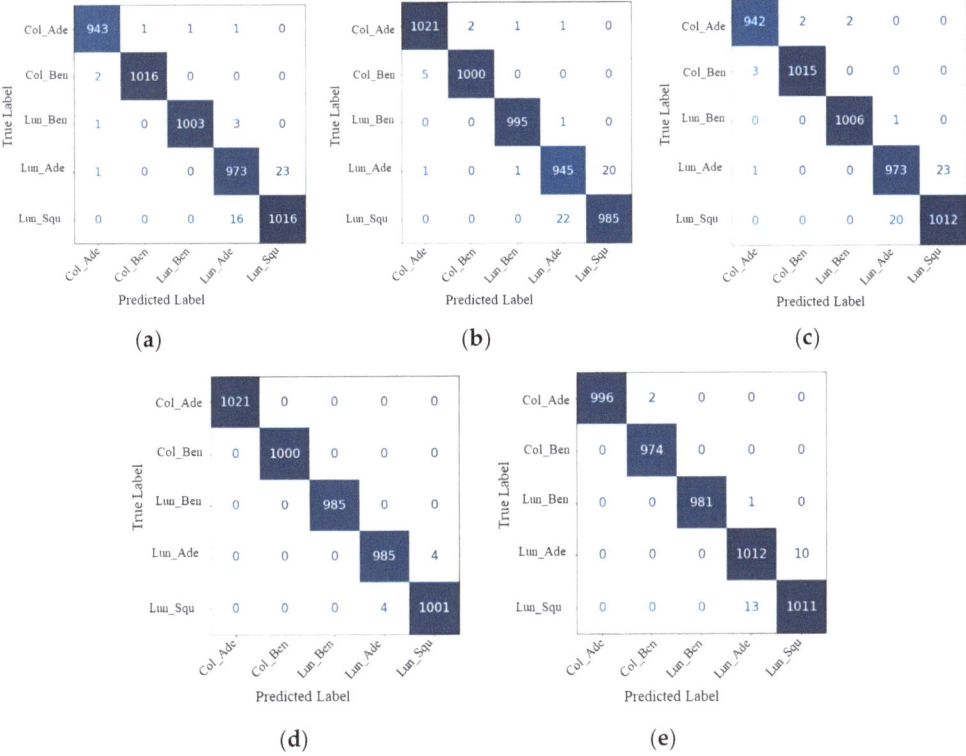

Figure 3. Confusion matrix of LW-MS-CCN model on (**a**) Fold 1, (**b**) Fold 2, (**c**) Fold 3, (**d**) Fold 4, and (**e**) Fold 5 for lung and colon cancer classifications.

4.4. XAI Visualization

The applied explainable DL algorithm Grad-CAM, which is explained in previous sections, can be observed to retrieve the information after the final convolution and transform it into a heatmap. This map displays the regions in which the verdict was concentrated to reach its decision. This heatmap is superimposed on the original image to help the medical practitioner recognize the regions that affect the outcome. Before using the softmax technique (activating the class with the greatest value and inhibiting the others), the numerical result of the classifier is also taken from the system's final layer.

The size of the original image is 180 × 180 pixels, whereas the resolution of the heatmap is 5 × 5 pixels (because of the final convolution layer before maximum pooling). As a result, the heatmap image needs to be over scaled before being overlaid on the original. This results in some portions of the heatmap not fitting completely with the original due to the decimals produced during this process of resolution improvement; nevertheless, when observing them, it is clear which parts of the image it refers to. When it comes to model prediction in Figure 7, the red color on the maps denotes greater attention paid to those locations, while the blue color denotes that less attention was paid to those regions. Each image belongs to a different class, so the red color as well as the blue color heatmap in each image are situated in different positions of that image.

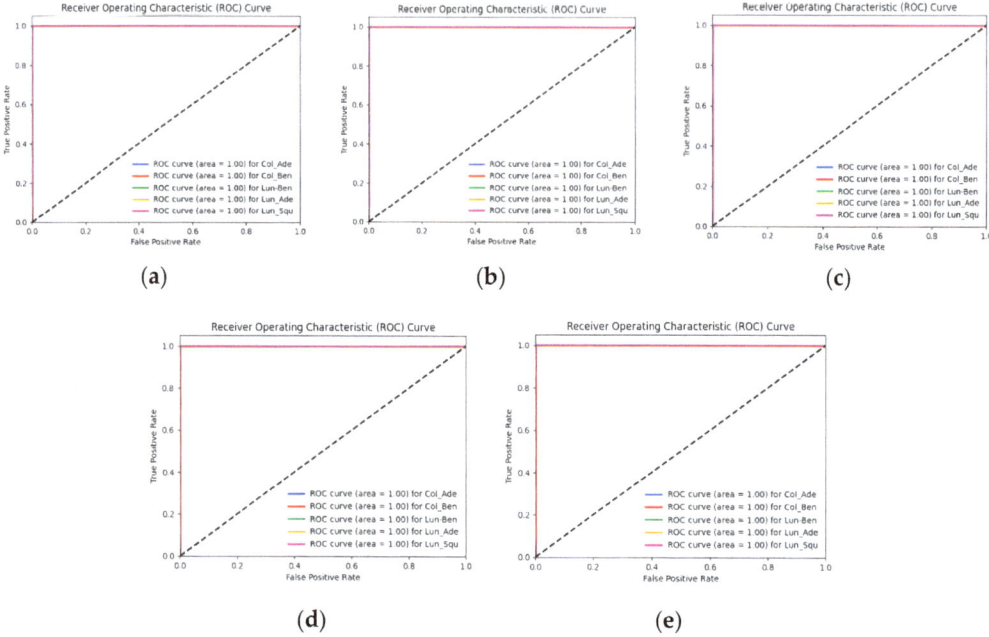

Figure 4. ROC curve of LW-MS-CCN model on (**a**) Fold 1, (**b**) Fold 2, (**c**) Fold 3, (**d**) Fold 4, and (**e**) Fold 5 for lung and colon cancer classifications.

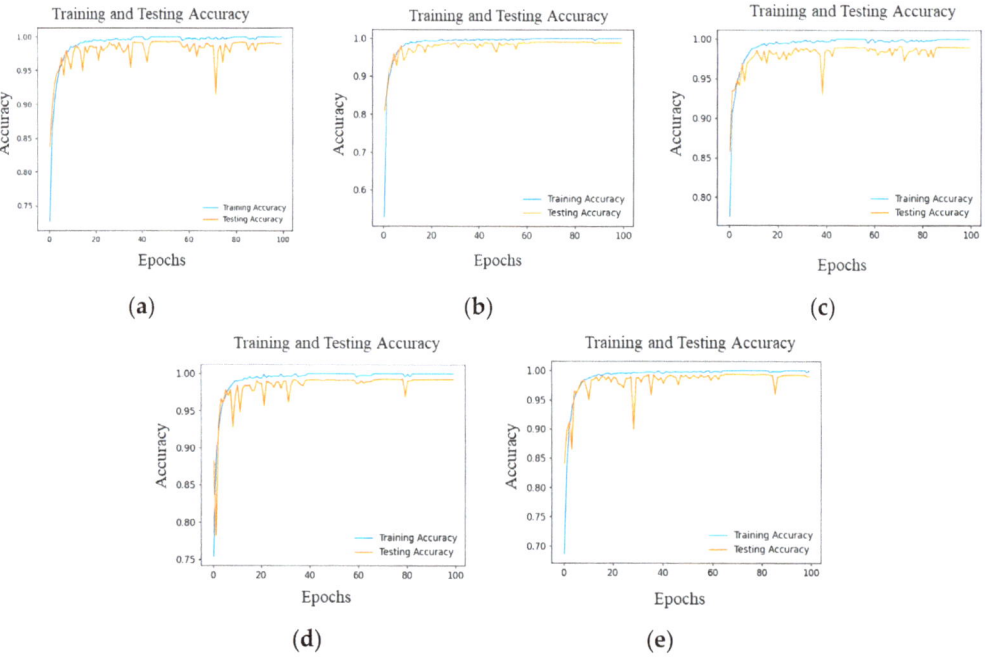

Figure 5. Accuracy curve of LW-MS-CCN model on (**a**) Fold 1, (**b**) Fold 2, (**c**) Fold 3, (**d**) Fold 4, and (**e**) Fold 5 for lung and colon cancer classifications.

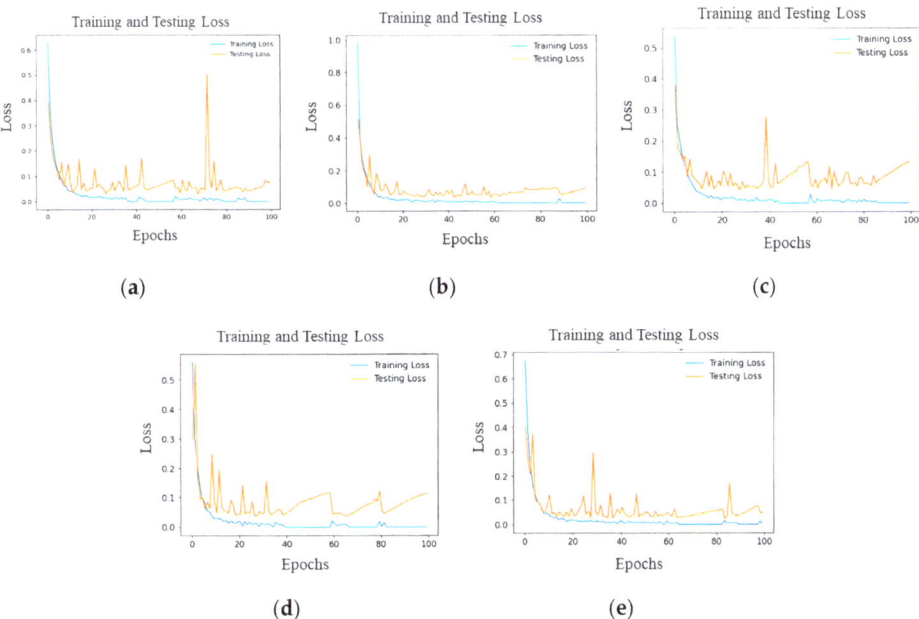

Figure 6. Loss curve of LW-MS-CCN model on (**a**) Fold 1, (**b**) Fold 2, (**c**) Fold 3, (**d**) Fold 4, and (**e**) Fold 5 for lung and colon cancer classifications.

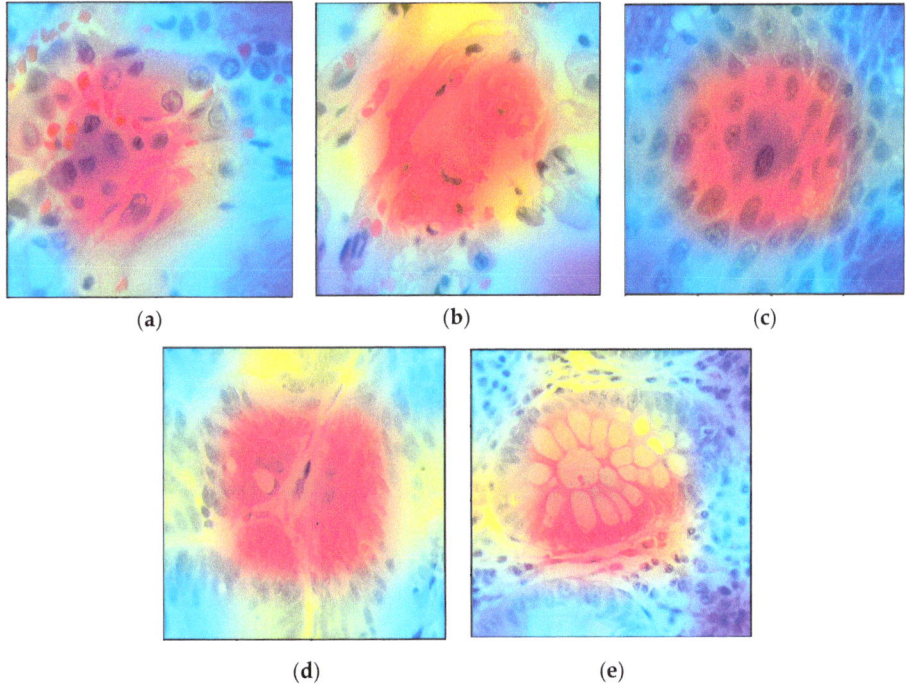

Figure 7. Grad-CAM visualization of (**a**) lung adenocarcinoma, (**b**) lung benign tissue, (**c**) lung squamous cell carcinoma, (**d**) colon adenocarcinoma, and (**e**) colon benign tissue.

This not only aids in model interpretability but also empowers healthcare practitioners to make informed decisions with additional information based on XAI-assisted analysis. By providing visual justification for the model's predictions, trust in the explainability and accuracy of the proposed model is the aim, ultimately facilitating its integration into clinical workflows for improved patient care.

Grad-CAM focuses on identifying the "class-discriminative" regions in the image, which are the areas that are most relevant to the predicted class. The visualization produced by Grad-CAM is specific to the model's prediction for a particular class. The SHAP results for each group explanation are set against a clear gray background. Here, the Shapley value represents the contribution of that feature to the model's prediction. SHAP provides a comprehensive explanation for individual predictions by quantifying the impact of each feature on the output.

In order for the model to determine the SHAP values for a particular set of instances, a SHAP explanation has to be first created. A customized SHAP partition explainer specifically made for deep learning models was made by using the SHAP—partition explainer function. For each instance in the dataset, the SHAP values show how much each pixel contributes to the model's output. The SHAP data are arranged in matrices, where columns stand for features and rows for instances. The features that push the prediction towards the positive class are shown by positive values, and those that push towards the negative class are indicated by negative values. Figure 8 shows an image plot of all five classes, generated by using the SHAP values. The plot shows the original image, with blue and red highlights in specific areas. Positive contributions to the class prediction are indicated by red areas, and negative contributions are indicated by blue areas. Blue zones reduce the likelihood of guessing a class, but red regions increase it. In Figure 8, a lower SHAP value to the left indicates a lower prediction value, while a higher SHAP value to the right indicates a greater prediction value. It can be seen in Figure 8 that for the Colon_Adenocarcinoma class, the prominence of red areas (positive SHAP values) in the plot signifies a tendency toward the prediction of the Colon_Adenocarcinoma class, indicating the correct prediction. In the second row, it has red pixels both in the Colon_Adenocarcinoma and the Colon_Bengin_Tissue classes, which is confusing. For Colon_Bengin_Tissue, all the pixels are red, whereas in Colon_Adenocarcinoma there are still some negative SHAP value. So, it is clear that the second row is Colon_Bengin_Tissue. The last row does not properly explain this, which is a limitation of the model.

Table 6 shows a full breakdown of how well the three explainability methods, Grad-CAM, and SHAP perform compared to a standard measure. The reference (Ref.) value column shows the optimal score for each parameter. This score is used to generate heatmaps that provide a clear and balanced representation of the data. It can be highlighted that a smaller value of nRMSE is preferable, and that higher values for SSIM and MS-SSIM indicate better similarity, with a value of 1 representing perfect similarity. Lower nRMSE values mean that the model is more accurate, and SHAP has the lowest number at 0.0678 ± 0.0245. Higher SSIM and MS-SSIM numbers indicate better structural similarity, and SHAP does very well in both, showing that it is good at capturing image features and is better than other methods.

Table 6. Performance metrics analysis of XAI methods.

Metric	Ref. Value	Grad-CAM	SHAP
nRMSE	0.0	0.0789 ± 0.0156	0.0678 ± 0.0245
SSIM	1.0	0.6198 ± 0.0259	0.7541 ± 0.0455
MS-SSIM	1.0	0.8934 ± 0.0754	0.8874 ± 0.0921

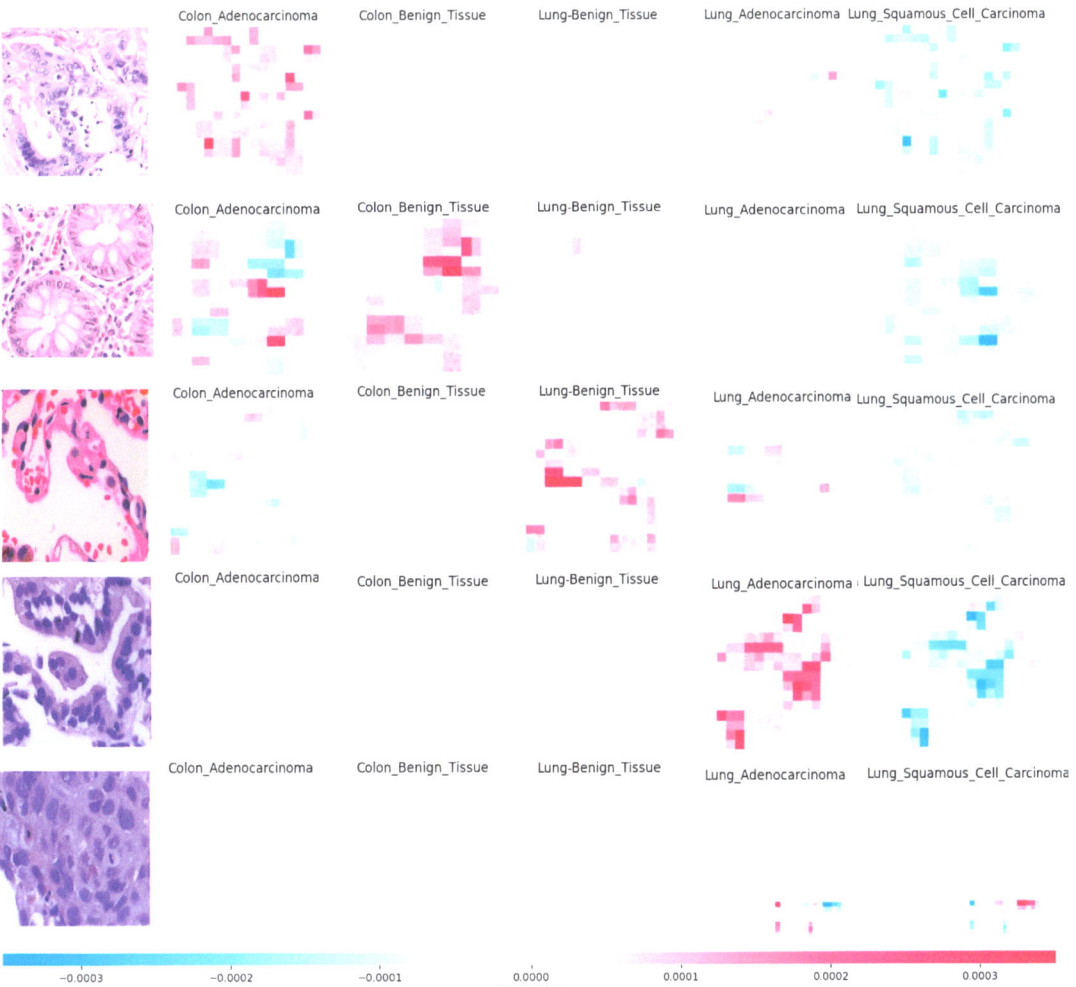

Figure 8. SHAP partition explainer of all five classes.

4.5. Web Application

In the context of LCC detection, the interpretability of the DL model is crucial for both medical professionals and patients. Gradio provides an intuitive and interactive platform that allows users, including non-technical stakeholders in the medical field, to comprehend and trust the predictions of the model. Gradio's user-friendly interfaces make it possible for oncologists, radiologists, and other healthcare professionals to interact with and understand the model without needing extensive technical expertise. Gradio simplifies the communication process between the model and the web interface. When a user interacts with the Gradio interface, the input data, the LCC image, are sent to the model. The model processes the input and generates predictions. Gradio receives the model's output and updates the web interface to display the results in a user-friendly format, which here is in text format, showing the predicted class. Utilizing Gradio's image input components allows for users to upload medical images for analysis, displaying the model's output and indicating the predicted class or probabilities for different cancer types.

In Figure 9, the web-application visualization can be seen, wherein the input images are classified correctly by the proposed model. So, in this way, from a user point of view, real-time prediction can be realized by the proposed model. The web application visualization demonstrates the accurate classification of input images by the proposed model, providing real-time predictions for different classes. Specifically:

(a) For colon adenocarcinomas, the proposed model correctly identifies and predicts this category.
(b) In the case of benign colon tissue, the proposed model accurately classifies the input images as such.
(c) Similarly, for benign lung tissue, the proposed model correctly predicts and categorizes the images.
(d) When it comes to lung adenocarcinomas, the proposed model reliably classifies the input images with precision.
(e) Finally, for lung squamous cell carcinoma, the proposed model consistently provides accurate real-time predictions.

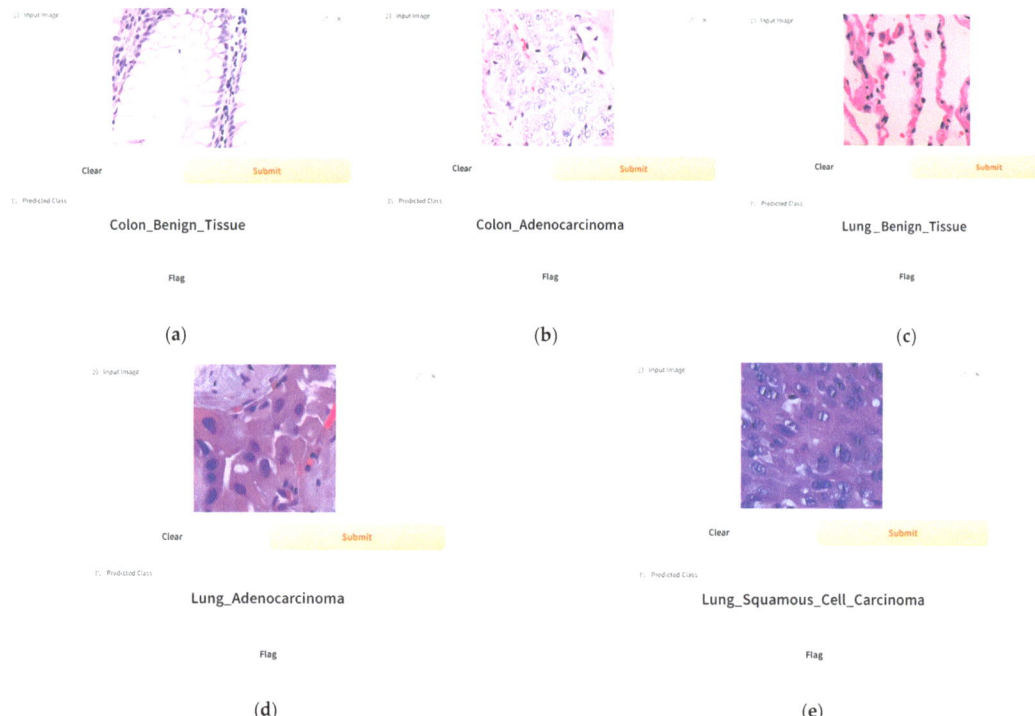

Figure 9. Web-application visualization of (**a**) benign colon tissue, (**b**) colon adenocarcinomas, (**c**) benign lung tissue, (**d**) lung adenocarcinomas, and (**e**) lung squamous cell carcinoma.

This web application, aided by Gradio, showcases the effectiveness of the proposed model from the user's perspective, ensuring reliable and precise predictions across various classes.

5. Discussion

To produce both quantitative as well as qualitative analyses, the suggested model was contrasted with other methods found in the literature. Table 7 indicates how well the proposed method performed on the lung and colon disease datasets.

Hasan et al. [3] have used custom CNN and PCA, and they achieved 99.80% accuracy for colon cancer only. XAI and end-to-end solutions were not used by the authors. On the other hand, this research paper provides the best solution for multi-class classification, provides an end-to-end pipeline solution, and uses explainable AI for visualization. Kumar et al. [30] have used DenseNet121 for feature extraction and an RF ML classifier to predict the actual class techniques. Mehmood S. et al. [50] have performed image enhancement and used AlexNet for training the data, achieving 98.40% accuracy. They used too many parameters. On the contrary, this research used 0.9 million parameters, which reduced the computational complexity. Masud M. et al. [21] have used traditional ML classifiers and achieved 96.33% accuracy, which is relatively low compared to other SOTA methods, whereas 99.20% accuracy was achieved in this paper. Hatuwal B. K. et al. [22] have also used a custom CNN, but it was only used for lung cancer. They achieved an accuracy of 97.20%. The hybrid ensemble learning technique was used by Talukder M. A. et al. [1], and it achieved 99.30% accuracy. Bukhari et al. [51] have used the pre-trained model ResNet50, which indicates that having more parameters also increases the computational complexity. The accuracy is also very low, at 93.13%. Balasundaram et al. [38] have made AdenoCanNet and AdenoCanSVM. They achieved 99% accuracy. The above-mentioned methods require different algorithms to detect ROI, but the model in this research article can detect ROI with the help of XAI. In comparison to [52], the proposed LW-MS CNN demonstrates superior efficiency with a parameter count of only 1.1 million, a substantial reduction from the 4.1 million parameters in the reference model. A model with fewer parameters requires less computational resources during training and inference. By incorporating convolutional layers with varying receptive field sizes, the model can capture both local and global features present in the input data. This multi-scale approach facilitates the detection of subtle abnormalities and distinctive characteristics across different scales, enhancing the model's sensitivity and discriminative power. Consequently, the model can provide a more comprehensive representation of the underlying pathology, leading to improved accuracy in cancer detection. This is especially beneficial for scenarios with limited computational power, such as edge devices or mobile applications. Training a model with fewer parameters is generally faster than training a larger model. This allows for quicker experimentation, faster model iteration, and reduced training time. Models with fewer parameters are less prone to overfitting, especially when dealing with limited data. The reduced parameter count makes the proposed model more suitable for deployment in resource-constrained environments, where memory and computation resources are limited.

Table 7. Comparison between the proposed model and other previous models.

References	Cancer Type	Methods	XAI	Accuracy	Precision	Recall	F1-Score
[34]	Lung and colon	Feature extraction	Yes	95.60%	95.8%	96.00%	95.90%
[21]	Lung and colon	CNN	No	96.33%	96.39%	96.37%	96.38%
[22]	Lung	CNN	No	97.20%	97.33%	97.33%	97.33%
[19]	Lung	CNN	No	97.89%	-	-	-
[19]	Colon	CNN	No	96.61%	-	-	-
[52]	Colon	CNN	No	99.50%	99.00%	100%	99.49%
[38]	Lung and colon	CNN	No	99.00%	-	-	-
[53]	Colon	CNN	No	99.21%	99.18%	98.23%	98.70%
[53]	Lung	CNN	No	98.30%	97.84%	98.16%	97.99%
Proposed	**Lung and colon**	**LW-MS-CCN**	**Yes**	**99.20%**	**99.16%**	**99.36%**	**99.16%**

The achievements and limitations of the proposed model can be highlighted as follows:

- The proposed model achieved an accuracy of 99.20% for the overall LCC class classification (five classes), indicating that it can detect LCC with greater accuracy than similar DL models.
- The suggested model is more appropriate for real-time applications, such as mobile or Internet of Medical Things (IoMT) devices, because it has fewer computationally expensive parameters (1.1 million) compared to existing DL models.
- The multi-scale aspect of the proposed model plays a pivotal role in extracting features at different hierarchical levels, thereby enriching its ability to discern intricate patterns inherent in LCC images.
- When compared to existing DL models, the suggested model is an end-to-end model since it can complete feature extraction and classification in a single pipeline. This reduces the system's complexity.
- The CV technique was employed to train and evaluate the suggested model, with the aim of reducing overfitting and enhancing the model's generalizability by applying it to three combinations of the LC25000 dataset.
- The integration of XAI algorithms, such as Grad-CAM and SHAP, enhances the model's interpretability by providing diverse and complementary insights into feature importance, enabling a more comprehensive understanding of the model's decision-making process.

Limitations:

- The proposed model has undergone testing on an LCC dataset using cross-validation methods. However, it has not yet undergone complete validation for application in real clinical scenarios. Additional clinical trials are necessary to validate the reliability and precision of the model in real-life scenarios.
- Despite the advancements in DL, the diagnosis of LCC still poses a difficult problem that requires a careful assessment of several parameters, such as the disease's location, shape, size, and the improvements observed following contrast enhancement. The suggested model may not comprehensively consider all of these parameters, suggesting a requirement for more enhancements to improve its accuracy in identifying LCC.
- Future work will focus on enhancing the model to minimize the margin of error in XAI.

In the realm of medical image analysis, the LW-MS CNN presents several advantages worthy of discussion. Firstly, its ability to efficiently process and analyze medical images while maintaining a relatively low computational footprint makes it highly suitable for real-time applications, offering timely diagnoses critical for patient care. Additionally, the incorporation of multi-scale features enables the model to capture intricate details across various levels of granularity, enhancing its sensitivity to the subtle abnormalities characteristic of LCC. This multi-scale architecture facilitates a more holistic understanding of the pathology present in the images, thereby potentially improving diagnostic accuracy. Moreover, the lightweight design of the model, with a modest parameter count of 1.1 million, not only ensures rapid inference but also makes it more accessible for deployment on resource-constrained environments, such as edge devices or low-power computing platforms. These combined attributes render the lightweight multi-scale CNN an attractive solution for addressing the pressing need for early and accurate cancer detection, ultimately contributing to improved patient outcomes and healthcare delivery. Finally, the suggested approach has the potential to increase the effectiveness and precision of LCC identification, particularly in real-world applications where computational power and speed are crucial considerations as well as to analyze the region of interest areas.

6. Conclusions

A novel end-to-end DL-based lung and colon detection model that is interpretable is proposed in this research. The proposed model demonstrates a high degree of accuracy in identifying the most prevalent types of cancer in the five-class classification of both LCC subtypes. The LW-MS CNN design of the suggested model, with 1.1 million trainable parameters, enables real-time applications, cutting down on processing time and boosting system effectiveness. The proposed model has less trainable parameters than other SOTA models, which indicates that the training and testing time are also less than in other SOTA models. Additionally, the CV strategy was utilized to address the overfitting issue and guarantee the generalizability of the model, providing an accuracy of 99.20% for the classification of LCC. Medical practitioners can use an inventive end-to-end application that was created to make use of the proposed model, which will provide precise forecasts and support decision making. As a result, the proposed model's capability to identify the type of LCC rapidly and accurately can help neurosurgeons and medical professionals make fast and correct clinical decisions about patients with LCC. In this study, interpretability approaches including Grad-CAM, and SHAP improve the understandability, dependability, and adaptability of lightweight CNN models to increase their efficacy. These methods assist users, developers, and data scientists in understanding model behavior, resolving problems, and improving the models' effectiveness and fairness.

However, more research is required to properly comprehend the potential and limitations of DL in LCC detection in the IoMT and to overcome the challenges of practical application. To prevent overfitting or incorrect diagnosis, it is crucial to utilize strong and proven DL models that have been trained on substantial and varied datasets. To ensure the generalizability of the model, it is also crucial to consider potential biases in the training data and to apply methods like CV. The proposed model can be used in clinics for the automated diagnosis of LCC. The model could have improved performance with more advanced image pre-processing and dataset segmentation, even though the architecture provides greater accuracy. Additionally, segmentation techniques improve performance results, and the region of interest of segmentation methods can be compared with the use of interpretability methods. The datasets on LCC that were recently made public will be investigated in the future to conduct an ablation study of the suggested model, aiming to demonstrate its reliability. In further study endeavors, it is important to contemplate the inclusion of comparisons with vision transformers to provide a more thorough perspective on the progressions within this domain.

Author Contributions: Conceptualization, M.A.H. and F.H.; funding acquisition, M.M.R. and S.R.S.; investigation, M.A.H., F.H., H.S. and M.O.F.G.; methodology, M.A.H., F.H. and F.R.; project administration, H.S. and M.O.F.G.; software, M.A.H. and F.H.; supervision, S.R.S., H.S., and M.O.F.G.; validation, M.M.R., H.S., M.O.F.G. and S.R.S.; writing—original draft, M.A.H. and F.H.; writing—review and editing, H.S., M.O.F.G., M.M.R. and S.R.S. All authors have read and agreed to the published version of the manuscript.

Funding: This research received no external funding.

Institutional Review Board Statement: Not applicable.

Informed Consent Statement: Not applicable.

Data Availability Statement: The data are contained within this article.

Acknowledgments: We express our gratitude to the editor and reviewer for their valuable feedback in enhancing the standard of our paper. We acknowledge the usage of language editing tools QuillBot [https://quillbot.com] which assisted in reducing grammatical errors.

Conflicts of Interest: The authors declare no conflicts of interest.

References

1. Talukder, M.A.; Islam, M.M.; Uddin, M.A.; Akhter, A.; Hasan, K.F.; Moni, M.A. Machine learning-based lung and colon cancer detection using deep feature extraction and ensemble learning. *Expert Syst. Appl.* **2022**, *205*, 117695. [CrossRef]
2. Dubey, R.S.; Goswami, P.; Baskonus, H.M.; Gomati, A.T. On the existence and uniqeness analysis of fractional blood glucose-insulin minimal model. *Int. J. Model. Simul. Sci. Comput.* **2022**, *14*, 2350008. [CrossRef]
3. Hasan, M.I.; Ali, M.S.; Rahman, M.H.; Islam, M.K. Automated Detection and Characterization of Colon Cancer with Deep Convolutional Neural Networks. *J. Healthc. Eng.* **2022**, *2022*, 5269913. [CrossRef] [PubMed]
4. Bawankar, B.U.; Chinnaiah, K. Implementation of ensemble method on DNA data using various cross validation techniques. *3c Tecnol. Glosas De Innovación Apl. A La Pyme* **2022**, *11*, 59–69. [CrossRef]

5. Godkhindi, A.M.; Gowda, R.M. Automated detection of polyps in CT colonography images using deep learning algorithms in colon cancer diagnosis. In Proceedings of the 2017 International Conference on Energy, Communication, Data Analytics and Soft Computing (ICECDS), Chennai, India, 1–2 August 2017; pp. 1722–1728.
6. Sarwinda, D.; Bustamam, A.; Paradisa, R.H.; Argyadiva, T.; Mangunwardoyo, W. Analysis of deep feature extraction for colorectal cancer detection. In Proceedings of the 2020 4th International Conference on Informatics and Computational Sciences (ICICoS), Semarang, Indonesia, 10–11 November 2020; pp. 1–5.
7. Attallah, O.; Abougharbia, J.; Tamazin, M.; Nasser, A.A. A BCI system based on motor imagery for assisting people with motor deficiencies in the limbs. *Brain Sci.* **2020**, *10*, 864. [CrossRef]
8. Ayman, A.; Attalah, O.; Shaban, H. An efficient human activity recognition framework based on wearable imu wrist sensors. In Proceedings of the 2019 IEEE International Conference on Imaging Systems and Techniques (IST), Abu Dhabi, United Arab Emirates, 9–10 December 2019; pp. 1–5.
9. Di Cataldo, S.; Ficarra, E. Mining textural knowledge in biological images: Applications, methods and trends. *Comput. Struct. Biotechnol. J.* **2017**, *15*, 56–67. [CrossRef] [PubMed]
10. Zhang, C.; Chen, T. From low level features to high level semantics. In *Handbook of Video Databases: Design and Applications*; CRC Press: Boca Raton, FL, USA, 2003.
11. Aslan, M.F.; Sabanci, K.; Durdu, A. A CNN-based novel solution for determining the survival status of heart failure patients with clinical record data: Numeric to image. *Biomed. Signal Process. Control* **2021**, *68*, 102716. [CrossRef]
12. Anwar, S.M.; Majid, M.; Qayyum, A.; Awais, M.; Alnowami, M.; Khan, M.K. Medical image analysis using convolutional neural networks: A review. *J. Med. Syst.* **2018**, *42*, 226. [CrossRef] [PubMed]
13. Alzubaidi, L.; Zhang, J.; Humaidi, A.J.; Al-Dujaili, A.; Duan, Y.; Al-Shamma, O.; Santamaría, J.; Fadhel, M.A.; Al-Amidie, M. Review of deep learning: Concepts, CNN architectures, challenges, applications, future directions. *J. Big Data* **2021**, *8*, 1–74.
14. Chaddad, A.; Peng, J.; Xu, J.; Bouridane, A. Survey of Explainable AI Techniques in Healthcare. *Sensors* **2023**, *23*, 634. [CrossRef]
15. Malafaia, M.; Silva, F.; Neves, I.; Pereira, T.; Oliveira, H.P. Robustness Analysis of Deep Learning-Based Lung Cancer Classification Using Explainable Methods. *IEEE Access* **2022**, *10*, 112731–112741. [CrossRef]
16. Selvaraju, R.R.; Cogswell, M.; Das, A.; Vedantam, R.; Parikh, D.; Batra, D. Grad-cam: Visual explanations from deep networks via gradient-based localization. In Proceedings of the 2017 IEEE International Conference on Computer Vision (ICCV), Venice, Italy, 22–29 October 2017; pp. 618–626.
17. Saranya, A.; Subhashini, R. A systematic review of Explainable Artificial Intelligence models and applications: Recent developments and future trends. *Decis. Anal. J.* **2023**, *7*, 100230.
18. Nishio, M.; Nishio, M.; Jimbo, N.; Nakane, K.J.C. Homology-based image processing for automatic classification of histopathological images of lung tissue. *Cancers* **2021**, *13*, 1192. [CrossRef]
19. Mangal, S.; Chaurasia, A.; Khajanchi, A. Convolution neural networks for diagnosing colon and lung cancer histopathological images. *arXiv* **2020**, arXiv:2009.03878.
20. Shandilya, S.; Nayak, S.R. Analysis of lung cancer by using deep neural network. In *Innovation in Electrical Power Engineering, Communication, and Computing Technology*; Springer: Berlin/Heidelberg, Germany, 2021; Volume 2022, pp. 427–436.
21. Masud, M.; Sikder, N.; Nahid, A.-A.; Bairagi, A.K.; AlZain, M.A.J.S. A machine learning approach to diagnosing lung and colon cancer using a deep learning-based classification framework. *Sensors* **2021**, *21*, 748. [CrossRef]
22. Hatuwal, B.K.; Thapa, H.C. Lung cancer detection using convolutional neural network on histopathological images. *Int. J. Comput. Trends Technol* **2020**, *68*, 21–24. [CrossRef]
23. Tasnim, Z. Deep learning predictive model for colon cancer patient using CNN-based classification. *Int. J. Adv. Comput. Sci. Appl.* **2021**, *12*, 687–696. [CrossRef]
24. Sikder, J.; Das, U.K.; Chakma, R.J. Supervised learning-based cancer detection. *Int. J. Adv. Comput. Sci. Appl.* **2021**, *863-869*, 863–869. [CrossRef]
25. Qasim, Y.; Al-Sameai, H.; Ali, O.; Hassan, A. Convolutional neural networks for automatic detection of colon adenocarcinoma based on histopathological images. In *International Conference of Reliable Information and Communication Technology*; Springer: Berlin/Heidelberg, Germany, 2020; pp. 19–28.
26. Mengash, H.A. Leveraging Marine Predators Algorithm with Deep Learning for Lung and Colon Cancer Diagnosis. *Cancers* **2023**, *15*, 1591. [CrossRef]
27. Attallah, O.; Aslan, M.F.; Sabanci, K. A Framework for Lung and Colon Cancer Diagnosis via Lightweight Deep Learning Models and Transformation Methods. *Diagnostics* **2022**, *12*, 2926. [CrossRef]
28. Al-Jabbar, M.; Alshahrani, M.; Senan, E.M.; Ahmed, I.A. Histopathological Analysis for Detecting Lung and Colon Cancer Malignancies Using Hybrid Systems with Fused Features. *Bioengineering* **2023**, *10*, 383. [CrossRef]
29. El-Ghany, S.A.; Azad, M.; Elmogy, M. Robustness Fine-Tuning Deep Learning Model for Cancers Diagnosis Based on Histopathology Image Analysis. *Diagnostics* **2023**, *13*, 699. [CrossRef]
30. Kumar, N.; Sharma, M.; Singh, V.P.; Madan, C.; Mehandia, S. An empirical study of handcrafted and dense feature extraction techniques for lung and colon cancer classification from histopathological images. *Biomed. Signal Process. Control* **2022**, *75*, 103596. [CrossRef]
31. Borkowski, A.A.; Bui, M.M.; Thomas, L.B.; Wilson, C.P.; DeLand, L.A.; Mastorides, S.M. Lung and colon cancer histopathological image dataset (lc25000). *arXiv* **2019**, arXiv:1912.12142.

32. Nahiduzzaman, M. Diabetic retinopathy identification using parallel convolutional neural network based feature extractor and ELM classifier. *Expert Syst. Appl.* **2023**, *217*, 119557. [CrossRef]
33. Ali, M.B. Domain mapping and deep learning from multiple MRI clinical datasets for prediction of molecular subtypes in low grade gliomas. *Brain Sci.* **2020**, *10*, 463. [CrossRef]
34. Chehade, A.H.; Abdallah, N.; Marion, J.-M.; Oueidat, M. Chauvet, Lung and colon cancer classification using medical imaging: A feature engineering approach. *Phys. Eng. Sci. Med.* **2022**, *45*, 729–746. [CrossRef] [PubMed]
35. Al-Zoghby, A.M.; Al-Awadly, E.M.K.; Moawad, A.; Yehia, N.; Ebada, A.I. Dual Deep CNN for Tumor Brain Classification. *Diagnostics* **2023**, *13*, 2050. [CrossRef]
36. Arrieta, A.B. Explainable Artificial Intelligence (XAI): Concepts, taxonomies, opportunities and challenges toward responsible AI. *Inf. Fusion* **2020**, *58*, 82–115. [CrossRef]
37. Hassija, V. Interpreting Black-Box Models: A Review on Explainable Artificial Intelligence. *Cogn. Comput.* **2023**, *16*, 45–74. [CrossRef]
38. Ananthakrishnan, B.; Shaik, A.; Chakrabarti, S.; Shukla, V.; Paul, D.; Kavitha, M.S.J.S. Smart Diagnosis of Adenocarcinoma Using Convolution Neural Networks and Support Vector Machines. *Sustainability* **2023**, *15*, 1399. [CrossRef]
39. Islam, M.R. Explainable transformer-based deep learning model for the detection of malaria parasites from blood cell images. *Sensors* **2022**, *22*, 4358. [CrossRef]
40. Asuncion, L.V.R.; De Mesa, J.X.P.; Juan, P.K.H.; Sayson, N.T.; Cruz, A.R.D. Thigh motion-based gait analysis for human identification using inertial measurement units (IMUs). In Proceedings of the 2018 IEEE 10th International Conference on Humanoid, Nanotechnology, Information Technology, Communication and Control, Environment and Management (HNICEM), Baguio City, Philippines, 29 November–2 December 2018; pp. 1–6.
41. Powers, D.M.W. What the F-measure doesn't measure: Features, Flaws, Fallacies and Fixes. *arXiv* **2015**, arXiv:1503.06410.
42. Powers, D.M.W. Evaluation: From precision, recall and F-measure to ROC, informedness, markedness and correlation. *arXiv* **2020**, arXiv:2010.16061.
43. Sasaki, Y. *The Truth of the F-Measure*; University of Manchester: Manchester, UK, 2007; p. 25.
44. Fawcett, T.J.M.L. ROC graphs: Notes and practical considerations for researchers. *Mach. Learn.* **2004**, *31*, 1–38.
45. Krzanowski, W.J.; Hand, D.J. *ROC Curves for Continuous Data*; Taylor & Francis Ltd.: London, UK, 2009.
46. Vergara, I.A.; Norambuena, T.; Ferrada, E.; Slater, A.W.; Melo, F. StAR: A simple tool for the statistical comparison of ROC curves. *BMC Bioinform.* **2008**, *9*, 265. [CrossRef] [PubMed]
47. Narkhede, S. *Understanding AUC-ROC Curve: Towards Data Science*; Toronto, ON, Canada, 2018; Available online: https://towardsdatascience.com/understanding-auc-roc-curve-68b2303cc9c5 (accessed on 14 February 2024).
48. Gorunescu, F. *Data Mining: Concepts, Models and Techniques*; Springer Science & Business Media: Berlin/Heidelberg, Germany, 2011.
49. Yulianto, A.; Sukarno, P.; Suwastika, N.A. Improving adaboost-based intrusion detection system (IDS) performance on CIC IDS 2017 dataset. *J. Phys.* **2019**, *1192*, 012018. [CrossRef]
50. Mehmood, S. Malignancy Detection in Lung and Colon Histopathology Images Using Transfer Learning with Class Selective Image Processing. *IEEE Access* **2022**, *10*, 25657–25668. [CrossRef]
51. Bukhari, S.U.K.; Syed, A.; Bokhari, S.K.A.; Hussain, S.S.; Armaghan, S.U.; Shah, S.S.H. The histological diagnosis of colonic adenocarcinoma by applying partial self supervised learning. *MedRxiv* **2020**. MedRxiv:15.20175760.
52. Sakr, A.S.; Soliman, N.F.; Al-Gaashani, M.S.; Pławiak, P.; Ateya, A.A.; Hammad, M. An Efficient Deep Learning Approach for Colon Cancer Detection. *Appl. Sci.* **2022**, *12*, 8450. [CrossRef]
53. Kumar, A.; Vishwakarma, A.; Bajaj, V. CRCCN-Net: Automated framework for classification of colorectal tissue using histopathological images. *Biomed. Signal Process. Control* **2023**, *79*, 104172. [CrossRef]

Disclaimer/Publisher's Note: The statements, opinions and data contained in all publications are solely those of the individual author(s) and contributor(s) and not of MDPI and/or the editor(s). MDPI and/or the editor(s) disclaim responsibility for any injury to people or property resulting from any ideas, methods, instructions or products referred to in the content.

Article

Blood Pressure Measurement Device Accuracy Evaluation: Statistical Considerations with an Implementation in R

Tanvi Chandel [1], Victor Miranda [2,*], Andrew Lowe [1] and Tet Chuan Lee [1]

[1] Institute of Biomedical Technologies, Auckland University of Technology, Auckland 1010, New Zealand; tanvi.chandel@aut.ac.nz (T.C.); tet.chuan.lee@aut.ac.nz (T.C.L.)
[2] Department of Mathematical Sciences, Auckland University of Technology, Auckland 1010, New Zealand
* Correspondence: victor.miranda@aut.ac.nz

Abstract: Inaccuracies from devices for non-invasive blood pressure measurements have been well reported with clinical consequences. International standards, such as ISO 81060-2 and the seminal AAMI/ANSI SP10, define protocols and acceptance criteria for these devices. Prior to applying these standards, a sample size of N >= 85 is mandatory, that is, the number of distinct subcjects used to calculate device inaccuracies. Often, it is not possible to gather such a large sample. Many studies apply these standards with a smaller sample. The objective of the paper is to introduce a methodology that broadens the method first developed by the AAMI Sphygmomanometer Committee for accepting a blood pressure measurement device. We study changes in the acceptance region for various sample sizes using the sampling distribution for proportions and introduce a methodology for estimating the exact probability of the acceptance of a device. This enables the comparison of the accuracies of existing device development techniques even if they were studied with a smaller sample size. The study is useful in assisting BP measurement device manufacturers. To assist clinicians, we present a newly developed "bpAcc" package in R to evaluate acceptance statistics for various sample sizes.

Keywords: blood pressure; ANSI/AAMI–SP10 standards; blood pressure measurement device; probability of acceptance

Citation: Chandel, T.; Miranda, V.; Lowe, A.; Lee, T.C. Blood Pressure Measurement Device Accuracy Evaluation: Statistical Considerations with an Implementation in R. *Technologies* 2024, 12, 44. https://doi.org/10.3390/technologies12040044

Academic Editors: Juvenal Rodriguez-Resendiz, Gerardo I. Pérez-Soto, Karla Anhel Camarillo-Gómez and Saul Tovar-Arriaga

Received: 19 February 2024
Revised: 22 March 2024
Accepted: 22 March 2024
Published: 25 March 2024

Copyright: © 2024 by the authors. Licensee MDPI, Basel, Switzerland. This article is an open access article distributed under the terms and conditions of the Creative Commons Attribution (CC BY) license (https://creativecommons.org/licenses/by/4.0/).

1. Introduction

Blood pressure (BP) is extensively used to assist health monitoring and diagnosis in healthcare settings. However, inaccuracies in BP measurement can result in misjudgments, potentially leading to severe consequences [1]. The clinical gold standard for BP measurement is BP measurement performed using arterial cannulation [2]; however, arterial cannulation is invasive and time-consuming and can only be performed by skilled personnel. It is also linked with cases of ischemia, lesions of nerves or vessels, embolism, and other complications [3]. In regular cases, BP is measured non-invasively [4], which yields measurement inaccuracies. Even slight measurement inaccuracies can result in misclassifying millions of individuals [5]. Hence, a precise measurement of blood pressure holds significant importance in public health. Underestimating true BP by merely 5 mmHg or less can have significant clinical consequences as several studies have inferred incorrect tagging of more than 20 million Americans as pre-hypertensive when, in fact, they are suffering from hypertension. Untreated hypertension can lead to a 25% increased risk of fatal strokes and fatal myocardial infarctions [1]. Conversely, if there is an overestimation of true BP by 5 mmHg, nearly 30 million Americans may receive inappropriate treatment with antihypertensive medications. This could result in exposure to potential side effects of the drugs, psychological distress due to misdiagnosis, and unnecessary financial liability [5]. In healthcare domains such as intensive care, accurate BP measurement is even more crucial. As a result, regulating BP measurement devices is a critical matter, and suitable processes must be used for clinical investigations to validate BP devices.

National regulators have made significant efforts towards global harmonization of the standards for medical devices. When designing a blood pressure (BP) measurement device, manufacturers must adhere to standardized protocols, ensuring that the device's inaccuracy falls within an acceptable range, typically expressed as mean error ± standard deviation of BP errors for non-invasive techniques. Even when within acceptable limits, continuous efforts are made to improve the accuracy using improved methods by adding parameters associated with blood pressure [6–8]. This pursuit aims to provide healthcare professionals with more reliable BP readings, reducing the likelihood of errors and supporting informed decision-making. The International Organization for Standardization (ISO), established in 1947, defines standards that are accepted worldwide. It comprises representatives from various national standards organizations. The ISO 81060-2:2018 standard defines the criterion for the clinical investigation of automated, non-invasive sphygmomanometers [9] and has been approved for use currently and recognized in whole or part by many national regulators. It supersedes region-specific standards such as EN 1060-4:2004 [10] and has been adopted in law, in contrast to validation protocols such as those recommended by the British Hypertension Society [11] and the European Society of Hypertension [12].

ISO 81060-2:2018 stipulates criteria for determining the acceptable accuracy of sphygmomanometers that originated from the initial work of the Committee of US Association for the Advancement of Medical Instrumentation (AAMI) in creating the American Standard for manual, electronic, or automated sphygmomanometers known as SP10 [13]. The standard also specifies safety, labeling, and performance requirements designed to ensure the safety and effectiveness of the device. ISO 81060-2, like SP10, mandates a minimum sample size (N) of 85 participants to be used to evaluate the BP device inaccuracy [9]. In addition to N >= 85, the standard requires the BP errors to be within −10 mmHg to 10 mmHg, also known as the tolerable error limit, and the estimated probability of tolerable error ($\hat{p}$) to be at least 85%. In practice, it is found that accuracy requirements are difficult to achieve, and process requirements are costly. Manufacturers attempt to adhere to this standard. However, only a small fraction of manufacturers can do so [14]. A study reports that less than 20% of the devices accessible today conform to an established guideline [2].

While compliance with this standard is appropriate for devices that are to be marketed, there are purposes other than regulation of medical devices for which studies involving fewer participants can still yield useful information. For instance, early evaluation of experimental devices would benefit from an earlier checkpoint, as it is often difficult for clinicians to gather 85 participants [15–17]. Currently, to our knowledge, there is no official method for evaluating studies with fewer participants. As a result, various research works in this field adopted potentially incorrect pass/fail criteria of the standard apparently without recognizing the difference between their research methods and those assumed by the standard. This paper aims to inform researchers and BP device manufacturers about the potential effects of employing different sample sizes for the validation of a BP measurement device.

We also offer recommendations to adjust the appropriate acceptance range (upper limit of acceptable standard deviation for a certain mean error) required for any study to adhere to criteria similar to the SP10 requirements. In addition to the different acceptance limits for different sample sizes, this paper provides a brief comparison of previous studies that investigated novel BP measurement methods with different sample sizes, and also assesses their adherence to the current standard.

2. SP10 Statistical Considerations

2.1. SP10 Acceptance Criteria

Multiple techniques are used for automated, non-invasive BP measurement. Most researchers/clinicians use an inflatable cuff to hinder the flow of blood in the upper arm. As the cuff is deflated, various methods can be employed to estimate the systolic and diastolic blood pressure (SBP; DBP) [18]. The error in estimation is the difference between the values obtained from the test device and the value obtained using a reference method, which is normally specified as auscultation by trained observers [19]. Acceptance criteria evolved from SP10's inception in 1987 to reflect a more defined statistical treatment that is currently adopted in ISO 81060-2.

Initially, the standard required that manufacturers should maintain the mean of errors within ±5 mmHg with a standard deviation no greater than 8 mmHg [20]. However, these static values did not consider the relation between the mean and standard deviation of errors. For the same standard deviation, $\hat{p}$ will be different if the sample mean is 0 mmHg and if the sample mean is 5 mmHg. Hence, SP10's criteria (reproduced as Table A1 in Appendix A) was introduced to span different values of the sample mean and the upper limit of standard deviation such that 85% of the errors are within the tolerable error range, where $\hat{p} = 0.85$. These values of acceptable standard deviation for a given sample mean represent the acceptance limit. For instance, for a mean error of 2 mmHg, the standard deviation must be less than or equal to 6.65 mmHg to accept the device. But this estimated probability of tolerable error ($\hat{p}$) is itself an estimate.

How far off it is from the true value depends on the sample size. As per SP10, a sample size of N = 85 yields a 90% chance or confidence that $\hat{p}$ will not differ by more than about 0.07 from its true probability of tolerable error (p) [20], given by

$$\hat{p} - p = 1.645 \times K, \quad (1)$$

where $K = \sqrt{\frac{1}{2\pi(N-1)}}$ and N is the sample size. We will refer to this difference as the "90% confidence between p and $\hat{p}$". In Equation (1), K is the standard deviation of the distribution of probability of tolerable error which is assumed to be asymptotically normal according to the SP10 standard, where the mean of the distribution is $\hat{p}$. Thus, for the device to be acceptable, $\hat{p}$ must be at least 85% for N = 85, because then one can be confident that p is at least 78%, as per the standard.

2.2. Brief Review of the Problem
2.2.1. Acceptance Criteria

According to SP10, Table A1 can be used with any number of participants, but it only considers the acceptance limit that is suitable for N ≥ 85. However, studies with fewer participants than the minimum of 85 specified by SP10 are not uncommon. For instance, one study proposes a novel BP estimation method based on Pulse Arrival Time (PAT) to estimate SBP and DBP [16]. Using 32 subjects, they report the BP error limit, mean error ± SD. These limits are 0.12 ± 6.15 (SBP, mmHg) and 1.31 ± 5.36 (DBP, mmHg). Another study validates a wireless BP monitor using 33 participants [6]. The estimated BP errors were −0.7 ± 6.9 mmHg for SBP and −1 ± 5.1 mmHg for DBP. A new calibration procedure that accounts for the Sympathetic Nervous System (SNS) on BP-PTT (Pulse Transit Time) was also proposed to estimate BP values using 10 subjects [21]. All these studies attain the $\hat{p} = 0.85$ criteria mentioned in SP10, but the sample size is less than 85. For smaller sample sizes, there is little guidance on how the acceptance limit should change such that one can be 95% certain that the true probability is at least 78%, which is recognized as the threshold for acceptability by SP10. While these studies are potentially valuable, it would be inappropriate to interpret results by making a comparison to the criteria in the standard which is just fixed for N = 85.

At present, the 90% confidence between $\hat{p}$ and p which is evaluated using Equation (1) only considers the sample size of a specific study. Using this 90% difference value, the standard makes some assumptions about the $\hat{p}$ such that $p \geq 0.78$. However, variations in the 90% difference are not only due to changes in sample size but also to the value of $\hat{p}$, obtained from the reported sample mean and standard deviation from the BP device. To tackle this issue, we propose a methodology that provides a more flexible approach to evaluating the 90% confidence between $\hat{p}$ and p with respect to the sample size, sample mean, and standard deviation from a statistical point of view. With this approach, the value of $\hat{p}$ can be evaluated for different sample sizes, which we can use to study the changes in the acceptance limit of different sample sizes such that the devices under test adhere to the SP10 criteria.

2.2.2. Probability of Acceptance (P_A)

As a result of the acceptance limit varying with sample size, the probability of acceptance, which essentially gives the probability of meeting the SP10 criterion for a particular sample mean, standard deviation, and sample size, will fluctuate. In this regard, this research also provides a mechanism for a more relevant comparison of the mean and standard deviation of the data from studies with varying sample sizes. To effectively compare studies with smaller sample sizes and distinguish the methodology (e.g., techniques and mathematical methods) being used to develop the BP measurement devices, a more robust statistical treatment is required to re-evaluate the literature less subjectively as per the international standards. We present aspects behind the computation of the probability of acceptance, denoted by P_A.

There are two key outcomes from this work. First, we study the changes in the acceptance limit for different sample sizes such that they adhere to the standards when the sample size is less than N = 85. Secondly, we provide a methodology for evaluating the probability of acceptance P_A, allowing comparison of different studies with varying sample sizes, assessing the accuracy of different methods and techniques being tested to build BP measurement devices. This work has a companion R package called "bpAcc" which implements the methodology introduced in this paper. This enables manufacturers and researchers to better judge their compliance with the accuracy criteria of ISO 81060-2 using a smaller sample size and more appropriately compare studies performed using different sample sizes.

3. Methodology

This section outlines the theoretical details of this research starting from the protocols currently in use by the SP10 standard. The parameters utilized in this section are also outlined in Table A2 within Appendix C.

3.1. Brief Review of the Statistical Components of SP10
3.1.1. Average Error and Tolerable Error

For each of the N participants, k = 3 pairs of blood pressure measurements are obtained: one measurement, $\delta^j_{k_1}$, produced by the usual auscultatory reference method, and the other, $\delta^j_{k_2}$, produced by the device being assessed. The difference, $\epsilon^j_k = \delta^j_{k_2} - \delta^j_{k_1}$, is called an error, and the average error for the jth participant is

$$\delta^j = \sum_{k=1}^{3} \frac{\epsilon^j_k}{3}, j = 1, \ldots, N \quad (2)$$

Statistically, we assume the average errors δ^j produced by the device D follow a θ-parameterized distribution $\mathbb{F} = \mathbb{F}(\theta)$. The maximum average error accepted, also known as the tolerable error, is denoted by Δ. The tolerable error is set to $\Delta = 10$ mmHg in this work, following the SP10 standard. Hence, the probability of tolerable error is given by

$$\mathbb{P}(|\delta j| \leq 10; \theta) = \mathbb{F}(10; \theta) - \mathbb{F}(-10; \theta). \quad (3)$$

The errors produced by any device are deemed acceptable if p is a minimum of γ_p, i.e.,

$$p = \mathbb{P}\left(\left|\delta^j\right| \leq 10; \theta\right) \geq \gamma_p, \ 0 \leq \gamma_p \leq 1 \quad (4)$$

Fundamentally, we assume the errors δ^j follow a normal distribution with parameters $\theta = (\mu_p, \sigma_p)^T, \mu_p \in \mathbb{R}, \sigma_p > 0$. μ_p and σ_p are the mean and standard deviation of the errors produced by the device readings and will be referred to as true mean error (or true bias) and true standard deviation, respectively.

3.1.2. σ_{γ_P} Acceptance Curve

Our interest focuses on $\sigma_{\gamma_P}{}^{MAX}$, a bivariate function of (μ_p, γ_P) given by

$$\sigma_{\gamma_P}{}^{MAX} = \sigma_{\gamma_P}(\mu_p, \gamma_P) = max\left\{\sigma_p; \mathbb{P}\left(\left|\delta^j\right| \leq 10; \theta = (\mu_p, \sigma_p)^T\right) \geq \gamma_P\right\}, \quad (5)$$

for $\mu_p \in (-10, 10)$ and fixed $\gamma_P \in (0, 1)$. The curve is called the σ_{γ_P} acceptance curve, or simply the acceptance curve. For every μ_p, $\sigma_{\gamma_P}{}^{MAX}$ is given by the maximum sample standard deviation producing a probability of tolerable error of at least γ_P. Figure 1 shows σ_{γ_P} acceptance curves represented by σ^{MAX} for $\gamma_P \in \{0.75, 0.80, 0.85, 0.90, 0.95\}$, and μ_p in (10, 10).

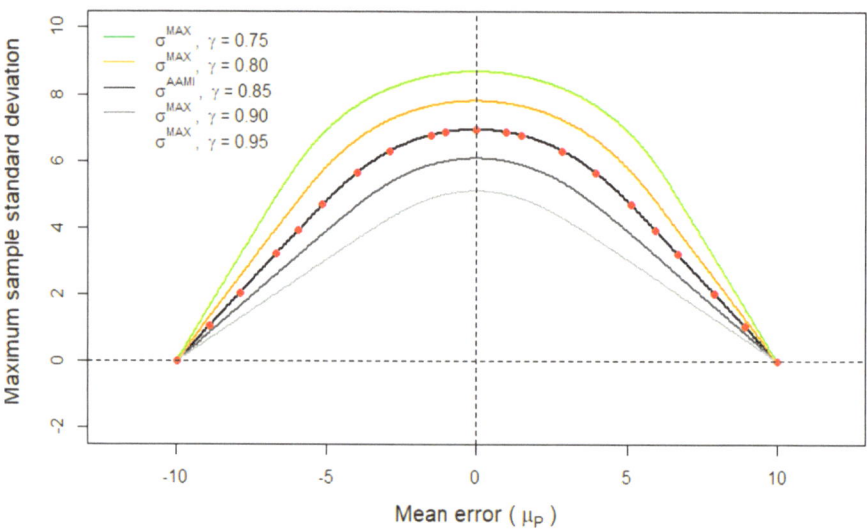

Figure 1. Acceptance curves obtained from Equation (5) for $\gamma_p \in \{0.75, 0.80, 0.85, 0.90, 0.95\}$; μ_p in (−10, 10).

3.1.3. ANSI/AAMISP10 Acceptance Criterion

As per the SP10 standard [20], a device D will be deemed acceptable if the estimated probability of tolerable error $\hat{p}$ is at least $\gamma_p = 0.85$, and the sample size is 85 subjects. From Equation (5), we define the σ^{AAMI} acceptance curve as

$$\sigma^{AAMI} = \sigma_{0.85}{}^{MAX}\left\{\sigma_p; \mathbb{P}\left(\left|\delta^j\right| \leq 10; \theta = (\mu_p, \sigma_p)^T\right) \geq 0.85\right\}, \quad (6)$$

with $\mu_p \in (-10, 10)$, $\theta = (\mu_p, \sigma_p)^T$. Since σ_{γ_P} acceptance curves narrow down as shown in Figure 1 as γ_P decreases, without loss of generality, we assume that the σ^{AAMI} acceptance curve is obtained at $\gamma_p = 0.85$, which is as follows:

$$\sigma^{AAMI} = max\left\{\sigma_p; \mathbb{P}\left(\left|\delta^j\right| \leq 10; \theta\right) \geq 0.85\right\} = \left\{\sigma_p; \mathbb{P}\left(\left|\delta^j\right| \leq 10; \theta\right) = 0.85\right\}. \quad (7)$$

The σ^{AAMI} acceptance curve is the thick black line and is given by the solution to

$$\mathbb{P}\left(\left|\delta^j\right| \leq \Delta; \theta = (\mu_p, \sigma_p)^T\right) = 0.85. \quad (8)$$

with $\Delta = 10$, and $\mu_0 \epsilon(-10, 10)$.

However, in practice, only size-limited samples of BP measurements are available for testing the device D. In the following, we will introduce the statistical assumptions required in this work. Let $S_t = \{\delta_t^1, \ldots, \delta_t^n\}$ be a size-n sample of BP average errors δ_t with $\delta_t \sim N(\mu_p, \sigma_p)$. The S_t-sample mean, and S_t-sample standard deviation are denoted by $\bar{x}_t$ and s_t, respectively. We will remove the superscripts $1, \ldots, n$ for simplicity.

Crucially, we replace μ_p with $\bar{x}_t$ in Equation (8). Then, the highest permissible value for s_t rendering the device $\sigma_{0.85}$-acceptable, denoted by $\sigma_{0.85}^{AAMI}$, is the function of $\bar{x}_t$ given by

$$\sigma_{0.85}^{AAMI} = \sigma_{0.85}^{AAMI}(\bar{x}_t) = \{\sigma_p; \mathbb{P}(|\delta_t| \leq 10; \bar{x}_t) = 0.85\}. \tag{9}$$

As a result, the device D is deemed acceptable under the SP10 acceptance criterion if and only if $s_t \leq \sigma_{0.85}^{AAMI}(\bar{x}_t)$.

The values $\sigma_{0.85}^{AAMI}$ for fixed $\bar{x}_t$ are obtained by directly applying the bisection method to Equation (7) as a function of σ_p. Table 1 gives $\sigma_{0.85}^{AAMI}$ to selected values of $\mu_p = \bar{x}_t$. The pairs $(\bar{x}_t, \sigma_{0.85}^{AAMI})$ mentioned in the table are the values displayed in Figure 1 as red dots.

Table 1. The upper limit of SD for selected values of $\bar{x}_t$.

$\bar{x}_t$	0	±1	±1.5	±2.87	±3.96	±5.12
$\sigma_{0.85}^{AAMI}$	6.947	6.874	6.782	6.311	5.664	4.696
$\bar{x}_t$	±5.93	±6.67	±7.15	±7.88	±8.9	±9.99
$\sigma_{0.85}^{AAMI}$	3.927	3.213	2.75	2.045	1.061	0.01

3.2. Sampling Distribution of Sampling Proportions

SP10's confidence limits for p, or true probability of tolerable error, rely on approximations of the Gaussian density using Taylor expansions around the mean and standard deviation [20], providing a biased, standard error depending just on N of the form $\frac{1}{2\pi(N-1)}$ and 95% confidence limits given by $\hat{p} \pm 1.645 \times \sqrt{\frac{1}{2\pi(N-1)}}$.

In this paper, as opposed to [20], we adopt a statistical standpoint to address the uncertainty attached to $\hat{p}$. Consider the binomial random variable, say Y, given by the number of errors falling in the interval $[-10, 10]$. The probability of "occurrence" or errors falling in $[-10, 10]$, denoted with p, is central to this paper. Essentially, we estimate p via maximum likelihood estimation (MLE) using the sampling distribution of proportions which results from the theoretical probability distribution of random-sampled proportions of fixed-size N from the population of errors. The MLE of p is given by $\frac{Y}{N}$. This method represents our main modeling framework allowing us to estimate p and compute probabilities associated with any sample. This framework has been implemented in the package "bpAcc" in R software, R version 4.2.0 [22].

The distribution of p, or proportion's sampling distribution, is asymptotically normal, based on the Central Limit Theorem, requiring a reasonably large sample size for estimation accuracy. Specifically, it requires $N\hat{p} \geq 5$ and $N(1 - \hat{p}) \geq 5$. Under such conditions, the distribution of p is approximately normal with mean $\hat{p}$ and standard deviation

$$sd = \sqrt{\frac{\hat{p}(1-\hat{p})}{N}}. \tag{10}$$

With the proposed approach, the 90% confidence between $\hat{p}$ and p is given by

$$p - \hat{p} = 1.645 \times \sqrt{\frac{\hat{p}(1-\hat{p})}{N}} \tag{11}$$

To comply with the SP10 standards, the 90% confidence between $\hat{p}$ and p should be such that $p \geq 0.78$. In this way, one can evaluate an updated value of $\hat{p}$ for any sample size (N) using Equation (12). This results in changes in the acceptance limits for different sample sizes which will be discussed in Section 5.

$$\hat{p} - 1.645 \times \sqrt{\frac{\hat{p}(1-\hat{p})}{N}} = 0.78 \tag{12}$$

3.3. Evaluation of the Probability of Acceptance (P_A)

3.3.1. Evaluating the Probability

According to the standard, there is 95% certainty that $p \geq 0.78$ with a sample size of N = 85, where 95% is the threshold for the probability of acceptance. This will serve as the benchmark for our proposed methodology, given by

$$P_A = 1 - \emptyset\left(0.78, \hat{p}, \sqrt{\frac{\hat{p}(1-\hat{p})}{N}}\right), \tag{13}$$

where $\emptyset$ is the cumulative density function of the normal distribution. Fundamentally, Equation (13) compares the probability of acceptance for previously published studies with reported mean and standard deviation for the BP errors under different sample sizes. For cases where $P_A \geq 0.95$, the device meets the SP10 standard. Currently, the acceptance region provided for N = 85 is used to validate devices that have used smaller sample sizes; however, with the proposed approach, we can now provide more insights on whether those devices are complying with the SP10 standards with fewer sample sizes or not.

For inference purposes, the proposed framework relies on reasonably large sample sizes, i.e., $N \geq 39$ such that $N\hat{p} \geq 5$ and $N(1 - \hat{p}) \geq 5$. However, the results provided by the simulation study described in Section 3.3.2 have shown closer approximations even for small sample sizes (N < 39), as shown in Table 2. For instance, $\mu_P = 2$ and $\sigma_P = 5.5$ are used to check for cases of samples that are less than 39 to compare the value of P_A. The selection of sample size for comparing the values of P_A in Table 2 is informed by some of the previous studies that have utilized smaller sample sizes to assess device inaccuracy through various evaluation methods [23–25].

Table 2. Simulated P_A and P_A obtained from normal approximation using proposed method vs. the method currently in use in the SP10 standard, for small sample sizes, with $\mu_P = 2$ and $\sigma_P = 5.5$.

	N = 10	N = 15	N = 20	N = 25
Simulated P_A	0.95	0.964	0.974	0.982
P_A using proposed framework	0.931	0.965	0.982	0.99
P_A using SP10 method	0.84	0.893	0.926	0.948

3.3.2. Simulation Study

The simulation study conducted to evaluate the probability of acceptance compares the results obtained with the proposed framework. We investigated a simple situation in which N random numbers from a normal distribution with a known mean and standard deviation were generated. The proportion of errors that fall within the tolerable error range is calculated, yielding the estimated probability of tolerable error, $\hat{p}$. The simulations are conducted for sim.count = 20,000 errors, and the proportion of $\hat{p} \geq 0.78$ is evaluated to determine the value of P_A. To obtain the probability values shown in Table 2, an R code has been provided in Appendix B.

To obtain an estimate of the proportion of instances that have $\hat{p} \geq 0.78$, this process was repeated 50 times. The proportion obtained in each of these repetitions is comparable with a maximum difference of 0.004. For mean = 2, standard deviation = 5.5, and sample size = 25, the median of these repetitions was 9.982. Future simulations would yield similar medians of proportions with only minute differences. We can demonstrate through simulations that our modeling framework is a better approximation than the present technique for N < 39.

4. Software Implementation

The concept of the acceptance region and the probability of acceptance have been implemented in the package "bpAcc" for the R statistical software. The function for evaluating the acceptance region for different sample sizes is AcceptR(), which directly computes Equation (6). Here, $\gamma_P = \hat{p}$ is evaluated for a given sample size, N, using Equation (12). The function PAccept() gives the probability of acceptance for a study that has reported a sample mean error and SD for a sample size to validate a BP measurement device. This function directly evaluates Equation (13). Arguments for both the functions from the package are provided in Tables 3 and 4. The Comprehensive R Archive Network contains concise documentation on user guidance, providing detailed descriptions of package functions and examples. Users can access this documentation when downloading the package in R.

Table 3. Arguments for AcceptR() from the R package bpAcc.

Argument	Comments
N	S_t—sample size.
distribution	Distribution the errors are pulled from. Default is "normal", i.e., normally distributed δ_i^k errors.
criteria	The underlying standard criteria for testing and data analysis. The default is "SP10:2006".

Table 4. Arguments for PAccept() from the R package bpAcc.

Argument	Comments
N	S_t—sample size.
Xbar, sd	Sample mean and sample standard deviation of δ_i^k-error distribution.
distribution	Distribution the errors are pulled from. Default is "normal", i.e., normally distributed δ_i^k errors.
criteria	The underlying standard criteria for testing and data analysis. The default is "SP10:2006".

4.1. AcceptR() Function

Figure 2 provides an upper limit on the sample standard deviation to make sure that $\hat{p}$ is at least 87.47% for N = 33. If the sample mean error is between two values in the table, linear interpolation is implemented. As an example, if the sample mean is −0.7 mmHg, this is $(-0.7 + 0.5)/(-1 + 0.5) = 0.40 = 40\%$ of the distance between −1.0 and −0.5, so one uses $0.40 \times 6.45 + (1 - 0.40) \times 6.50 = 6.48$. The sample standard deviation would have to be 6.48 or less to accept the device.

```
> ## n = 33, xbar = -0.7, sd = 6.9

> AcceptR(n = 33)
---------------------------------------------------------------
For 33 samples, 87.47% of errors must be within -10 mmHg to 10
mmHg.
---------------------------------------------------------------
xbar        sd

    0.0     6.522419

    0.5     6.503214

    1.0     6.445114

    1.5     6.346654

    2.0     6.205400

    2.5     6.018054

    3.0     5.780903

    3.5     5.491068

    4.0     5.149097

    4.5     4.762372

    5.0     4.345659
```

Figure 2. Sample output from AcceptR() for sample size N = 33.

4.2. PAccept() Fuction

During the initial research and development phase of a BP measurement device, different methods can be compared by evaluating P_A which gives the probability of a device meeting the standards using the "PAccept()" function. Usually, different sample sizes are used to evaluate the device. This function can be directly used to determine how far existing studies or devices are from the acceptable standard.

For instance, when two methods to develop a device are compared, where Method 1 provides a device inaccuracy with sample mean error $\pm$ SD = 4 $\pm$ 5.1 and Method 2 provides sample mean error $\pm$ SD = 3 $\pm$ 6.2 for N = 33, to validate which method provides better accuracy and is acceptable as per the standards, the R code chunk provided in Figure 3 is used.

```
> ## n = 33, xbar = 4, sd = 5.1

> PAccept(xbar=4,sd=5.1,N=33)

--------------------------------------------------

The probability of acceptance as per SP10 is
0.9557126

The device is meeting the SP10 criteria.

--------------------------------------------------

> ## n = 33, xbar = 3, sd = 6.2

> PAccept(xbar=3,sd=6.2,N=33)

--------------------------------------------------

The probability of acceptance as per SP10 is 0.8801012
 The device is not meeting the SP10 criteria.
--------------------------------------------------
```

Figure 3. Sample output from PAccept() for two different cases with different device inaccuracies.

5. Applications

5.1. Acceptance Region for Different Sample Sizes

Using the proposed methodology, we can obtain the value of $\hat{p}$ such that the standard criteria are also met using Equation (12). This adjusted value of $\hat{p}$ for a given N is termed as the revised estimated probability of tolerable error or revised $\hat{p}$ as illustrated in Figure 4. The figure indicates the changes in the revised $\hat{p}$ for different sample sizes. This implies that there will be different acceptance regions per sample size as opposed to a single acceptance region in the SP10 standard [20], which is illustrated in Figure 5. The figure shows the acceptance region for a range of sample sizes between 5 and 85, each showing the upper limit of SD for a given mean error that must be followed such that the true probability is at least 0.78 95% of the time to adhere to the SP10 standards.

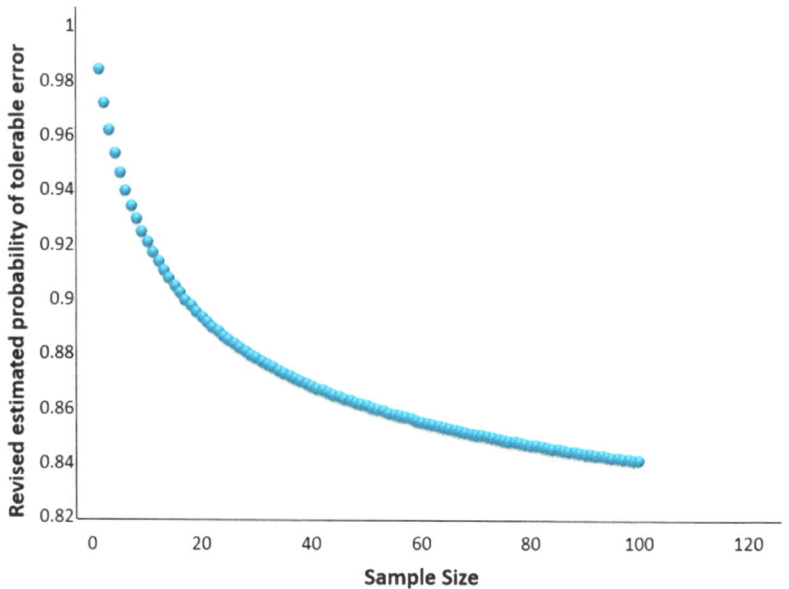

Figure 4. Relationship between the revised estimated probability of tolerable error for different sample sizes (N).

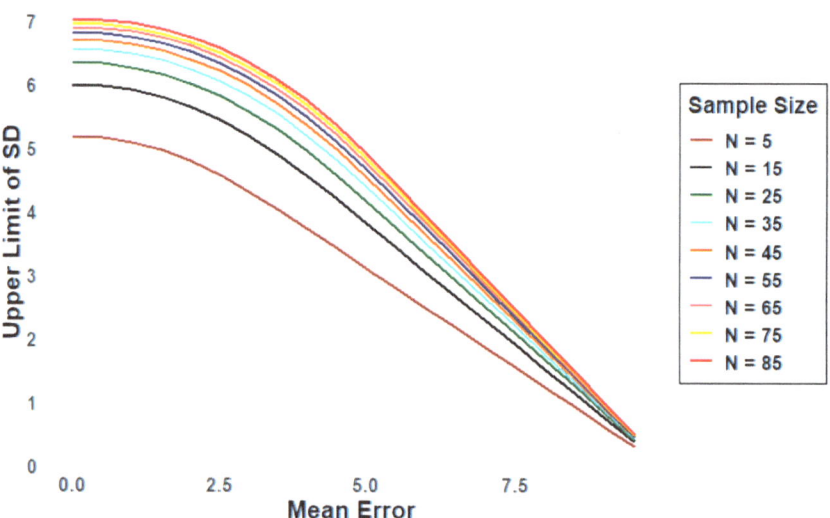

Figure 5. Changes in the acceptance region (upper limit of SD for a given sample mean, $\bar{x}$) as per SP10 for different sample sizes.

5.2. BP Technologies: Comparison of Different Methods

Since the acceptance region varies for different sample sizes based on the revised $\hat{p}$, the value of P_A will also vary for any reported mean error and SD. The P_A values can be directly evaluated using Equation (13). Tables 5 and 6 provide a list of different studies that have reported device inaccuracy based on their development techniques or research methods. The techniques or methods outlined in the tables represent only a subset of the diverse range of technologies employed in the development of blood pressure (BP) measurement devices. While the table highlights specific studies utilizing various methods, it is important to recognize that numerous other technologies and approaches are also being explored within the field of device development. The device inaccuracy in both the tables signifies the error ($\bar{x} \pm SD$) associated with the BP device. These BP estimation errors are measured in mmHg. The proposed methods allow us to evaluate the probability of acceptance of devices reported by these studies and hence also provide a comparison between different BP development techniques/methods. For instance, a study conducted to develop a BP device using an oscillometric method has reported a mean of BP errors $\bar{x}$ as -0.7 and SD as 6.9 for a sample size of 33 [6]. The SP10 standard states that 85 samples should be used, and for $\bar{x} = -0.7$ mean BP error, the SD should not be more than 6.95. Even though the SD reported by the study with 33 samples is less than 6.95, it would be incorrect to interpret this as compliance with the SP10 criteria because the smaller sample size will also influence the acceptability. By evaluating P_A, we can analyze that effect. For this study, $P_A \sim 0.87$, which is less than the threshold of acceptability upon which SP10 is based, i.e., $P_A \geq 0.95$. Hence, the device does not meet the criteria of acceptability.

Table 5. Comparison statistics of the previous clinical studies that have reported device inaccuracy based on SBP values.

Study	Method/Techniques	Sample Size	Device Inaccuracy ($\bar{x} \pm SD$)	Probability, P_A ($p \geq 0.78$)
[6]	Oscillometry	33	-0.7 ± 6.9	0.873
[21]	PTT	10	1.04 ± 6.88	0.730
[7]	PTT-PPG	33	1.17 ± 5.72	0.997
[23]	Standing	25	-0.462 ± 8	0.539
[16]	PAT	32	0.12 ± 6.15	0.984
[26]	PTT	33	-0.06 ± 6.63	0.934
[24]	PTT-linear	20	0 ± 6.73	0.859
	PTT-nonlinear		0 ± 5.56	0.995
[15]	ML	45	4.53 ± 2.68	0.999

Table 6. Comparison statistics of the previous clinical studies that have reported device inaccuracy based on DBP values.

Study	Method/Techniques	Sample Size	Device Inaccuracy ($\bar{x} \pm SD$)	Probability, P_A ($p \geq 0.78$)
[21]	PTT	10	-2.16 ± 6.60	0.732
[7]	PTT-PPG	33	0.40 ± 7.11	0.825
[25]	PTT-IPG	15	-0.5 ± 5.07	0.999
[27]	PWV-	15	-0.06 ± 5.46	0.991
[6]	Oscillometry	33	-1.0 ± 5.1	0.999
[16]	PAT	32	1.31 ± 5.36	0.999
[26]	PTT	33	-0.25 ± 5.63	0.999

6. Discussion and Conclusions

International standards such as ISO 86010-2 serve an important purpose in providing clarity to consumers, manufacturers, and regulators that medical devices (at least with respect to the scope of the standard) are safe and effective. With this purpose, standards provide clear pass/fail criteria, which reflect the level of a device's performance and acceptability. In this regard, the pass/fail criteria set out in ISO 86010-2:2018, inherited from SP10, are broadly recognized and represent an implicit definition of what constitutes acceptable errors in blood pressure measurement. In SP10, the mathematical translation of this definition into pass/fail criteria utilizes an approximate approach that results in formulas for confidence intervals that are functions only of the sample size, disregarding the sampling errors in the form of estimated probability of tolerable error. In this work, we have proposed a method using a solid statistical theory to determine confidence intervals. The proportion's sampling distribution is a more accurate statistical approach for studying the random errors producing p since it additionally takes the mean and the standard deviation of measurement errors into consideration.

By detailing the expected changes in device acceptability, the paper contributes valuable knowledge to the existing research in this field. This work also provides an adjusted acceptance limit of BP errors based on the same definition of acceptable performance underlying the SP10 standard for studies that use a sample size of less than 85. The adjusted limits are expected to be useful in the initial validation of BP technologies. Device manufacturers can use these adjusted acceptance limits to estimate compliance with the standards using smaller sample sizes, reducing the cost of development and/or allowing faster iterative development.

An important use case for this research is the ability to compare reported results, for example, when there are several technologies/methods/algorithms being used for estimating BP, and the individual reports utilize different sample sizes, as shown in Table 5. In each case, the reported device inaccuracy ($\bar{x}$ and SD) appears to be within the SP10 criteria. However, the sample sizes are significantly smaller. Using the methods presented here, the calculation of the probability of acceptance, P_A, allows a quantitative comparison of the existing literature and also to the SP10 criteria. For most of the presented studies, it is apparent that the reported results do not reach an equivalent level of confidence to SP10. P_A allows direct comparison of results with different means and standard deviations, for example, a study with a high mean and low SD [15] and a study with a much lower mean and higher SD [26]. Many hypertension societies now offer clinicians a comprehensive list of blood pressure measurement devices, facilitating informed decision-making for clinical trials. Currently, the list can only assess whether the device is recommended or not based on its performance, utilizing the acceptance range specified in the standards [28]. With the introduction of the proposed method, clinicians can now more appropriately compare the reported inaccuracies across varying sample sizes. This functionality empowers clinicians when evaluating devices for their specific research needs.

We demonstrate that no more than 70 samples are required to maintain the 85% estimated probability of tolerable error, as opposed to the N = 85 stated in the standards, as illustrated in Figure 4. There are cases where being able to correctly interpret the results of a study with a smaller sample would be beneficial, for example, with a population subset with only infants. Our framework still makes a statistical assumption of reasonably large sample sizes, ideally $N \geq 39$. Although the proposed method is optimized for at least 39 samples, it is instructive to see how the framework performs for fewer samples. For this, we performed a simulation study in which we varied the mean and the standard deviation of the error distribution. We experimented with high variance but not exceeding the SP10 mandate for the standard deviation, that is, $\sigma \leq 6.9$. The approximations with the proposed framework demonstrated closer results compared to the approximations with the existing framework. Table 2 presents one such result, while additional scenarios are elaborated in detail in Appendix D. The results become more distant for both frameworks as the sample size decreases. We witness this relation because with smaller sample sizes, the tendency of the sampling distribution to approximate normal distribution decreases. These results are further confirmation that caution should be applied when using smaller sample sizes, particularly when N < 39. To extend the proposed framework for less than 39 samples, further research is required.

This study presents a formal statistical evaluation of the device's conformity with international standards, primarily through the evaluation of the probability of acceptance P_A depending on the mean error, the standard deviation of the error, and the sample size. While evaluating a device's inaccuracy, international standards also mandate that the device follows guidelines in regard to selecting the cuff size, providing the subject with a resting period prior to measurement, etc. Any deviation from these protocols has the potential to introduce bias, though the specific impact remains inconclusive as most studies do not explicitly state whether these protocols were adhered to. Additionally, to ensure enough samples in varied categories, including covering high and low blood pressure groups, ranges, and distributions of arm sizes, a minimum device sample size of N = 85 is necessary. In such cases, a smaller sample size will not be representative of the population.

With significant modifications for the standard SP10, we introduce a mathematical framework to accommodate different underlying definitions of acceptable error and confidence. This is of relevance to those developing new BP measurement technologies which are often tested initially in smaller samples, such as cuffless, wearable BP measurement devices that perform continuous readings to gather trends for a long period of time. These technologies help in assessing real-time fluctuations which might be useful for clinical trials that aim to gather longitudinal data on blood pressure trends and responses to interventions.

Finally, to assist in the calculations presented, this paper also introduces the companion R package "bpAcc", an implementation of this methodology involving functions to directly compute the acceptance limit and P_A without having to deal with the mathematical complexities. At present, our framework has the infrastructure to afford normally distributed errors, as stated by the argument "distribution" from "PAccept()" and "AcceptR()". Future work includes upgrading both functions to handle errors other than normal. Initial steps have been taken in this direction with both functions being currently trained and tested using the one-parameter (λ, or degrees of freedom) Student-t distribution. We are focused on selecting real-valued distributions with practical benefits for clinicians and manufacturers, rather than a theory-based selection of choices. Essentially, more data are useful, but experiments are often expensive. We aim to provide choices spanning various sample sizes by providing statistical infrastructure to maximize the user's ability to identify faulty devices (e.g., Type I error) for BP measurement. Over time, both functions will be enhanced with further arguments to handle, e.g., criteria other than "SP10:2006".

Author Contributions: Conceptualization, T.C. and V.M.; methodology, T.C. and V.M.; validation, V.M. and A.L.; formal analysis, T.C.; writing—original draft preparation, T.C.; writing—review and editing, A.L. and T.C.L.; visualization, V.M. and T.C.; supervision, V.M., A.L. and T.C.L.; funding acquisition, A.L. All authors have read and agreed to the published version of the manuscript.

Funding: This research was funded by NZ MBIE Endeavour Smart Ideas grant, grant number "AUTX1904".

Institutional Review Board Statement: Not applicable.

Informed Consent Statement: Not applicable.

Data Availability Statement: No new data were created or analyzed in this study. Data sharing is not applicable to this article.

Conflicts of Interest: The authors declare no conflicts of interest.

Appendix A

Table A1. This is Table F.1 in [20]. It shows the upper limit on the sample standard deviation to yield at least 85% estimated probability of tolerable error.

Sample Mean Error	Standard Deviation
0	6.95
±0.5	6.93
±1.0	6.87
±1.5	6.78
±2	6.65
±2.5	6.93
±3.0	6.87
±3.5	6.78

Table A1. *Cont.*

Sample Mean Error	Standard Deviation
±4	6.65
±4.5	6.93
±5.0	6.87

Appendix B

This section contains Algorithm 1 that is useful to generate the probability values present in Table 2. The algorithm is used to evaluate the probability of acceptance using the simulation study.

Algorithm 1

1. function simulate (μ, σ, n)
2. $\quad \delta := 10$
3. $\quad$ Set $\hat{p}$ as the probability of tolerable error using (μ, σ, δ)
4. $\quad \kappa := 20000$
5. $\quad$ Set P, Γ, Υ to φ
6. $\quad \epsilon := -10$
7. $\quad E := 10$
8. $\quad$ for i = 1 to κ do
9. $\quad\quad$ Set Θ as a randomly normalized array using (n, μ, σ)
10. $\quad\quad \varkappa := 0$
11. $\quad\quad$ for each θ in Θ do
12. $\quad\quad\quad$ if $\theta > \epsilon$ and $\theta < E$ then
13. $\quad\quad\quad\quad \varkappa := \varkappa + 1$
14. $\quad\quad\quad$ end if
15. $\quad\quad$ end for
16. $\quad\quad \varkappa := \varkappa / n$
17. $\quad\quad$ Insert $\varkappa$ in P
18. $\quad\quad$ Insert mean(Θ) in Γ
19. $\quad\quad$ Insert stddev(Θ) in Υ
20. $\quad$ end for
21. $\quad \Omega := 0$
22. $\quad$ for each ρ in P do
23. $\quad\quad$ if $\rho > 0.78$ then
24. $\quad\quad\quad \Omega := \Omega + 1$
25. $\quad\quad$ end if
26. $\quad$ end for
27. $\quad \Omega := \Omega / \kappa$
28. $\quad$ return Ω
29. end function

Appendix C

Table A2. List of parameters used in Section 3 with their description.

Parameter	Description
ϵ_k^j	BP error for the jth participant, which is the difference between the test device measurement, $\delta_{k_2}^j$, and the reference device, $\delta_{k_1}^j$
δ^j	Average of the errors for three measurements for the jth participant
Δ	Tolerable error, i.e., errors within -10 mmHg and 10 mmHg
$\hat{p}$	Estimated probability of tolerable error

Table A2. Cont.

Parameter	Description
p	True probability of tolerable error
γ_p	Maximum value of the probability of a tolerable error
θ	Parameterized distribution for BP errors
μ_p, σ_p	True mean and true standard deviation, respectively
σ^{MAX}	Maximum standard deviation for a certain value of mean and γ_p
σ^{AAMI}	Maximum standard deviation for a certain value of mean as per SP10 where $\gamma_p = 0.85$
$\bar{x}_t, s_t$	Sample mean and sample standard deviation of BP errors, respectively
$\emptyset$	Cumulative density function of the standard normal distribution

Appendix D

Table A3. Simulated P_A and P_A obtained from normal approximation using proposed method vs. the method currently in use in the SP10 standard, for small sample sizes, with $\mu_P = 0$ and $\sigma_P = 6.9$.

	N = 10	N = 15	N = 20	N = 25
Simulated P_A	0.83	0.83	0.84	0.85
P_A using proposed framework	0.74	0.79	0.82	0.85
P_A using SP10 method	0.71	0.75	0.79	0.81

Table A4. Simulated P_A and P_A obtained from normal approximation using proposed method vs. the method currently in use in the SP10 standard, for small sample sizes, with $\mu_P = 2.5$ and $\sigma_P = 6.9$.

	N = 10	N = 15	N = 20	N = 25
Simulated P_A	0.75	0.74	0.74	0.74
P_A using proposed framework	0.65	0.68	0.71	0.73
P_A using SP10 method	0.64	0.67	0.69	0.72

Table A5. Simulated P_A and P_A obtained from normal approximation using proposed method vs. the method currently in use in the SP10 standard, for small sample sizes, with $\mu_P = 5$ and $\sigma_P = 6.9$.

	N = 10	N = 15	N = 20	N = 25
Simulated P_A	0.52	0.46	0.42	0.38
P_A using proposed framework	0.42	0.40	0.38	0.37
P_A using SP10 method	0.41	0.39	0.37	0.36

References

1. Handler, J. The Importance of Accurate Blood Pressure Measurement. *Perm. J.* 2009, *13*, 51–54. [CrossRef]
2. Stergiou, G.S.; Alpert, B.S.; Mieke, S.; Wang, J.; O'Brien, E. Validation protocols for blood pressure measuring devices in the 21st century. *J. Clin. Hypertens.* 2018, *20*, 1096–1099. [CrossRef]
3. Scheer, B.V.; Perel, A.; Pfeiffer, U.J. Clinical review: Complications and risk factors of peripheral arterial catheters used for haemodynamic monitoring in anaesthesia and intensive care medicine. *Crit. Care* 2002, *6*, 198–204. [CrossRef]
4. Meidert, A.S.; Saugel, B. Techniques for non-invasive monitoring of arterial blood pressure. *Front. Med.* 2017, *4*, 231. [CrossRef] [PubMed]
5. Jones, D.W.; Appel, L.J.; Sheps, S.G.; Roccella, E.J.; Lenfant, C. Measuring Blood Pressure Accurately New and Persistent Challenges. *JAMA* 2003, *289*, 1027–1030. [CrossRef] [PubMed]
6. Wang, Q.; Zhao, H.; Chen, W.; Li, N.; Wan, Y. Validation of the iHealth BP7 wrist blood pressure monitor, for self-measurement, according to the European Society of Hypertension International Protocol revision 2010. *Blood Press. Monit.* 2014, *19*, 54–57. [CrossRef] [PubMed]

7. Ding, X.; Yan, B.P.; Zhang, Y.T.; Liu, J.; Zhao, N.; Tsang, H.K. Pulse Transit Time Based Continuous Cuffless Blood Pressure Estimation: A New Extension and A Comprehensive Evaluation. *Sci. Rep.* **2017**, *7*, 1–11. [CrossRef] [PubMed]
8. Imai, Y.; Sasaki, S.; Minami, N.; Munakata, M.; Hashimoto, J.; Sakuma, H.; Sakuma, M.; Watanabe, N.; Imai, K.; Sekino, H.; et al. The accuracy and the performance of the A&D TM 2421, a new ambulatory blood pressure monitoring device based on the cuff-oscillometric method and the Korotkoff sound technique. *Am. J. Hypertens.* **1992**, *5*, 719–726.
9. ISO. *Non-Invasive Sphygmomanometers—Part 2: Clinical Investigation of Intermittent*; ISO: Geneva, Switzerland, 2018.
10. *EN 1060-4:2004*; Non-Invasive sphygmomanometers—Part 4: Test Procedures to Determine the Overall System Accuracy of Automated non-invasive Sphygmomanometers. European Committee for Standardization; BSI: London, UK, 2004.
11. O'Brien, N.A.E.; Petrie, J.; Littler, W.; de Swiet, M.; Padfield, P.L.; O'Malley, K.; Jamieson, M.; Altman, D.; Bland, M. The British Hypertension Society protocol for the evaluation of automated and semi-automated blood pressure measuring devices with special reference to ambulatory systems. *J. Hypertens.* **1990**, *8*, 607–619. [CrossRef] [PubMed]
12. O'brien, E.; Pickering, T.; Asmar, R.; Myers, M.; Parati, G.; Staessen, J.; Mengden, T.; Imai, Y.; Waeber, B.; Palatini, P.; et al. Working Group on Blood Pressure Monitoring of the European Society of Hypertension International Protocol for validation of blood pressure measuring devices in adults. *Blood Press. Monit.* **2002**, *7*, 3–17. [CrossRef] [PubMed]
13. Stergiou, G.S.; Alpert, B.; Mieke, S.; Asmar, R.; Atkins, N.; Eckert, S.; Frick, G.; Friedman, B.; Graßl, T.; Ichikawa, T.; et al. A universal standard for the validation of blood pressure measuring devices: Association for the Advancement of Medical Instrumentation/European Society of Hypertension/International Organization for Standardization (AAMI/ESH/ISO) Collaboration Statement. *Hypertension* **2018**, *71*, 368–374. [CrossRef]
14. Alpert, B.; Friedman, B.; Osborn, D. AAMI blood pressure device standard targets home use issues. *Biomed. Instrum. Technol.Assoc. Adv. Med. Instrum.* **2010**, 69–72.
15. Jung, Y.K.; Baek, H.C.; Soo, M.I.; Myoung, J.J.; In, Y.K.; Sun, I.K. Comparative study on artificial neural network with multiple regressions for continuous estimation of blood pressure. In Proceedings of the Annual International Conference of the IEEE Engineering in Medicine and Biology—Proceedings, Shanghai, China, 17–18 January 2005; Volume 7, pp. 6942–6945. [CrossRef]
16. Esmaili, A.; Kachuee, M.; Shabany, M. Nonlinear Cuffless Blood Pressure Estimation of Healthy Subjects Using Pulse Transit Time and Arrival Time. *IEEE Trans. Instrum. Meas.* **2017**, *66*, 3299–3308. [CrossRef]
17. Dong, Y.; Kang, J.; Yu, Y.; Zhang, K.; Li, Z.; Zhai, Y. A Novel Model for Continuous Cuff-less Blood Pressure Estimation. In Proceedings of the 2018 11th International Symposium on Communication Systems, Networks & Digital Signal Processing (CSNDSP), Budapest, Hungary, 18–20 July 2018.
18. Kallioinen, N.; Hill, A.; Horswill, M.S.; Ward, H.E.; Watson, M.O. Sources of inaccuracy in the measurement of adult patients' resting blood pressure in clinical settings: A systematic review. *J. Hypertens.* **2017**, *35*, 421–441. [CrossRef]
19. Bonnafoux, P. Auscultatory and oscillometric methods of ambulatory blood pressure monitoring, advantages and limits: A technical point of view. *Blood Press. Monit.* **1996**, *1*, 181–185.
20. *ANSI/AAMI SP10:2002/(R)2008*; American National Standard for Manual, Electronic, or Automated Sphygmomanometers. American National Standards Institute: New York, NY, USA, 2008.
21. Lui, H.-W.; Chow, K.-L. A Novel Calibration Procedure of Pulse Transit Time based Blood Pressure measurement with Heart Rate and Respiratory Rate. In Proceedings of the Annual International Conference of the IEEE Engineering in Medicine and Biology Society, EMBS, Honolulu, HI, USA, 18–21 July 2018; Volume 2018, pp. 4318–4322. [CrossRef]
22. Victor, M.; Tanvi, C.; Andrew, L.; Tet, L.C. *_bpAcc: Blood Pressure Device Accuracy Evaluation: Statistical Considerations.*; R package Version 0.0-2; CRAN: Auckland, New Zealand, 2024.
23. Pandian, P.; Mohanavelu, K.; Safeer, K.; Kotresh, T.; Shakunthala, D.; Gopal, P.; Padaki, V. Smart Vest: Wearable multi-parameter remote physiological monitoring system. *Med. Eng. Phys.* **2008**, *30*, 466–477. [CrossRef]
24. Wibmer, T.; Doering, K.; Kropf-Sanchen, C.; Rüdiger, S.; Blanta, I.; Stoiber, K.M.; Rottbauer, W.; Schumann, C. Pulse Transit Time and Blood Pressure During Cardiopulmonary Exercise Tests. *Physiol. Res.* **2014**, *63*, 287–296. [CrossRef] [PubMed]
25. Huynh, T.H.; Jafari, R.; Chung, W.Y. Noninvasive cuffless blood pressure estimation using pulse transit time and impedance plethysmography. *IEEE Trans. Biomed. Eng.* **2019**, *66*, 967–976. [CrossRef] [PubMed]
26. Masè, M.; Mattei, W.; Cucino, R.; Faes, L.; Nollo, G. Feasibility of cuff-free measurement of systolic and diastolic arterial blood pressure. *J. Electrocardiol.* **2011**, *44*, 201–207. [CrossRef] [PubMed]
27. Huynh, T.H.; Jafari, R.; Chung, W.Y. A robust bioimpedance structure for smartwatch-based blood pressure monitoring. *Sensors* **2018**, *18*, 2095. [CrossRef]
28. Padwal, R.; Berg, A.; Gelfer, M.; Tran, K.; Ringrose, J.; Ruzicka, M.; Hiremath, S.; Accuracy in Measurement of Blood Pressure (AIM-BP) Collaborative. The Hypertension Canada blood pressure device recommendation listing: Empowering use of clinically validated devices in Canada. *J. Clin. Hypertens.* **2020**, *22*, 933–936. [CrossRef] [PubMed]

Disclaimer/Publisher's Note: The statements, opinions and data contained in all publications are solely those of the individual author(s) and contributor(s) and not of MDPI and/or the editor(s). MDPI and/or the editor(s) disclaim responsibility for any injury to people or property resulting from any ideas, methods, instructions or products referred to in the content.

Article

Machine Learning Approaches to Predict Major Adverse Cardiovascular Events in Atrial Fibrillation

Pedro Moltó-Balado [1,2], Silvia Reverté-Villarroya [3,*], Victor Alonso-Barberán [4], Cinta Monclús-Arasa [1], Maria Teresa Balado-Albiol [5], Josep Clua-Queralt [6] and Josep-Lluis Clua-Espuny [6,7,*]

[1] Primary Health-Care Center Tortosa Oest, Institut Català de la Salut, Primary Care Service (SAP) Terres de l'Ebre, CAP Baix Ebre Avda de Colom, 16-20, 43500 Tortosa, Spain; pemolto@gmail.com (P.M.-B.)
[2] Biomedicine Doctoral Programme, Universitat Rovira I Virgili, 43500 Tortosa, Spain
[3] Nursing Department, Advanced Nursing Research Group at Rovira I Virgili University, Biomedicine Doctoral Programme Campus Terres de l'Ebre, Av. De Remolins, 13, 43500 Tortosa, Spain
[4] Institut d'Educació Secundària El Caminàs, C/Pintor Soler Blasco, 3, Conselleria d'Educació, 12003 Castellón, Spain
[5] Primary Health-Care Center CS Borriana I, Conselleria de Sanitat, Avinguda Nules, 31, 12530 Borriana, Spain
[6] Primary Health-Care Center EAP Tortosa Est, Institut Català de la Salut, CAP El Temple Plaça Carrilet, s/n, 43500 Tortosa, Spain; jcluaq@gmail.com
[7] Research Support Unit Terres de l'Ebre, Institut Universitari d'Investigació en Atenció Primària Jordi Gol (IDIAPJGol) (Barcelona), Ebrictus Research Group, Terres de l'Ebre, 43500 Tortosa, Spain
* Correspondence: silvia.reverte@urv.cat (S.R.-V.); jlclua@telefonica.net (J.-L.C.-E.); Tel.: +34-977510018 (J.-L.C.-E.)

Citation: Moltó-Balado, P.; Reverté-Villarroya, S.; Alonso-Barberán, V.; Monclús-Arasa, C.; Balado-Albiol, M.T.; Clua-Queralt, J.; Clua-Espuny, J.-L. Machine Learning Approaches to Predict Major Adverse Cardiovascular Events in Atrial Fibrillation. *Technologies* 2024, 12, 13. https://doi.org/10.3390/technologies12020013

Academic Editors: Juvenal Rodriguez-Resendiz, Gerardo I. Pérez-Soto, Karla Anhel Camarillo-Gómez, Saul Tovar-Arriaga and Jeffrey W. Jutai

Received: 16 December 2023
Revised: 20 January 2024
Accepted: 21 January 2024
Published: 23 January 2024

Copyright: © 2024 by the authors. Licensee MDPI, Basel, Switzerland. This article is an open access article distributed under the terms and conditions of the Creative Commons Attribution (CC BY) license (https://creativecommons.org/licenses/by/4.0/).

Abstract: The increasing prevalence of atrial fibrillation (AF) and its association with Major Adverse Cardiovascular Events (MACE) presents challenges in early identification and treatment. Although existing risk factors, biomarkers, genetic variants, and imaging parameters predict MACE, emerging factors may be more decisive. Artificial intelligence and machine learning techniques (ML) offer a promising avenue for more effective AF evolution prediction. Five ML models were developed to obtain predictors of MACE in AF patients. Two-thirds of the data were used for training, employing diverse approaches and optimizing to minimize prediction errors, while the remaining third was reserved for testing and validation. AdaBoost emerged as the top-performing model (accuracy: 0.9999; recall: 1; F1 score: 0.9997). Noteworthy features influencing predictions included the Charlson Comorbidity Index (CCI), diabetes mellitus, cancer, the Wells scale, and CHA_2DS_2-VASc, with specific associations identified. Elevated MACE risk was observed, with a CCI score exceeding 2.67 ± 1.31 ($p < 0.001$), CHA_2DS_2-VASc score of 4.62 ± 1.02 ($p < 0.001$), and an intermediate-risk Wells scale classification. Overall, the AdaBoost ML offers an alternative predictive approach to facilitate the early identification of MACE risk in the assessment of patients with AF.

Keywords: atrial fibrillation; major adverse cardiovascular events (MACE); machine learning; artificial intelligence

1. Introduction

Despite being the most prevalent cardiac arrhythmia, the early identification, diagnosis, and treatment of atrial fibrillation (AF) remain challenging. AF affects millions of individuals globally and is linked to a heightened risk of stroke, heart failure, and mortality [1–4]. These medical conditions collectively fall under the term Major Adverse Cardiovascular Events (MACE) and are subject to extensive research [5]. The diagnosis of AF is associated with a fourfold increase in heart failure incidence and an eightfold increase in MACE occurrence [6].

Risk factors for MACE in AF patients have been identified as age, gender, hypertension, diabetes (known as "traditional"), biomarkers, genetic variants, imaging parameters, and left atrial function [7–10]. In recent years, there has been growing interest in identifying new predictors of MACE in AF patients [11] beyond traditional ones such as obesity, chronic obstructive pulmonary disease (COPD), or chronic renal failure [7,8,12]; this novel approach is associated with a reduced risk of MACE, including mortality and thromboembolism [13].

Several proposals for stroke risk assessment in AF have been developed, such as CHA_2DS_2-VASc [14], the Framingham score [15], Anticoagulation and Risk Factors in Atrial Fibrillation (ATRIA) [16], Cohorts for Heart and Aging Research in Genomic Epidemiology for Atrial Fibrillation (CHARGE-AF) [17,18], and Atrial Fibrillation Research In CATalonia (AFRICAT) [19]. However, there are still challenges and limitations with clinical risk scores that restrict their applicability to certain populations. Moreover, the discriminatory ability of clinical risk scores in predicting stroke risk for an individual is at best moderate [20]. For MACE risk specifically, some studies [21,22] have proposed additional scoring systems or modifications to existing scores to better predict cardiovascular events in patients with AF. Leveraging artificial intelligence (AI) and machine learning (ML) techniques on electronic health record (EHR) data offers a potential avenue to further refine these risk prediction models. However, it is important to note that the extent of performance improvement achieved through AI and ML approaches can vary [23,24].

Therefore, more comprehensive risk prediction models incorporating a wider range of predictors or with more prognostic value are needed. Such models can be achieved using ML algorithms, which offer a promising approach in AF patients [2], as they can integrate large amounts of data from multiple sources and identify complex patterns and correlations that may not be evident using traditional statistical methods.

The heterogeneous mechanisms and risk factors associated with AF make it necessary to target personalized treatment approaches, requiring extensive patient data to identify specific patterns. AI algorithms are particularly suitable for handling high-dimensional data, predicting outcomes, and ultimately optimizing strategies for patient management [25]. Recent advances in ML have resulted in great success and have also been utilized to analyze electrocardiogram (ECG) data and predict the future occurrence of arrhythmias. Future Innovations in Novel Detection for Atrial Fibrillation (FIND-AF), an extensively scalable ML algorithm, is capable of analyzing routinely collected primary care data to identify individuals with an elevated risk of short-term AF [26]. Other studies have demonstrated the utility of machine learning-based models in AF for real-time identification of a variety of rhythms using 12-lead or single-lead ECG recordings, as well as for diagnosis, outcome prediction, disease characterization, and treatment assessment [2,27–33]. However, they do not address the discrimination of cardioembolic from noncardioembolic stroke among individuals with AF with high accuracy and surpassing traditional risk scores. These methods provide precise and efficient algorithms for data analysis, improving prediction accuracy, pattern identification, and task automation. If patients at higher risk of MACE could be identified, treatment strategies could be developed to potentially reduce incidence and associated complications.

The primary objectives of this study encompassed the identification of noteworthy clinical indicators associated with MACE in patients with new AF. It further aimed to assess the prognostic impact of these predictors within a community cohort, aged 65–95 years, tracked from 2015 to 2021.

2. Materials and Methods
2.1. Study Design

This was an observational study, and the data were retrospectively collected where possible, or manually collected otherwise. The specific codes of the International Classification of Diseases (ICD-10) were used. The project encompassed the broader demographic of individuals aged 65–95 years ($n = 40,297$) who did not have AF as part of their inclusion criteria and was conducted within the Primary Care facilities of Terres de l'Ebre, located in Catalonia, Spain, during the period spanning from 1 January 2015 to 31 December 2021.

The data were available from the electronic medical datasets (E-CAP and SAP) managed by the Catalan Health Institute (ICS), which collect information from primary care centers and hospitals in the health region anonymously and without contact with the cases included, as follows:

1. The Health Plan [33] outlines healthcare priorities in the "Terres de l'Ebre" Healthcare Region (Catalonia, Spain) from 2021 to 2025.
2. The HC3 Patient Episode Dataset provides clinical information of care on inpatient and outpatient care in Catalan hospitals.
3. The clinical database of 11 primary care teams includes comprehensive health data for 97.7% of residents, covering symptoms, tests, diagnoses, comorbidities, prescribed medication, and referrals.
4. The Integrated System of Electronic Prescription (SIRE) captures information on prescribed medications.
5. The Statistics Institute of Catalonia includes demographic information [34–36].

The datasets generated, used, and analyzed during the current study are available from the corresponding author on reasonable request.

2.2. Eligibility Criteria

All patients over 65 years of age from Terres de l'Ebre (N 55,459) without AF or MACE in their clinical history were considered, and the following criteria were defined:

1. Outcomes: AF patients who had a MACE.
2. Inclusion criteria: Subjects aged 65–95 years who met the inclusion criteria: high risk-AF (according to the risk model and belonging to Q4) [19], active clinical history in any of the health centers of the territory with information accessible through the shared history (HC3), without previous AF or MACE, residing in the territory, and attached to any of the Primary Care Teams (EAP) of the territory.
3. Exclusion criteria: under 65 years of age or over 95 years of age, living outside Terres de l'Ebre, a previous diagnosis of AF, treatment with anticoagulants, impaired cognitive status, Barthel score < 55 points, or pacemaker or defibrillator wearer. Non-availability or loss of accessibility to the information necessary for the study was considered a reason for exclusion.

2.3. Data and Preprocessing

The overall composition of the dataset for MACE prediction is given in Table 1 Numerical calculations and data analysis were performed using Python library version 3. Code and models used for the analysis are available online (https://github.com/vmalonsobarberan/MACE) (accessed on 15 December 2023).

Table 1. Comparison of the performance of different models.

Machine Learning Model	Accuracy	Precision	Recall	F1 Score	Sensitivity	Specificity	PPV	NPV	AUC
Random Forest	96.78%	0.8456	0.9263	0.8841	0.9885	0.8456	0.9741	0.9263	96.78%
Extra Trees	98.82%	0.9641	0.9554	0.9597	0.9923	0.9641	0.9938	0.9554	98.82%
AdaBoost	99.99%	0.9994	1	0.9997	1	0.9994	0.9999	1	99.99%
XGBoost	99.95%	1	0.9971	0.9985	0.9995	1	1	0.997	99.95%
LightGBM	99.96%	1	0.9977	0.9988	0.9996	1	1	0.9977	99.96%

2.4. Model Development

To develop ML models for estimation, we took features of the individuals with newly diagnosed AF who developed MACE, following the eligibility criteria. ML model development was performed using the SKLearn and TensorFlow libraries due to their versatility and ease of programming. For each fold, hyperparameters were tuned on training data using a randomized search after the determination of a candidate hyperparameter set. Evaluation of validation data was performed using the metrics described in the next section.

Five different ML models were implemented based on the following algorithms: Random Forest, Extra Trees, AdaBoost, XGBoost, and LightGBM. They were trained on all the features (variables) used in the study to predict the development of MACE within one year as well as to predict the development of AF.

A fundamental part of the study, prior to the construction of the learning models, consisted of "Feature Engineering", which consists of the analysis and selection of the variables, as well as the processing of the data they contain. To this end, those that only contribute noise and/or are correlated with others that have a greater influence on the objective we aimed to predict were eliminated.

The performance of MACE prediction was quantified using the following metrics: precision, recall, accuracy, and F1 score (combination of precision and recall). Two thirds of the data (36,973) were randomly selected for training and model building using different approaches and optimized to reduce the prediction error. The remaining 1/3 (18,486) was used for testing and validation. The models underwent testing using this separate test data to assess their performance on data that had not been utilized during the training phase. This evaluation aimed to determine whether the models could effectively generalize and make accurate predictions on unseen data.

2.5. Model Performance Analysis

Several metrics were used to evaluate the algorithms, including prediction robustness, completeness, sensitivity, specificity, precision, recall, accuracy, and F1 score (combination of precision and recall). Evaluation of these metrics allowed us to adjust the hyperparameters of the model to improve the most desirable aspect of the model. The model with the highest and most robust performance was chosen after evaluating the performance of the different models using the mean value of the area under the ROC curve. The assessment of our models included consideration of the standard deviation of the results to evaluate their stability, along with an analysis of sensitivity, specificity, and accuracy. After fitting and evaluating different models, the best model was selected, and the hyperparameters were adjusted to obtain the optimal results.

2.6. Model Interpretability

The Shapley Additive exPlanations (SHAP) method was used to analyze which factors were the most important and to what extent they contribute to the model's predictions. An individual automatic explainability model was also created to allow an analysis to be made for each individual patient. The latter allows, after analyzing a patient's variables, to explain how likely a patient with AF is to have a MACE and which factors contribute to this prediction and to what extent.

2.7. Statistical Analysis

The traditional statistical analysis of the baseline data was previously documented [6].

3. Results

3.1. Study Population Patient Characteristics

The study encompassed a cohort of 2574 individuals devoid of prior MACE incidents, with a mean age of 81.22 ± 7.91 years and a gender distribution of 52.01% women. A detailed analysis of baseline characteristics, as outlined previously [6], revealed notable distinctions among the study groups. Notably, women who experienced MACE exhibited a higher mean age (82.23 ± 7.59 years, compared to 80.53 ± 8.05 years for males, $p < 0.001$) and a higher prevalence of cardiovascular risk factors and comorbidities. Refer to Table 2 for a comprehensive overview of the selected variables instrumental in model construction.

Table 2. Distribution of AF patients according to the presence of MACE.

Variables	No MACE	(%)	MACE	(%)	p	All
All	1527	59.32%	1047	40.68%		2574
Woman	785	51.41%	558	53.30%	0.356	1343
Age average	80.53 ± 8.05		82.23 ± 7.59		<0.001	81.22 ± 7.91
Hypertension, arterial	1112	72.82%	833	79.56%	<0.001	1945
Diabetes mellitus	406	26.59%	363	34.67%	<0.001	769
Dyslipemia	692	45.32%	524	50.05%	0.020	1216
Vascular disease	59	3.86%	286	27.32%	<0.001	345
Dementia/cognitive impairment	174	11.39%	136	12.99%	<0.001	310
Liver disease	6	0.39%	4	0.38%	1.000	10
Renal failure	339	22.20%	337	32.19%	<0.001	676
Cancer	516	33.79%	340	32.47%	0.496	856
Thyroid disease	109	7.14%	106	10.12%	0.018	215
OSAHS [1]	60	3.93%	66	6.30%	0.007	126
COPD [2]	225	14.73%	222	21.20%	<0.001	447
Inflammatory disease (Crohn's and Colitis)	9	0.59%	7	0.67%	0.804	16
Deep vein thrombosis	20	1.31%	17	1.62%	0.506	37
Weight (kg)	77.47 ± 5.7		78.03 ± 16.51		0.038	77.69 ± 16.04
BMI [3]	29.32 ± 5.28		29.75 ± 5.51		0.041	29.49 ± 5.38
Heart rate/min	76.05 ± 1847		75.71 ± 18.47		0.625	75.91 ± 18.47
Cholesterol mg/dL	184.23 ± 38.07		164.98 ± 38.14		<0.001	176.4 ± 39.24
ProBNP (pg/mL)	1550		3301.75 ± 2882.7		0.625	2951.4 ± 2616.52
Dimer D (ng/mL)	1753.59 ± 2714.47		1319.72 ± 2954.13		0.337	1532.56 ± 2838.47
Glomerular filtration rate (mL/min/1.73 m^2)	66.11 ± 19.8		59.85 ± 20.74		<0.001	63.48 ± 20.43
Serum albumin (g/dL)	4.94 ± 5.43		5.04 ± 14.85		0.835	4.98 ± 10.68
Lymphocytes (×10^3/µL)	2.12 ± 1.11		2.02 ± 1.62		0.072	2.08 ± 1.34
Statins	505	33.07%	607	57.98%	<0.001	945
Anticoagulation	1207	79.04%	787	75.16%	0.021	1994
Antivitamin-K	613	40.14%	331	31.61%	<0.001	944
NOAC [4]	595	38.96%	458	43.74%	0.015	1053
Anti-aggregants	67	4.38%	74	7.06%	0.003	141
Pfeiffer score ± SD	2.91 ± 3.1		2.61 ± 2.8		0.218	2.75 ± 2.94
CHA$_2$DS$_2$-VASc ± SD	3.26 ± 0.95		4.62 ± 1.02		<0.001	3.81 ± 1.20
CCI [5] ± SD	1.24 ± 1.19		2.67 ± 1.31		<0.001	1.82 ± 1.43
CONUT score ± SD	1.31 ± 0.54		1.48 ± 0.61		<0.001	1.38 ± 0.58
Wells score ± SD	1.35 ± 0.48		1.33 ± 0.47		0.415	1.34 ± 0.47
COVID-19	150	9.82%	110	10.51%	0.573	260
Death	1279	83.76%	777	74.21%	<0.001	2056

[1]. OSAHS: obstructive sleep apnea-hypopnea syndrome; [2]. COPD: chronic obstructive pulmonary disease; [3]. BMI: Body Mass Index; [4]. NOAC: new oral anticoagulants; [5]. CCI: Charlson Comorbidity Index.

3.2. Machine Learning Model

3.2.1. Comparison between the Different Models

In the comparative analysis of various pre-trained models, AdaBoost emerged as the top-performing model, showcasing exceptional metrics, with an accuracy of 0.9999, recall of 1, and an F1 score of 0.9997. This marked superiority was evident, making AdaBoost the optimal choice, balancing both sensitivity and specificity (Figure 1).

Following closely behind, XGBoost (accuracy: 0.9995; recall: 0.9971; F1: 0.9985) and LightGBM (accuracy: 0.9996; recall: 0.9977; F1: 0.9988) emerged as the second-best models in our evaluation (Table 1). Notably, Random Forest and Extra Trees, while achieving commendable Area Under the Curve values (Figure 2), did not match the performance levels achieved by AdaBoost.

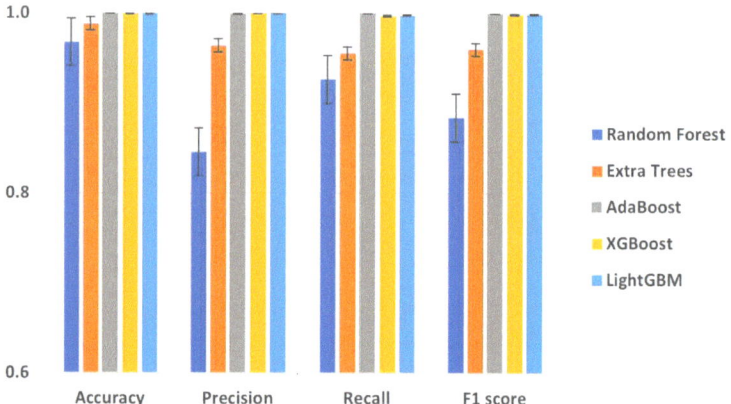

Figure 1. Comparison of the performance of different models.

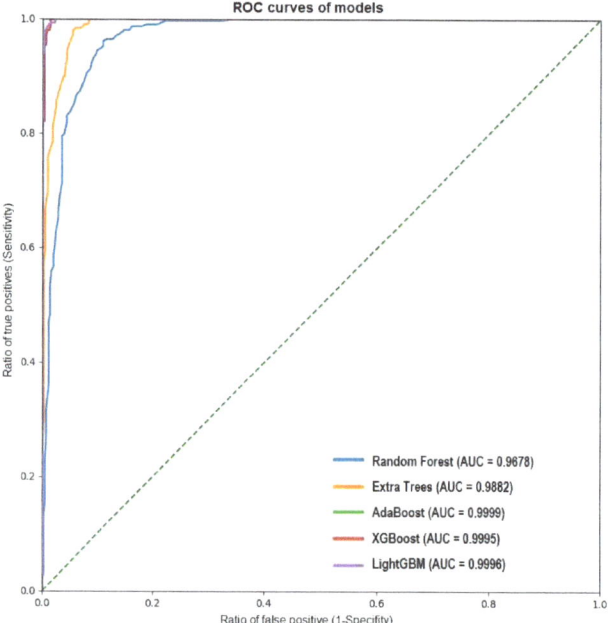

Figure 2. Comparison of AUC results between the machine learning models.

The confusion matrices of the different models and cross-validation were calculated. Each model has a confusion matrix. The models were ranked by true positive rates (Table 2).

3.2.2. Predictors by Outcomes

Figure 3 shows the main prognostic factors for MACE in AF patients. From most to least important were an elevated CCI, cancer, diabetes mellitus, COPD/asthma/bronchitis, cognitive impairment, vascular disease, high values of the CHA_2DS_2-VASc, and Wells scale.

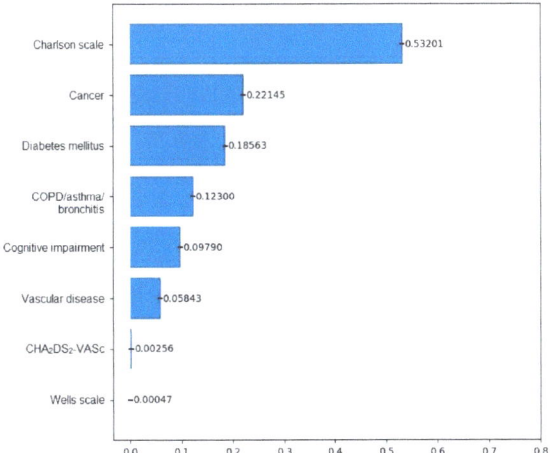

Figure 3. Strength of the main prognostic factors for MACE.

3.2.3. Model Interpretation

Figure 4 encapsulates a comprehensive overview of the feature contributions within the optimal model, AdaBoost. The SHAP (SHapley Additive exPlanations) summary chart delineates the significance of various characteristics, with the following five features emerging as the most influential: CCI, diabetes mellitus, cancer, Wells scale, and CHA_2DS_2-VASc. This SHAP summary chart not only identifies the primary features impacting the prediction but also quantifies their respective magnitudes through the SHAP values. The figures provide valuable insights into the relative importance of each feature, aiding in a nuanced understanding of the predictive dynamics within the AdaBoost model.

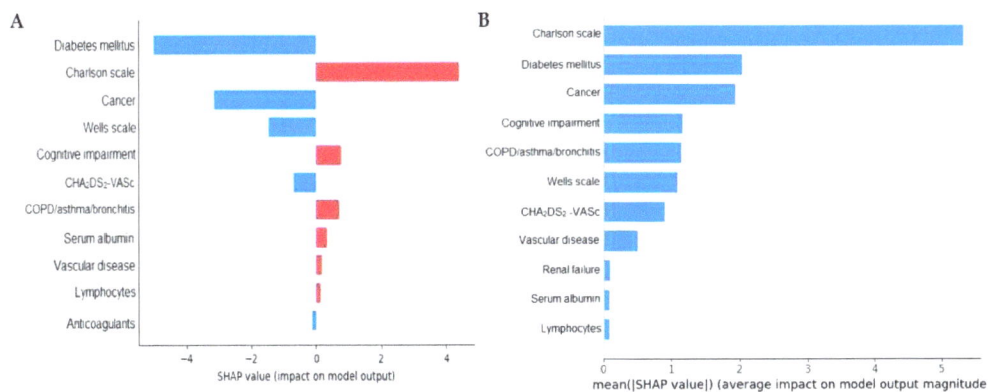

Figure 4. SHAP summary plot of optimum model. (**A**) The warm SHAP plot shows the distribution of SHAP values for each characteristic. (**B**) Bar chart according to feature importance.

Figure 5, the SHAP bar plot, serves as a visual representation elucidating the overall significance of each feature in predicting the occurrence of MACE. The height of the bars directly correlates with the importance of each feature to the model—higher bars denote greater importance. This graphical representation offers a clear and straightforward depiction of the overall magnitude and relevance of individual features in influencing the predictive outcome of MACE within the model. The visual emphasis on bar height facilitates an immediate understanding of the relative contributions of different features, enhancing the interpretability of the model's decision-making process.

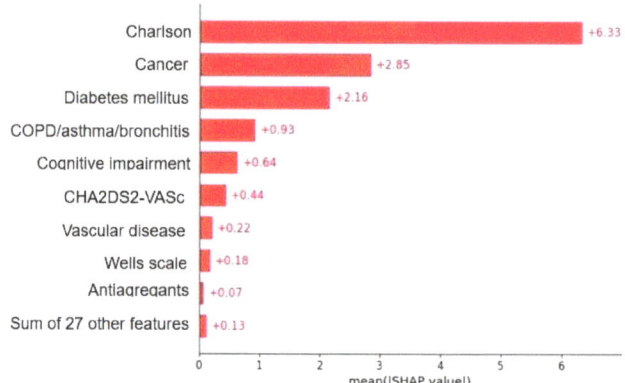

Figure 5. SHAP bar plot showing the overall magnitude and importance of the features.

The analysis delved into the influence of specific diseases, as outlined in the model, as predictors of MACE in AF patients. A CCI score exceeding 2.67 ± 1.31 ($p < 0.001$), a CHA_2DS_2-VASc score of 4.62 ± 1.02 ($p < 0.001$), and an intermediate-risk classification in the Wells scale were all observed to significantly elevate the risk of MACE. These findings underscore the nuanced interplay of individual patient characteristics, providing valuable insights into the factors contributing to the heightened risk of MACE in AF patients.

In Figure 6, the force chart dynamically illustrates the contributions of each feature in directing the model prediction from the base value to the ultimate result. The length of the colored bars within the chart serves as a visual indicator of the magnitude of each feature's contribution. This graphical representation offers a dynamic and insightful portrayal of how individual features influence the model's predictions, emphasizing the varying degrees of impact that contribute to the final outcome. The length of each bar provides a quick and intuitive assessment of the relative importance of each feature in shaping the model's decision-making process.

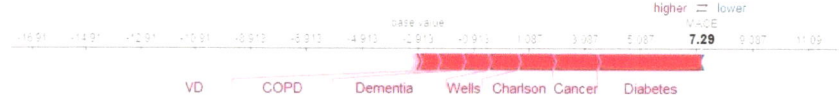

Figure 6. Dynamic force chart.

4. Discussion

The study identified AdaBoost as the best-performing model for MACE prediction in AF patients. Additionally, the CCI, concurrent cancer diagnosis, diabetes mellitus, and Wells and CHA_2DS_2-VASc scores emerged as primary predictors of MACE among patients newly diagnosed with AF. In a previous investigation [6], subsequent adjustments for age, gender, body mass index, cardiovascular risk factors, antiplatelets, and anticoagulants revealed that only the CHA_2DS_2-VASc, CCI, and CONUT scores remained as independent prognostic factors for MACE in individuals with a recent diagnosis of AF [6].

The various potential benefits of the results can be described in the different sections included in the flowchart for the approach and treatment of AF [14] as risk stratification, the prevention of thromboembolism among patients with silent AF and stroke without a previous diagnosis of AF, and for specific comorbidities such as chronic coronary disease, peripheral artery disease, heart failure, chronic kidney disease, and cognitive impairment.

AF almost quintuples the risk of MACE [6], especially ischemic stroke and heart failure. The 23.5% with known AF were not receiving oral anticoagulant therapy [37]. The AF was associated with more severity, disability, and a 20% increase in stroke-related costs. The clinical benefits of appropriate anticoagulation are widely recognized, and clinicians should be aware of the importance of anticoagulation therapies in stroke prophylaxis, the occurrence of stroke, and the downstream economic burden on an increasingly aging population [38]. Patients with AF may benefit from evaluating factors such as the AdaBoost model. This information can assist in making informed decisions about treatment.

The decision to prescribe oral anticoagulants for preventing MACE in patients with intermediate annual risk of thromboembolic events, as determined by classic risk scores like CHA_2DS_2-VASc or an equivalent, and who are uncertain about the benefits of anticoagulation, may require additional discussion. This is due to the diverse magnitude of risk associated with each factor across different populations. Managing specific patient groups, particularly those with risk factors for MACE, can improve risk discrimination by incorporating additional factors, as seen with the AdaBoost model.

Moreover, it addresses the optimization of treatment decisions concerning the burden of AF in relation to the associated risks of thromboembolism and ischemic stroke. This involves assessing the need for anticoagulant treatment decisions in individuals experiencing either paroxysmal or persistent AF because of the predictive significance of the AF burden [39,40]. A pioneering aspect of this approach involves the comprehensive analysis of large patient cohorts and the integration of diverse data sources, including blood biomarkers, electrical signals, and medical images [41]. The significance of this research extends into the domain of Personalized Risk Assessment, providing a promising approach for the early non-invasive detection of AF. This extends to optimizing treatment approaches and anticipating long-term clinical trajectories.

The algorithm emphasizes the CCI as the primary predictor, a widely utilized tool in the medical field for predicting the risk of mortality linked to chronic health conditions. It encompasses various factors such as heart disease, diabetes, and cancer and assigns specific weights to each based on their impact on mortality. The cumulative score is then employed to estimate an individual's overall health status and prognosis. A higher CCI score correlates with an elevated risk of adverse outcomes or mortality. Remarkably, until now, the CCI has not been previously associated with the risk of thromboembolism in patients recently diagnosed with AF. Notably, there have been instances where the use of anticoagulant therapy was linked to a lower CCI score [42]. While the CCI has undergone extensive validation and widespread use in predicting outcomes across various medical contexts, its application in specific situations, such as predicting outcomes in patients with AF [6], may not have been as comprehensively explored.

The presence of cancer emerges as the second-ranking predictor of MACE. While the algorithm does not specify the type of cancer, numerous studies have explored the connection between cancer and thromboembolism in patients with AF. Some of these studies not only identify cancer as a significant predictor of MACE, encompassing thromboembolic events [43], but also suggest that the onset of new AF is associated with an elevated risk of developing cancer [44,45]. These findings underscore the intricate interplay between AF, cancer, and thromboembolic complications, as well as the importance of considering both conditions in clinical assessment and management [46].

The Wells score has not been widely recognized as a prognostic factor for thromboembolism among patients with AF; it is typically used to assess the likelihood of deep vein thrombosis and pulmonary embolism. AF and venous thromboembolism share several common risk factors. Moreover, the presence of AF may be linked to a higher risk of developing VTE, and individuals with a high risk of experiencing VTE may also face an elevated risk of developing AF [47]. This bidirectional association highlights the potential interplay between these two conditions, suggesting that they may influence each other's occurrence and progression. Further research is warranted to fully understand the complex relationship between AF and VTE and its implications for clinical management and preventive strategies.

Diabetes mellitus and peripheral artery disease play an important role as a predictor of MACE [7,48]. Although they are also variables included in the $CHA_2DS_2\text{-}VAS_c$ and CCI scales, they alone are also an important variable for the development of MACE, and the significance of $CHA_2DS_2\text{-}VAS_c$ is widely recognized among patients with nonvalvular AF receiving oral anticoagulants [6,14,49,50]. In a recent study [51], machine learning models demonstrated satisfactory performance in forecasting MACE among patients with Type 2 diabetes mellitus. Notably, these models exhibited a higher accuracy in predicting strokes than myocardial infarction and heart failure.

Eventually, the study shed light on the significant role of COPD in the development of MACE among patients with AF, in alignment with existing evidence [8,12,50,52]. Prolonged P-wave duration acts as a potent precursor to AF, a condition that may be triggered by obstructive sleep apnea [53]. The presence of COPD in AF patients may contribute to an increased risk of MACE, emphasizing the importance of considering and managing this comorbidity when evaluating cardiovascular outcomes in this patient population.

While simpler models, such as logistic regression and decision trees, are more straightforward to interpret, they frequently exhibit inferior predictive performance compared to more sophisticated algorithms, including ensembles of decision trees like XGBoost and random forests [54]. Harnessing ML [53] algorithms facilitates the early identification of subtle indicators of thromboembolism risk from intricate datasets, thereby uncovering latent relationships among the risk factors associated with AF. The LightGBM model revealed associations between ischemic stroke and various peripheral blood biomarkers (such as creatinine, glycated hemoglobin, and monocytes) not considered by $CHA_2DS_2\text{-}VASc$ and demonstrated significance in predicting ischemic stroke among AF patients [55,56]. These algorithms not only facilitate the analysis and correction of potential confounding factors but also serve as powerful tools to identify and mitigate bias in the AI system. Continuous monitoring using ML algorithms offers ongoing assessment of thromboembolic risk among AF patients, contributing to the tracking of disease progression, monitoring treatment response, and promptly detecting any sudden changes in health status. Additionally, by enhancing follow-up through the prediction of patient-specific risks, these algorithms can prioritize follow-up visits and interventions, ultimately leading to improved patient outcomes.

Using the Deep Learning methodology, the results were slightly inferior to those achieved with Machine Learning (accuracy of 0.9678). The primary reason for this discrepancy may be the fact that neural networks demand a substantial amount of data to effectively learn. They are characterized by an abundance of parameters that require tuning, allowing them to grasp intricate, high-dimensional patterns. However, this proves to be a disadvantage when the dataset is limited. In instances of small datasets, these models become prone to overfitting, essentially 'memorizing' the training data rather than 'learning' the underlying pattern. Consequently, this results in suboptimal generalization performance when applied to unseen data.

The strengths of the study include the models of prediction, the high-quality datasets, and strict adherence to data privacy regulations, as well as clinical context and domain knowledge, making it easy to interpret the reasons behind their predictions. In summary, incorporating machine learning algorithms into the clinical management of individuals at high risk of AF and those with AF yields potential benefits, including personalized risk assessment, data-driven decision support, and improved patient care. However, further validation in independent studies is required.

Some limitations should be considered, as external validation is essential before effectively adopting and integrating AI systems into patient care. One crucial factor that largely determines the efficiency and accuracy of these models is the quantity of data available. For small datasets, like in our case, traditional machine learning models tend to outperform their deep learning counterparts, contrary to popular belief. AI models trained on specific datasets might not generalize well to different populations or healthcare settings, and overfitting could limit their applicability. Additionally, it is important to note that correlation does not necessarily imply causation. Establishing causal relationships between risk factors for AF and thromboembolism requires further research and experimentation. By addressing these limitations and maintaining responsible and effective AI use, we can enhance our understanding beyond not only the early detection of AF but also the risk associated with the incidence of MACE, providing opportunities to intervene in modifiable risk factors, and including aspects such as monitoring methods, detection technologies, and biomarkers linked to the association between AF and thromboembolism, ultimately leading to enhanced patient care outcomes.

Artificial intelligence-based clinical decision support systems may improve the outcomes among patients who have AF, but the efficacy of the tool in the real world is seldom reported. Future research could explore additional advantages, such as personalized risk assessment. By analyzing extensive datasets, including social determinants of health [18,57,58], biomarkers [59], multimodality imaging parameters [60,61], and nutritional status associated with AF risk [57,62], a comprehensive assessment can be made. This integration facilitates a more comprehensive and personalized risk assessment for each individual, allowing the identification of distinctive patterns and factors specific to the patient. This approach leads to more accurate risk predictions compared to traditional statistical models [6,23,63] and, consequently, may improve treatment decision making.

5. Conclusions

The application of Machine Learning, employing multiple models, indicates that the AdaBoost model is the most effective in predicting MACE in patients with newly diagnosed AF, with an accuracy of 0.9999, recall of 1, and an F1 score of 0.9997. The primary prognostic factors identified included an elevated Charlson Comorbidity Index, cancer, diabetes mellitus, COPD, cognitive impairment, vascular disease, and high values on the CHA_2DS_2-VASc and Wells scale. This finding contributes to the optimization of treatment decisions concerning the burden of AF in relation to the associated risks of thromboembolism and ischemic events.

Author Contributions: Conceptualization, P.M.-B., S.R.-V. and J.-L.C.-E.; methodology, P.M.-B., S.R.-V. and J.-L.C.-E.; software P.M.-B. and V.A.-B.; validation, P.M.-B., S.R.-V. and J.-L.C.-E.; formal analysis, P.M.-B., V.A.-B. and J.-L.C.-E.; investigation, P.M.-B. and J.-L.C.-E.; resources, P.M.-B. and J.-L.C.-E.; data curation, P.M.-B.; writing—original draft preparation, P.M.-B., S.R.-V. and J.-L.C.-E.; writing—review and editing, P.M.-B., S.R.-V., C.M.-A., M.T.B.-A., J.C.-Q. and J.-L.C.-E.; visualization, P.M.-B. and C.M.-A.; supervision, S.R.-V. and J.-L.C.-E.; project administration, P.M.-B.; funding acquisition, P.M.-B. All authors have read and agreed to the published version of the manuscript.

Funding: This research received no external funding.

Institutional Review Board Statement: This retrospective chart review study involving human participants was in accordance with the ethical standards of the institutional and national research committee and with the 1964 Helsinki Declaration and its later amendments or comparable ethical standards. Approval was obtained from the Ethics Committee of the Institut Universitari d'Investigació en Atenció Primària Jordi Gol with the registration number 22/243-P (30 November 2022). Registry information was collected from the government-run healthcare provider responsible for all inpatient care in the county, without contact with participants, in order to gather data from the study. The manuscript does not contain clinical studies or patient data that might disclose the identity of the people under study.

Informed Consent Statement: Not applicable. For this type of study, formal consent is not required, and the requirement for the informed consent of patients was waived prior to the inclusion of their medical data in this study.

Data Availability Statement: Numerical calculations and data analysis were performed using Python library version 3. The code and models used for the analysis are available online (https://github.com/

vmalonsobarberan/MACE) (15 December 2023). The datasets generated, used, and analyzed during the current study are available from the corresponding author (P.M.-B.) upon reasonable request.

Acknowledgments: The authors would like to thank the teachers and students of the specialization course in Artificial Intelligence and BigData 2022/2023 of the IES El Caminàs (Castellón, Spain) for their work, help, and support, especially A.N. Molina-Gutiérrez, the main programmer, and L.D. Taciulet, under the supervision of V. Alonso-Barberán. *ChatGPT* AI system, as a free-to-use tool, was used as a supporting tool for reviewing or considering options in the English translation of the original language.

Conflicts of Interest: The authors declare no conflict of interest.

References

1. Rosenstrøm, S.; Risom, S.S.; Hove, J.D.; Brødsgaard, A. Living with Atrial Fibrillation: A Family Perspective. *Nurs. Res. Pract.* **2022**, *2022*, 7394445. [CrossRef] [PubMed]
2. Sánchez de la Nava, A.M.; Atienza, F.; Bermejo, J.; Fernández-Avilés, F. Artificial intelligence for a personalized diagnosis and treatment of atrial fibrillation. *Am. J. Physiol. Heart Circ. Physiol.* **2021**, *320*, H1337–H1347. [CrossRef] [PubMed]
3. Odutayo, A.; Wong, C.X.; Hsiao, A.J.; Hopewell, S.; Altman, D.G.; Emdin, C.A. Atrial fibrillation and risks of cardiovascular disease, renal disease, and death: Systematic review and meta-analysis. *BMJ* **2016**, *354*, i4482. [CrossRef] [PubMed]
4. Blum, S.; Aeschbacher, S.; Coslovsky, M.; Meyre, P.B.; Reddiess, P.; Ammann, P.; Erne, P.; Moschovitis, G.; Di Valentino, M.; Shah, D.; et al. Long-term risk of adverse outcomes according to atrial fibrillation type. *Sci. Rep.* **2022**, *12*, 2208. [CrossRef]
5. Bosco, E.; Hsueh, L.; McConeghy, K.W.; Gravenstein, S.; Saade, E. Major adverse cardiovascular event definitions used in observational analysis of administrative databases: A systematic review. *BMC Med. Res. Methodol.* **2021**, *21*, 241. [CrossRef]
6. Moltó-Balado, P.; Reverté-Villarroya, S.; Monclús-Arasa, C.; Balado-Albiol, M.T.; Baset-Martínez, S.; Carot-Domenech, J.; Clua-Espuny, J.L. Heart Failure and Major Adverse Cardiovascular Events in Atrial Fibrillation Patients: A Retrospective Primary Care Cohort Study. *Biomedicines* **2023**, *11*, 1825. [CrossRef] [PubMed]
7. Boriani, G.; Vitolo, M.; Diemberger, I.; Proietti, M.; Valenti, A.C.; Malavasi, V.L.; Lip, G.Y.H. Optimizing indices of atrial fibrillation susceptibility and burden to evaluate atrial fibrillation severity, risk and outcomes. *Cardiovasc. Res.* **2021**, *117*, 1–21. [CrossRef] [PubMed]
8. Chen, X.; Lin, M.; Wang, W. The progression in atrial fibrillation patients with COPD: A systematic review and meta-analysis. *Oncotarget* **2017**, *8*, 102420–102427. [CrossRef]
9. Zhang, J.; Johnsen, S.P.; Guo, Y.; Lip, G.Y.H. Epidemiology of Atrial Fibrillation: Geographic/Ecological Risk Factors, Age, Sex, Genetics. *Card. Electrophysiol. Clin.* **2021**, *13*, 1–23. [CrossRef]
10. Ardashev, A.V.; Belenkov, Y.N.; Matsiukevich, M.C.; Snezhitskiy, V.A. Atrial Fibrillation and Mortality: Prognostic Factors and Direction of Prevention. *Kardiologiia* **2021**, *61*, 91–98. [CrossRef]
11. Pastori, D.; Menichelli, D.; Lip, G.Y.H.; Sciacqua, A.; Violi, F.; Pignatelli, P.; ATHERO-AF study group†. Family History of Atrial Fibrillation and Risk of Cardiovascular Events: A Multicenter Prospective Cohort Study. *Circ. Arrhythm. Electrophysiol.* **2020**, *13*, e008477. [CrossRef] [PubMed]
12. Raparelli, V.; Pastori, D.; Pignataro, S.F.; Vestri, A.R.; Pignatelli, P.; Cangemi, R.; Proietti, M.; Davì, G.; Hiatt, W.R.; Lip, G.Y.H.; et al. Major adverse cardiovascular events in non-valvular atrial fibrillation with chronic obstructive pulmonary disease: The ARAPACIS study. *Intern. Emerg. Med.* **2018**, *13*, 651–660. [CrossRef] [PubMed]
13. Romiti, G.F.; Pastori, D.; Rivera-Caravaca, J.M.; Ding, W.Y.; Gue, Y.X.; Menichelli, D.; Gumprecht, J.; Kozieł, M.; Yang, P.S.; Guo, Y.; et al. Adherence to the 'Atrial Fibrillation Better Care' Pathway in Patients with Atrial Fibrillation: Impact on Clinical Outcomes-A Systematic Review and Meta-Analysis of 285,000 Patients. *Thromb. Haemost.* **2022**, *122*, 406–414. [CrossRef]
14. Joglar, J.A.; Chung, M.K.; Armbruster, A.L.; Benjamin, E.J.; Chyou, J.Y.; Cronin, E.M.; Deswal, A.; Eckhardt, L.L.; Goldberger, Z.D.; Gopinathannair, R.; et al. 2023 ACC/AHA/ACCP/HRS Guideline for the Diagnosis and Management of Atrial Fibrillation: A Report of the American College of Cardiology/American Heart Association Joint Committee on Clinical Practice Guidelines. *Circulation* **2023**, *149*, e1–e156. [CrossRef] [PubMed]
15. D'Agostino, R.B., Sr.; Vasan, R.S.; Pencina, M.J.; Wolf, P.A.; Cobain, M.; Massaro, J.M.; Kannel, W.B. General cardiovascular risk profile for use in primary care: The Framingham Heart Study. *Circulation* **2008**, *117*, 743–753. [CrossRef]
16. Fang, M.C.; Go, A.S.; Chang, Y.; Borowsky, L.H.; Pomernacki, N.K.; Udaltsova, N.; Singer, D.E. A new risk scheme to predict warfarin-associated hemorrhage: The ATRIA (Anticoagulation and Risk Factors in Atrial Fibrillation) Study. *J. Am. Coll. Cardiol.* **2011**, *58*, 395–401. [CrossRef]
17. Lip, G.Y.; Nieuwlaat, R.; Pisters, R.; Lane, D.A.; Crijns, H.J. Refining clinical risk stratification for predicting stroke and thromboembolism in atrial fibrillation using a novel risk factor-based approach: The euro heart survey on atrial fibrillation. *Chest* **2010**, *137*, 263–272. [CrossRef] [PubMed]
18. Jung, S.; Song, M.K.; Lee, E.; Bae, S.; Kim, Y.Y.; Lee, D.; Lee, M.J.; Yoo, S. Predicting Ischemic Stroke in Patients with Atrial Fibrillation Using Machine Learning. *Front. Biosci. (Landmark Ed)* **2022**, *27*, 80. [CrossRef]

19. Muria-Subirats, E.; Clua-Espuny, J.L.; Ballesta-Ors, J.; Lorman-Carbo, B.; Lechuga-Duran, I.; Fernández-Saez, J.; Pla-Farnos, R.; On Behalf Members of Africat Group. Incidence and Risk Assessment for Atrial Fibrillation at 5 Years: Hypertensive Diabetic Retrospective Cohort. *Int. J. Environ. Res. Public Health* **2020**, *17*, 3491. [CrossRef]
20. Wan, D.; Andrade, J.; Laksman, Z. Thromboembolic risk stratification in atrial fibrillation-beyond clinical risk scores. *Rev. Cardiovasc. Med.* **2021**, *22*, 353–363. [CrossRef]
21. Rivera-Caravaca, J.M.; Marín, F.; Esteve-Pastor, M.A.; Raña-Míguez, P.; Anguita, M.; Muñiz, J.; Cequier, Á.; Bertomeu-Martínez, V.; Valdés, M.; Vicente, V.; et al. Usefulness of the 2MACE Score to Predicts Adverse Cardiovascular Events in Patients with Atrial Fibrillation. *Am. J. Cardiol.* **2017**, *120*, 2176–2181. [CrossRef] [PubMed]
22. Froehlich, L.; Meyre, P.; Aeschbacher, S.; Blum, S.; Djokic, D.; Kuehne, M.; Osswald, S.; Kaufmann, B.A.; Conen, D. Left atrial dimension and cardiovascular outcomes in patients with and without atrial fibrillation: A systematic review and meta-analysis. *Heart* **2019**, *105*, 1884–1891. [CrossRef] [PubMed]
23. Black, J.E.; Kueper, J.K.; Williamson, T.S. An introduction to machine learning for classification and prediction. *Fam. Pract.* **2023**, *40*, 200–204. [CrossRef] [PubMed]
24. Wang, Y.C.; Xu, X.; Hajra, A.; Apple, S.; Kharawala, A.; Duarte, G.; Liaqat, W.; Fu, Y.; Li, W.; Chen, Y.; et al. Current Advancement in Diagnosing Atrial Fibrillation by Utilizing Wearable Devices and Artificial Intelligence: A Review Study. *Diagnostics* **2022**, *12*, 689. [CrossRef]
25. Attia, Z.I.; Noseworthy, P.A.; Lopez-Jimenez, F.; Asirvatham, S.J.; Deshmukh, A.J.; Gersh, B.J.; Carter, R.E.; Yao, X.; Rabinstein, A.A.; Erickson, B.J.; et al. An artificial intelligence-enabled ECG algorithm for the identification of patients with atrial fibrillation during sinus rhythm: A retrospective analysis of outcome prediction. *Lancet* **2019**, *394*, 861–867. [CrossRef] [PubMed]
26. Nadarajah, R.; Wu, J.; Hogg, D.; Raveendra, K.; Nakao, Y.M.; Nakao, K.; Arbel, R.; Haim, M.; Zahger, D.; Parry, J.; et al. Prediction of short-term atrial fibrillation risk using primary care electronic health records. *Heart* **2023**, *109*, 1072–1079. [CrossRef] [PubMed]
27. Xia, Y.; Wulan, N.; Wang, K.; Zhang, H. Detecting atrial fibrillation by deep convolutional neural networks. *Comput. Biol. Med.* **2018**, *93*, 84–92. [CrossRef] [PubMed]
28. Hill, N.R.; Ayoubkhani, D.; McEwan, P.; Sugrue, D.M.; Farooqui, U.; Lister, S.; Lumley, M.; Bakhai, A.; Cohen, A.T.; O'Neill, M.; et al. Predicting atrial fibrillation in primary care using machine learning. *PLoS ONE* **2019**, *14*, e0224582. [CrossRef]
29. Mamoshina, P.; Bueno-Orovio, A.; Rodriguez, B. Dual Transcriptomic and Molecular Machine Learning Predicts all Major Clinical Forms of Drug Cardiotoxicity. *Front. Pharmacol.* **2020**, *11*, 639. [CrossRef]
30. Wu, X.; Zheng, Y.; Chu, C.H.; He, Z. Extracting deep features from short ECG signals for early atrial fibrillation detection. *Artif. Intell. Med.* **2020**, *109*, 101896. [CrossRef]
31. Jahan, M.S.; Mansourvar, M.; Puthusserypady, S.; Wiil, U.K.; Peimankar, A. Short-term atrial fibrillation detection using electrocardiograms: A comparison of machine learning approaches. *Int. J. Med. Inform.* **2022**, *163*, 104790. [CrossRef] [PubMed]
32. Wesselius, F.J.; van Schie, M.S.; De Groot, N.M.S.; Hendriks, R.C. Digital biomarkers and algorithms for detection of atrial fibrillation using surface electrocardiograms: A systematic review. *Comput. Biol. Med.* **2021**, *133*, 104404. [CrossRef] [PubMed]
33. Yue, Y.; Chen, C.; Liu, P.; Xing, Y.; Zhou, X. Automatic Detection of Short-Term Atrial Fibrillation Segments Based on Frequency Slice Wavelet Transform and Machine Learning Techniques. *Sensors* **2021**, *21*, 5302. [CrossRef]
34. Pla de salut de la Regió Sanitària Terres de l'Ebre 2021–2025. Direcció General de Planificació i Recerca en Salut: Tortosa, Spain, 2022. Available online: https://scientiasalut.gencat.cat/handle/11351/7964 (accessed on 12 September 2023).
35. Idescat. Anuario Estadístico de Cataluña. Densidad de Población. Comarcas y Aran, Ámbitos y Provincias. Available online: https://www.idescat.cat/indicators/?id=aec&n=15227&lang=es (accessed on 12 September 2023).
36. Idescat. Indicadors Demogràfics i de Territoris. Estructura Per Edats, Envelliment i Dependència. Comarques i Aran. Available online: https://www.idescat.cat/pub/?id=inddt&n=915&lang=en (accessed on 12 September 2023).
37. Sociedad Española de Cardiología. Atlas del Mal Control de la Anticoagulación en Pacientes con Fibrilación Auricular No Valvular. Available online: https://secardiologia.es/images/secciones/clinica/atlas-mal-control-anticoagulacion-INFOGRAFIA.pdf (accessed on 23 June 2023).
38. Li, X.; Tse, V.C.; Au-Doung, L.W.; Wong, I.C.; Chan, E.W. The impact of ischaemic stroke on atrial fibrillation-related healthcare cost: A systematic review. *Europace* **2017**, *19*, 937–947. [CrossRef] [PubMed]
39. Papakonstantinou, P.E.; Tsioufis, K. Optimizing Anticoagulation Management in Atrial Fibrillation: Beyond the Guidelines. How and for Whom? *J. Cardiovasc. Pharmacol.* **2023**, *81*, 397–399. [CrossRef] [PubMed]
40. Verheugt, F.W.A.; Fox, K.A.A.; Virdone, S.; Ambrosio, G.; Gersh, B.J.; Haas, S.; Pieper, K.S.; Kayani, G.; Camm, A.J.; Parkhomenko, A.; et al. Outcomes of Oral Anticoagulation in Atrial Fibrillation Patients With or Without Comorbid Vascular Disease: Insights From the GARFIELD-AF Registry. *Am. J. Med.* **2023**, *136*, 1187–1195. [CrossRef]
41. Haq, I.U.; Chhatwal, K.; Sanaka, K.; Xu, B. Artificial Intelligence in Cardiovascular Medicine: Current Insights and Future Prospects. *Vasc. Health Risk Manag.* **2022**, *18*, 517–528. [CrossRef]
42. Lahoz, C.; Cardenas, J.; Salinero-Fort, M.Á.; Mostaza, J.M. Prevalence of atrial fibrillation and associated anticoagulant therapy in the nonagenarian population of the Community of Madrid, Spain. *Geriatr. Gerontol. Int.* **2019**, *19*, 203–207. [CrossRef]
43. Khorana, A.A.; Mackman, N.; Falanga, A.; Pabinger, I.; Noble, S.; Ageno, W.; Moik, F.; Lee, A.Y.Y. Cancer-associated venous thromboembolism. *Nat. Rev. Dis. Primers.* **2022**, *8*, 11–24. [CrossRef]
44. Guha, A.; Fradley, M.G.; Dent, S.F.; Weintraub, N.L.; Lustberg, M.B.; Alonso, A.; Addison, D. Incidence, risk factors, and mortality of atrial fibrillation in breast cancer: A SEER-Medicare analysis. *Eur. Heart J.* **2022**, *43*, 300–312. [CrossRef]

45. Zhang, M.; Li, L.L.; Zhao, Q.Q.; Peng, X.D.; Wu, K.; Li, X.; Ruan, Y.F.; Bai, R.; Liu, N.; Ma, C.S. The Association of New-Onset Atrial Fibrillation and Risk of Cancer: A Systematic Review and Meta-Analysis. *Cardiol. Res. Pract.* **2020**, *27*, 2372067. [CrossRef] [PubMed]
46. Sorigue, M.; Miljkovic, M.D. Atrial Fibrillation and Stroke Risk in Patients with Cancer: A Primer for Oncologists. *J. Oncol. Pract.* **2019**, *15*, 641–650. [CrossRef] [PubMed]
47. Lutsey, P.L.; Norby, F.L.; Alonso, A.; Cushman, M.; Chen, L.Y.; Michos, E.D.; Folsom, A.R. Atrial fibrillation and venous thromboembolism: Evidence of bidirectionality in the Atherosclerosis Risk in Communities Study. *J. Thromb. Haemost.* **2018**, *16*, 670–679. [CrossRef] [PubMed]
48. Anandasundaram, B.; Lane, D.A.; Apostolakis, S.; Lip, G.Y. The impact of atherosclerotic vascular disease in predicting a stroke, thromboembolism and mortality in atrial fibrillation patients: A systematic review. *J. Thromb. Haemost.* **2013**, *11*, 975–987. [CrossRef] [PubMed]
49. Nabauer, M.; Oeff, M.; Gerth, A.; Wegscheider, K.; Buchholz, A.; Haeusler, K.G.; Hanrath, P.; Meinertz, T.; Ravens, U.; Sprenger, C.; et al. Prognostic markers of all-cause mortality in patients with atrial fibrillation: Data from the prospective long-term registry of the German Atrial Fibrillation NETwork (AFNET). *Europace* **2021**, *23*, 1903–1912. [CrossRef] [PubMed]
50. Lip, G.Y.H.; Murphy, R.R.; Sahiar, F.; Ingall, T.J.; Dhamane, A.D.; Ferri, M.; Hlavacek, P.; Preib, M.T.; Keshishian, A.; Russ, C.; et al. Risk Levels and Adverse Clinical Outcomes Among Patients with Nonvalvular Atrial Fibrillation Receiving Oral Anticoagulants. *JAMA Netw. Open* **2022**, *5*, e2229333. [CrossRef]
51. Abegaz, T.M.; Baljoon, A.; Kilanko, O.; Sherbeny, F.; Ali, A.A. Machine learning algorithms to predict major adverse cardiovascular events in patients with diabetes. *Comput. Biol. Med.* **2023**, *164*, 107289. [CrossRef]
52. Shao, X.H.; Yang, Y.M.; Zhu, J.; Zhang, H.; Liu, Y.; Gao, X.; Yu, L.T.; Liu, L.S.; Zhao, L.; Yu, P.F.; et al. Comparison of the clinical features and outcomes in two age-groups of elderly patients with atrial fibrillation. *Clin. Interv. Aging* **2014**, *9*, 1335–1342. [CrossRef]
53. Ueda, A.; Kasagi, S.; Maeno, K.I.; Naito, R.; Kumagai, T.; Kimura, Y.; Kato, M.; Kawana, F.; Tomita, Y.; Narui, K.; et al. Cross-Sectional Relationship Between Atrial Conduction Delay and Arterial Stiffness in Patients with Obstructive Sleep Apnea. *Vasc. Health Risk Manag.* **2023**, *19*, 733–740. [CrossRef]
54. Silva, G.F.S.; Fagundes, T.P.; Teixeira, B.C.; Chiavegatto Filho, A.D.P. Machine Learning for Hypertension Prediction: A Systematic Review. *Curr. Hypertens. Rep.* **2022**, *24*, 523–533. [CrossRef]
55. Parikh, R.B.; Teeple, S.; Navathe, A.S. Addressing Bias in Artificial Intelligence in Health Care. *JAMA* **2019**, *322*, 2377–2378. [CrossRef] [PubMed]
56. Areti, P.; Daniel, H.; Greg, S.; Eirini, M.; Panos, D. Prediction of Atrial Fibrillation and Stroke Using Machine Learning Models in UK Biobank. (Available at SSRN). Available online: https://www.medrxiv.org/content/10.1101/2022.10.28.22281669v1 (accessed on 10 November 2023). [CrossRef]
57. Arero, G.; Arero, A.G.; Mohammed, S.H.; Vasheghani-Farahani, A. Prognostic Potential of the Controlling Nutritional Status (CONUT) Score in Predicting All-Cause Mortality and Major Adverse Cardiovascular Events in Patients with Coronary Artery Disease: A Meta-Analysis. *Front. Nutr.* **2022**, *9*, 850641. [CrossRef]
58. Essien, U.R.; Kornej, J.; Johnson, A.E. Social determinants of atrial fibrillation. *Nat. Rev. Cardiol.* **2021**, *118*, 763–773. [CrossRef] [PubMed]
59. Palà, E.; Bustamante, A.; Clúa-Espuny, J.L.; Acosta, J.; González-Loyola, F.; Santos, S.D.; Ribas-Segui, D.; Ballesta-Ors, J.; Penalba, A.; Giralt, M.; et al. Blood-biomarkers and devices for atrial fibrillation screening: Lessons learned from the AFRICAT (Atrial Fibrillation Research in CATalonia) study. *PLoS ONE* **2022**, *17*, e0273571. [CrossRef]
60. Gentille-Lorente, D.; Hernández-Pinilla, A.; Satue-Gracia, E.; Muria-Subirats, E.; Forcadell-Peris, M.J.; Gentille-Lorente, J.; Ballesta-Ors, J.; Martín-Lujan, F.M.; Clua-Espuny, J.L. Echocardiography and Electrocardiography in Detecting Atrial Cardiomyopathy: A Promising Path to Predicting Cardioembolic Strokes and Atrial Fibrillation. *J. Clin. Med.* **2023**, *12*, 7315. [CrossRef] [PubMed]
61. López-Galvez, R.; Rivera-Caravaca, J.M.; Roldán, V.; Orenes-Piñero, E.; Esteve-Pastor, M.A.; López-García, C.; Saura, D.; González, J.; Lip, G.Y.H.; Marín, F. Imaging in atrial fibrillation: A way to assess atrial fibrosis and remodeling to assist decision-making. *Am. Heart J.* **2023**, *258*, 1–16. [CrossRef]
62. Zhang, S.; Stubbendorff, A.; Ericson, U. The EAT-Lancet diet, genetic susceptibility and risk of atrial fibrillation in a population-based cohort. *BMC Med.* **2023**, *21*, 280. [CrossRef]
63. Li, Z.; Zhang, X.; Ding, L.; Jing, J.; Gu, H.-Q.; Jiang, Y.; Meng, X.; Du, C.; Wang, C.; Wang, M.; et al. Rationale and design of the GOLDEN BRIDGE II: A cluster-randomised multifaceted intervention trial of an artificial intelligence-based cerebrovascular disease clinical decision support system to improve stroke outcomes and care quality in China. *Stroke Vasc. Neurol.* **2023**, svn-2023. [CrossRef]

Disclaimer/Publisher's Note: The statements, opinions and data contained in all publications are solely those of the individual author(s) and contributor(s) and not of MDPI and/or the editor(s). MDPI and/or the editor(s) disclaim responsibility for any injury to people or property resulting from any ideas, methods, instructions or products referred to in the content.

Review

Level of Technological Maturity of Telemonitoring Systems Focused on Patients with Chronic Kidney Disease Undergoing Peritoneal Dialysis Treatment: A Systematic Literature Review

Alejandro Villanueva Cerón [1], Eduardo López Domínguez [2,*], Saúl Domínguez Isidro [3], María Auxilio Medina Nieto [4], Jorge De La Calleja [4] and Saúl Eduardo Pomares Hernández [5]

1. Department of Computer Systems Engineering, Tecnológico Nacional de México Campus Álamo Temapache, Veracruz 92750, Mexico
2. Department of Computer Science, Centro de Investigación y de Estudios Avanzados del Instituto Politécnico Nacional, Mexico City 07360, Mexico
3. Faculty of Statistics and Informatics, Universidad Veracruzana, Veracruz 91020, Mexico
4. Postgraduate Department, Universidad Politécnica de Puebla, Puebla 72640, Mexico
5. Department of Computer Science, Instituto Nacional de Astrofísica, Óptica y Electrónica, Puebla 72840, Mexico
* Correspondence: eduardo.lopez.dom@cinvestav.mx

Citation: Villanueva Cerón, A.; López Domínguez, E.; Domínguez Isidro, S.; Medina Nieto, M.A.; De La Calleja, J.; Pomares Hernández, S.E. Level of Technological Maturity of Telemonitoring Systems Focused on Patients with Chronic Kidney Disease Undergoing Peritoneal Dialysis Treatment: A Systematic Literature Review. *Technologies* 2023, 11, 129. https://doi.org/10.3390/technologies11050129

Academic Editors: Juvenal Rodriguez-Resendiz, Gerardo I. Pérez-Soto, Karla Anhel Camarillo Gómez and Saul Tovar-Arriaga

Received: 13 July 2023
Revised: 5 September 2023
Accepted: 11 September 2023
Published: 18 September 2023

Copyright: © 2023 by the authors. Licensee MDPI, Basel, Switzerland. This article is an open access article distributed under the terms and conditions of the Creative Commons Attribution (CC BY) license (https://creativecommons.org/licenses/by/4.0/).

Abstract: In the field of eHealth, several works have proposed telemonitoring systems focused on patients with chronic kidney disease (CKD) undergoing peritoneal dialysis (PD) treatment. Nevertheless, no secondary study presents a comparative analysis of these works regarding the technology readiness level (TRL) framework. The TRL scale goes from 1 to 9, with 1 being the lowest level of readiness and 9 being the highest. This paper analyzes works that propose telemonitoring systems focused on patients with CKD undergoing PD treatment to determine their TRL. We also analyzed the requirements and parameters that the systems of the selected works provide to the users to perform telemonitoring of the patient's treatment undergoing PD. Fourteen works were relevant to the present study. Of these works, eight were classified within TRL 9, two were categorized within TRL 7, three were identified within TRL 6, and one within TRL 4. The works reported with the highest TRL partially cover the requirements for appropriate telemonitoring of patients based on the specialized literature; in addition, those works are focused on the treatment of patients in the automated peritoneal dialysis (APD) modality, which limits the care of patients undergoing the continuous ambulatory peritoneal dialysis (CAPD) modality.

Keywords: chronic kidney disease; peritoneal dialysis; technology readiness level; telemonitoring system

1. Introduction

eHealth is the term coined as the set of information and communication technologies (ICT) used as a tool in the health field [1]. For the World Health Organization (WHO), this concept of eHealth is related to the safe and cost-effective use of ICT in different settings [1]. In the eHealth context, the telemedicine paradigm has emerged [2] and is defined as the remote assistance of medical services. Its implementation requires information and communication technologies, as well as human resources specialized in implementing such systems [2]. There are several concepts associated with telemedicine [3]:

- Remote patient monitoring or control allows patients with chronic and degenerative diseases to be monitored from their homes, work environments, etc.
- Storage or forwarding technology stores clinical data and information to be forwarded to another clinical facility for interpretation, e.g., examination or X-ray imaging.

- Interactive telemedicine allows doctors and patients to be connected in real time through video conferencing [4]. Telemedicine consists of health professionals using ICT to diagnose, treat, and prevent diseases.
- The telemonitoring or remote monitoring of biomedical parameters seeks the patient's participation in managing their disease, promoting prevention and self-care.

In the telemonitoring field, several works [5–20] have proposed systems focused on telemonitoring patients with chronic and degenerative diseases, such as diabetes mellitus, heart failure, chronic obstructive pulmonary disease, and chronic kidney disease (CKD). CKD is a condition in which the kidneys are damaged and cannot filter blood as well as they should. Because of this, excess fluid and waste from blood remain in the body and may cause other health problems. When the loss of the kidney's ability to filter blood in chronic kidney disease is severe, kidney function must be replaced either by hemodialysis, dialysis peritoneal treatment, or kidney transplantation. This refers to a chronic renal disease stage 5 (CRDS5). CRDS5 generates multiple health complications to patients, such as hypertension, anemia, cardiovascular complications, chronic kidney-disease-related mineral bone disorder, salt and water retention, metabolic acidosis, and electrolyte disorders, as well as uremic symptoms. Therefore, CKD is considered catastrophic because of high morbidity and mortality rates and poor quality of life [21,22]. Compared to hemodialysis treatment, peritoneal dialysis (PD) treatment has several potential advantages, such as the fact that it is a technique that is easy to learn and apply, it has greater feasibility of use in remote communities and lower costs, and fewer specialized health personnel are required. In addition, PD treatment allows greater survival in the first years, permission to patients to travel, fewer dietary restrictions, and better preservation of residual renal function. Moreover, it is reported that patients have greater satisfaction with PD treatment and better quality of life, and the treatment can be performed by patients themselves at home, among others [23]. In this context, some works [24,25] have reviewed the specialized literature on different aspects of telemonitoring systems oriented to patients with CKD on PD. However, no work has been undertaken that presents a systematic review regarding the level of technological maturity of such telemonitoring systems from the perspective of the technology readiness level (TRL) [26]. TRL measures how far a technology has progressed along its development path from basic research (TRL 1) to mature technologies ready for commercialization (TRL 9). The TRL [26] is a method developed by the National Aeronautics and Space Administration (NASA) in the 1970s as a tool that measures the degree of maturity of a technology. Among the main advantages of using technology maturity levels in ICT projects related to telemonitoring systems are the following: they generate a standardized analysis of the project's status, they allow the identification and management of risks in projects, and they help to classify the proposed work in terms of research, development, and innovation, which contributes to decision-making in terms of funding and project transition between technology maturity levels [26–28]. Therefore, this paper analyzes works that propose telemonitoring systems focused on patients with CKD undergoing PD treatment to determine their level of technological maturity based on the NASA TRL framework. The results of our work show the main contributions, gaps, and opportunities of the proposed work concerning the TRL framework, which motivated the discussion of important future research directions.

2. Materials and Methods

The approach used in our systematic literature review (SLR) embraced the strategies and rules depicted by Kitchenham et al. [29] and Ali et al. [30]. In the first instance, the research questions were defined to identify the purpose of the review and the interest under study.

- Q1: What kinds of services and parameters are used in telemonitoring systems focused on patients with CKD on PD proposed in the specialized literature? This question is motivated by identifying the functional requirements and desirable parameters to monitor in a system-oriented telemonitoring of CKD on PD.

- Q2: Is it possible to identify the level of technological maturity of telemonitoring systems for patients with CKD on PD proposed in the specialized literature? This question aims to analyze the level of technological development achieved in these types of solutions by adopting a measurement framework used worldwide.
- Q3: How has the publication rate of studies related to telemonitoring systems for patients with CKD on PD proposed in the literature changed over time? This question aims to examine the period and frequency of publication of such works.
- Q4: Can we identify opportunity areas and challenges still pending in the telemonitoring systems for patients with CKD on PD implemented to date? This question aims to identify gaps in care that remain open in this treatment area to contribute to its solution.

2.1. Search Strategy

As a comprehensive search strategy, each research question was formulated as individual search terms, highlighting:

- Peritoneal dialysis;
- Telemonitoring system;
- Telehealth;
- Telemedicine;
- CKD;
- Telenephrology.

In the same way, some synonyms, abbreviations, and alternative spellings of the terms described were used. The search strategy was developed using these terms in com-bination with the logical operators AND and OR to generate more specific search sequences to query in the search engines. As can be seen in Table 1, the scope of the search query was defined based on three topics:

1. Telemonitoring via a device or service;
2. Applications or the scope of the software of such applications;
3. Type of disability or chronic illness targeted by the applications.

Table 1. Context and example search query.

Context	Quantity
Telemonitoring	(telemonitoring OR "telemonitoring system" OR tele-monitoring OR "tele monitoring") AND
Application	(apps or software or device(s) or application(s) or service(s)) AND
Disease	(CKD OR "chronic kidney disease" OR "peritoneal dialysis")

The electronic database sources used in this SLR included those relevant to our research's aim. Thus, the SLR was based on the following digital databases: Springer, Elsevier, EBSCO, SCOPUS, Cochrane Library, MEDLINE via PubMed, IEEE Xplore, and Google Scholar. The selection of studies is described in Section 2.4.

2.2. Inclusion Criteria

The purpose of the selection criteria is to identify primary studies that directly respond to the research questions [29]. In our case, each study found was evaluated using the following inclusion criteria (Cr_In):

- Cr_In1: Systems focused on telemonitoring of CKD patients on PD treatment were considered for the study.
- Cr_In2: The study includes evaluations of the described systems considering end users in clearly identified scenarios (real or simulated).
- Cr_In3: The article was published in an indexed, refereed, peer-reviewed journal or conference proceedings in the specialty.

- Cr_In4: We considered studies with experimental or research-oriented designs, i.e., randomized control trials (RCTs), non-RCTs, and pre- and post-experimental studies.
- Cr_In5: Studies include eHealth interventions with digital information or any communication technology component using any type of device.

2.3. Exclusion Criteria

The exclusion criteria (Cr_Ex) used in our SLR were as follows:

- Cr_Ex1: The study is not written in English or Spanish.
- Cr_Ex2: The article was published before 2014.
- Cr_Ex3: The study considers systems focused on the telemonitoring of patients with CKD but treated exclusively by hemodialysis.

2.4. Selection of Studies

We obtained 91 articles from PubMed, 36 articles from IEEE Xplore, 31 from Cochrane Library, 100 from Google Scholar, and 40 from EBSCO. The sum of these results produced 298 articles. After removing duplicate articles and those written in languages other than English and Spanish, 223 articles remained. Later, the title, keywords, and abstracts of these 223 articles were reviewed and 20 articles were selected. Those articles were read and checked in detail, and only 14 articles were included in the review because 6 articles did not meet the Cr_In2, Cr_In4, and Cr_In5 criteria. Figure 1 shows this process using the guidelines defined by Kitchenham et al. [29] and Ali et al. [30].

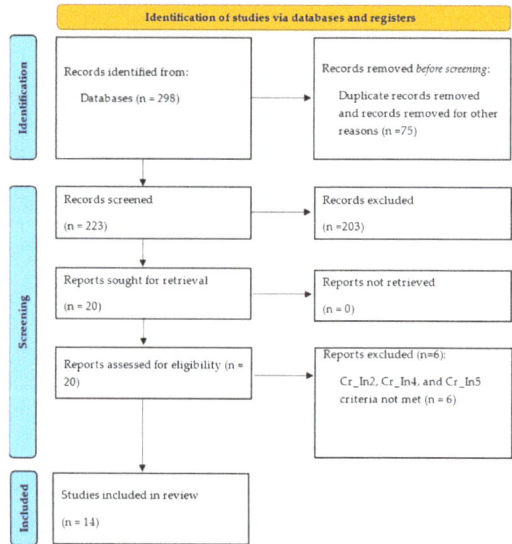

Figure 1. Paper selection process of our literature review.

Studies from different publication sources focused on implementing various types of systems aimed at patients with CKD on PD. From 10 different publication sources (journals), 14 articles relevant to the present study were found [7–20]. Based on the research scopes covered by those journals, the studies were divided into three main categories: computer science, health, and technology. Twelve studies were published in journals with a health focus [8,9,11–20], one in a journal with an informatics focus [10], and one in a journal with a technology focus [7], as shown in Table 2. According to the type of study in question, these publications provided the most relevant and necessary information for the analysis of each telemonitoring system, demonstrating that relevant research efforts are being made in this area.

Table 2. Publication sources of selected studies.

Reference	Name of the Conference/Journal	Research Domain
[7]	PLos One	Technology
[8]	American Journal of Nephrology	Health
[9]	BMC Nephrology	Health
[10]	Applied Clinical Informatics	Informatics
[11]	Brazilian Journal of Nephrology	Health
[12]	Peritoneal Dialysis International	Health
[13]	BMC Nephrology	Health
[14]	Blood Putification: Official Journal of the International Society of Hemofiltration	Health
[15]	Peritoneal Dialysis International	Health
[16]	Journal of Nephrology: Official Journal of the Italian Society of Nephrology	Health
[17]	Blood Putification: Official Journal of the International Society of Hemofiltration	Health
[18]	Peritoneal Dialysis International	Health
[19]	Nefrología Latinoamericana: Official Journal of the Sociedad Latinoamericana de Nefrología e Hipertensión	Health
[20]	Peritoneal Dialysis International	Health

2.5. Risk of Bias Control

We considered various digital databases to ensure a comprehensive search, achieve higher sensitivity levels, and reduce publication bias of sources [31]. The inclusion of papers published only in indexed, refereed, and peer-reviewed journals or conferences ensures a certain degree of conceptual and methodological rigor at the scientific level [32]. In our case, we did not include papers published before 2014 because the technologies used in the proposed systems before this year are obsolete today. The inclusion of two pairs of evaluators during the selection process of relevant works, and of a third evaluator in case of disagreement, substantially reduced the bias of the evaluators that could have arisen from the subjective nature of applying the inclusion and exclusion criteria. In our work, the inclusion criteria Cr_In2, Cr_In4, and Cr_In5 were defined because we were interested in determining the actual levels of implementation, use, acceptance, and effectiveness of technologies used in the remote monitoring of people with CKD.

2.6. Requirements and Parameters to Be Monitored in PD Treatment

We also assessed selected works in terms of requirements and parameters to be monitored in PD treatment based on the study proposed by Nayat et al. [33]. This study states that a telemonitoring system focused on peritoneal dialysis must meet the following requirements:

- Allow the user flexibility in movements and activities. Some of the systems analyzed provide total flexibility at any time to patients in terms of movements and activities, while other systems provide such flexibility only "out of treatment time", i.e., while the treatment is being developed, it restricts the patient movements and activities.
- Allow bidirectional communications through image capture or high-definition video. Some systems fully complied with this item, allowing bidirectional communication in "real-time" between the patient and medical staff; however, others only allowed an "asynchronous" communication between actors or some kind of restricted communication, such as audio transmissions.
- Provide intuitive and straightforward alarm systems. The systems that met this criterion have implemented various alarm systems.
- Incorporate modifiable and customizable mechanisms. Flexibility is analyzed in terms of modification, adaptation, and customization of various functional aspects ("customizable concerning treatment").

- Generate useful reports. The systems that complied with this feature have incorporated several mechanisms to generate reports or treatment reports.
- Are nonintrusive and portable. In general, in the analyzed studies, some systems complied with both aspects, being nonintrusive and having portability in their operation. Only a couple of systems complied only with being "nonintrusive" in their operation but omitted portability.

On the other hand, the parameters of PD exchanges to be monitored must be the following [33]: filling and draining volumes, filling and draining times, blood pressure, pulse, oxygen saturation, weight or bioimpedance, time/duration of treatment stay, number of exchanges, dialysis prescription, symptoms during therapy, alarms and patient response to alarms, and activity during the day.

These characteristics allow a system to provide a required level of virtual assistance, leading to greater patient satisfaction, improved comfort, and eventually, higher acceptance levels of peritoneal dialysis as a preferred form of renal replacement therapy [33].

3. Results

Our SLR aimed to find as many primary studies as possible that respond to the research questions. The results obtained for each question are as follows:

RQ1. Telemonitoring requirements and parameters used in telemonitoring systems focused on patients with CKD on PD.

For each of the 14 selected studies, Tables 3 and 4 present the analysis result considering the characteristics previously described and supported by Nayak et al. [33]. In Table 3, the first column (from left to right) lists each of the studies analyzed, the following six columns describe the requirements for PD telemonitoring in each system, and the last column shows the peritoneal dialysis modality to which each system is oriented (continuous ambulatory peritoneal dialysis—CAPD, automated peritoneal dialysis—APD, or both modalities). In Table 4, the first column (from left to right) lists each of the works analyzed; the following 12 columns describe the consideration or not of each parameter to be monitored in PD.

Table 3. Requirements for telemonitoring of PD [33].

Studies Included	R1	R2	R3	R4	R5	R6	PSZ	DP Modality
[7]	✓	✓	✓	✓	✓	✓	24	APD/CAPD
[8]	✓	✓	✗	✗	✗	✓	112	CAPD
[9]	✓	✓	✗	✓	✓	✓	✗	APD/CAPD
[10]	✓	✓	✓	✓	✗	✓	300	CAPD
[11]	✓	✓	✓	✓	✓	✓	✗	APD/CAPD
[12]	✓	✓	✗	✗	✓	✓	69	CAPD
[13]	✓	✓	✗	✗	✓	✓	✗	CAPD
[14]	✓	✓	✓	✓	✓	✓	6	APD/CAPD
[15]	✓	✓	✓	✓	✓	✓	100	APD
[16]	✓	✓	✓	✓	✓	✓	35	APD
[17]	✓	✓	✓	✓	✓	✓	1023	APD
[18]	✓	✓	✓	✓	✓	✓	65	APD
[19]	✓	✓	✓	✓	✓	✓	396	APD
[20]	✓	✓	✓	✓	✓	✓	1	APD

Where R1: Allows user flexibility in movement and activities, R2: Two-way communications with high-definition video or image capture, R3: Simple and intuitive alarm systems with a high degree of specificity, R4: Modifiable and customizable, R5: Generate useful reports, R6: Nonintrusive and portable, and PSZ: Patient sample size.

Table 4. Parameters of PD exchanges to be monitored [33].

Studies Included	P1	P2	P3	P4	P5	P6	P7	P8	P9	P10	P11	P12
[7]	✓	✓	✓	✓	✓	✓	✓	✓	✓	✓	✓	✗
[8]	✗	✗	✓	✓	✗	✗	✓	✗	✓	✓	✗	✗
[9]	✓	✓	✓	✓	✗	✓	✓	✓	✓	✗	✓	✗
[10]	✗	✗	✓	✗	✗	✓	✓	✗	✓	✗	✓	✗

Table 4. *Cont.*

Studies Included	P1	P2	P3	P4	P5	P6	P7	P8	P9	P10	P11	P12
[11]	✓	✓	✓	✗	✓	✓	✓	✗	✗	✗	✓	✗
[12]	✗	✗	✓	✗	✓	✗	✗	✗	✓	✓	✗	✓
[13]	✗	✗	✗	✗	✗	✗	✗	✓	✓	✓	✗	✗
[14]	✓	✓	✗	✗	✗	✗	✗	✓	✓	✗	✗	✗
[15]	✓	✓	✓	✓	✓	✓	✓	✓	✓	✓	✓	✗
[16]	✓	✓	✓	✓	✓	✓	✓	✓	✓	✓	✓	✗
[17]	✓	✓	✓	✓	✓	✓	✓	✓	✓	✓	✓	✗
[18]	✓	✓	✓	✓	✓	✓	✓	✓	✓	✓	✓	✗
[19]	✓	✓	✓	✓	✓	✓	✓	✓	✓	✓	✓	✗
[20]	✓	✓	✓	✓	✓	✓	✓	✓	✓	✓	✓	✗

Where P1: Fill and drain volumes, P2: Fill and drain times, P3: Blood pressure, P4: Pulse, P5: Oxygen saturation, P6: Weight or bioimpedance, P7: Time/duration of treatment dwell, P8: Number of exchanges, P9: Prescription of dialysis, P10: Symptoms during therapy, P11: Alarms and patient response to alarms, and P12: Activity during the day.

The results in Tables 3 and 4 show that the requirements and parameters described by Nayak et al. [33] are partially considered in all the selected studies. Regarding the requirements for telemonitoring of people with CKD on PD, only three studies [7,11,14] cover entirely all the requirements proposed by Nayak et al. [33]. Concerning the parameters to be monitored in peritoneal dialysis treatments, seven studies [7,15–20] considered 11 of the 12 parameters proposed by Nayak et al. [33] (see Table 4). The results also show that several of the studies privilege the implementation of requirements, such as ensuring flexibility in patient movements and activities [7–20], allowing bidirectional communication among patients and medical teams [7–20], and granting portability and nonintrusiveness to the systems [7–20], rather than the consideration of other requirements also described by Nayak et al. [33] (see Table 3). In the selected studies, we found implementations using proprietary systems from Baxter Healthcare Corporation [15–20]. These implementations mostly consider the characteristics defined by Nayak et al. [33]; however, they are oriented only to peritoneal dialysis treatment under the APD modality, excluding patients in the CAPD modality (see Table 3).

RQ2. Degree of technological development of telemonitoring systems for patients with CKD on PD proposed in the specialized literature

Given the detailed analysis executed for each paper, it was also possible to identify the distinctive characteristics of the levels of technology maturity described by Ibáñez de Aldecoa [34]. In addition, as a complement to the process of identifying the levels of technological development in which the analyzed projects are located, the Guide for the Diagnosis of the Level of Technological Maturity implemented by the National Council of Humanities, Science, and Technology (CONAHCyT) of Mexico in various calls for proposals were used to systemically present the process of technological development and innovation [35]. The results of the ranking process were as follows (see Table 5):

- Eight telemonitoring systems described within the selected studies [8,10,15–20] were identified within level 9 of implemented technology maturity;
- Two selected telemonitoring systems [7,14] were identified within the maturity level 7 of the implemented technology;
- Three selected telemonitoring systems [9,12,13] were identified within the maturity level 6 of the implemented technology;
- One selected system [11] was identified within the maturity level 4 of the implemented technology.

Table 5. Classification of selected studies.

Studies Included	TRL-1	TRL-2	TRL-3	TRL-4	TRL-5	TRL-6	TRL-7	TRL-8	TRL-9
[7]							✓		
[8]									✓
[9]					✓				
[10]									✓
[11]				✓					
[12]						✓			
[13]						✓			
[14]							✓		
[15]									✓
[16]									✓
[17]									✓
[18]									✓
[19]									✓
[20]									✓

RQ3. Changes over time in the publication rate of studies related to telemonitoring systems for patients with CKD on PD.

Concerning the publication date, the frequency of publication has changed over time. In 2014, three articles were published [12–14]. One study was published in 2015 [11] and one in 2019 [18]. In 2017, two studies were registered [9,10]. Four articles were published in 2018 [7,8,19,20]. Finally in 2020, three studies were published [15–17]. Twelve studies were published in journals with a health research scope [8,9,11–20], one in a journal with an informatics research scope [10], and one in a journal with a technology research scope [7]. Approximately 30% of the studies published in health research journals were published in *Peritoneal Dialysis International*. On the other hand, it is noted that there is limited research on telemonitoring systems focused on patients in peritoneal dialysis treatment from a technological perspective.

RQ4. Areas of opportunity and pending challenges of telemonitoring systems focused on patients with CKD on PD proposed in the specialized literature.

After the analysis, it was also possible to identify some opportunity areas, such as:

1. Most studies limit their treatment approach to peritoneal dialysis care modalities, either APD or CAPD, which restrict patient care and treatment options (see Table 3).
2. It is important to consider the advantages of certification in the early stages of development. This concept provides compliance with regulations applicable to telemonitoring systems oriented to health care. Thus, some of the selected systems declare such registration neither from the beginning of the process nor any description during the system design, which limits the possibility of obtaining a greater assessment of their degree of technological development.
3. The telemonitoring systems proposed are still in the process of evaluating and validating their usefulness from the patient perspective with CKD on PD.

In general, it is also possible to observe the need to promote the development of new implementations of telemonitoring systems, both software and hardware, specifically oriented to the treatment of patients with CKD on PD. Within the studies analyzed, four of them [7,9,11,14] refer to implementations of software systems, two studies [8,10] refer to implementations of hardware devices for monitoring, six studies [15–20] deal with implementations of systems that work together, both software and hardware, and two studies [12,13] refer to implementations of support systems via telephone, through a scheme of periodic interviews with patients.

4. Discussion

The telemonitoring systems focused on patients with CKD ongoing PD must have the ability to have bidirectional, fast, and real-time communication to help solve treatment problems, according to the study presented by Nayak et al. [33]. Nevertheless, it is observed that in many of the systems analyzed (7 of the 14), the communication between the patient and the medical staff is asynchronous and not necessarily in real time, which could hinder the rapid detection and attention of incidents present in PD treatment. In addition, the telemonitoring systems must obtain and analyze the treatment information in an automatized form. However, the review shows that many analyzed systems require the manual recording of biometric data or additional compatible devices to capture this information. Another point to highlight is that many analyzed systems do not consider the main parameters that need to be monitored in a system focused on peritoneal dialysis treatment. For example, the parameter that is generally not considered in the selected studies is related to monitoring patient activity during the day (a parameter considered only in the description of the system referred to by Juan Li et al. [12]). It was also observed that it is necessary to consider, within the selected systems, the nutritional intake of the patients, since accurate monitoring of the nutritional parameters can help to prevent and/or treat early the deterioration of the nutritional status, body composition, and functional capacity of the patients [36,37].

On the other hand, we have identified that all the systems with the highest level [8,10,15–20] (8 of the 14 studies), according to the TRL framework, correspond to implementations made with the use of proprietary systems of international companies specialized in telemedicine treatment of CKD such as Baxter (www.baxter.com, accessed on 1 July 2023), eQOL (https://eqol.ca, accessed on 1 July 2023), and GlobalMed (www.globalmed.com, accessed on 1 July 2023). It is important to note that despite the advantages of the proprietary systems described, there are also limitations in their use, derived from their status as proprietary systems, such as incompatibility with other devices or systems developed by different companies, difficulty in the migration of patient information between systems, and user dependency in general regarding cost, information, and functionalities. In addition, it was observed that all the implementations that make use of proprietary systems of the company Baxter [15–20] are focused on the care of patients with chronic CKD treated by peritoneal dialysis under the APD modality, which limits the care of patients who are under the CAPD modality. Regarding the two selected telemonitoring systems [7,14], which were identified within the maturity level 7 of the implemented technology, it was observed that they do not declare the beginning of the process related to the registration of the certifications required in these types of systems for their commercialization, which limits their classification in a higher level; however, the effort shown in the development of these systems is remarkable because of their level of development through a process of independent research and open architecture.

Finally, and based on the results obtained, we identify the following findings of the telemonitoring systems for patients with CKD on PD:

- The telemonitoring systems proposed are still in the process of validating their efficacy and effectiveness in the patient's treatment with CKD on PD.
- There is an orientation to discuss usability, functionality, and cost–benefit.
- There is an evident need to provide telemonitoring treatment options for both APD and CAPD peritoneal dialysis patients.
- The requirements and parameters that a PD system must monitor have already been defined and accepted as a reference in the literature, but many of the implementations analyzed do not consider them or comply with them in a limited way.
- Aspects such as the use, management, and ownership of personal data become relevant concerning telemonitoring systems for patients since these systems imply having the ability to capture treatment information and can even integrate clinical records of each patient. In the case of Mexico, there are various regulations in this regard whose observance must be met, such as the Mexican Official Standard NOM-004-SSA3-2012, which confirms the criteria of ownership and confidentiality (among others) for the management of information and the clinical record itself.
- The telemonitoring systems proposed must comply with specific standards and requirements that vary from country to country, which increases the complexity of the technological development of this type of system.

5. Conclusions

This paper presented a systematic literature review of telemonitoring systems focused on patients with CKD undergoing PD based on the TRL framework. The results of this SLR describe the main contributions and limitations of selected works concerning the TRL framework. The implementations of telemonitoring systems that reached the highest level of technological maturity correspond to studies developed with the use of proprietary devices and services of international companies specialized in telemedicine treatment of CKD with some limitations regarding their status as proprietary systems incompatible with other devices or systems. Their main limitation is that they are oriented only to treating patients in the APD modality, which limits the care of patients undergoing the CAPD modality. The level of technological maturity is highly relevant for telemonitoring systems. Therefore, this work can serve as a reference point for researchers and technologists focused on developing telemonitoring systems for patients with CKD undergoing PD. Future work will extend to analyzing the level of technological maturity of cyber–physical systems aimed at telemonitoring CKD patients undergoing PD.

Author Contributions: Study conception and design: A.V.C. and E.L.D.; acquisition of data: A.V.C. and E.L.D.; analysis and interpretation of data: A.V.C., E.L.D., S.E.P.H., S.D.I., M.A.M.N. and J.D.L.C.; drafting of manuscript: A.V.C., E.L.D., S.D.I., M.A.M.N., J.D.L.C. and S.E.P.H.; revising the manuscript critically for important intellectual content: A.V.C., E.L.D., S.D.I., M.A.M.N., J.D.L.C. and S.E.P.H.; approval of the version of the manuscript to be published: A.V.C., E.L.D., S.D.I., M.A.M.N., J.D.L.C. and S.E.P.H. All authors have read and agreed to the published version of the manuscript.

Funding: This research received no external funding, and the APC was funded by Centro de Investigación y de Estudios Avanzados del Instituto Politécnico Nacional.

Conflicts of Interest: The authors declare no conflict of interest.

References

1. World Health Organization. Available online: http://www.emro.who.int/health-topics/ehealth/ (accessed on 26 June 2023).
2. Leite, H.; Gruber, T.; Hodgkinson, I.R. Flattening the infection curve- understanding the role of telehealth in managing COVID-19. *Leadersh. Health Serv.* **2020**, *33*, 221–226. [CrossRef]
3. Norris, A.C. *Essentials of Telemedicine and Telecare*; John Wiley & Sons Ltd.: West Sussex, UK, 2002; p. 188.
4. Vidal, A.J.; Acosta, R.R.; Pastor, N.; Sánchez, U.; Morrison, D.; Narejos, S.; Salvador, A.; López-Segui, F. Telemedicine in the face of the COVID-19 pandemic. *Atención Primaria* **2020**, *52*, 418–422. [CrossRef] [PubMed]
5. Nazish, S.; Mirfa, M.; Pouria, K. An exploration of usability issues in telecare monitoring systems and possible solutions: A systematic literature review. *Disabil. Rehabil. Assist. Technol.* **2020**, *15*, 271–281. [CrossRef]
6. Gómez, E.J.; Del-Pozo, F.; Hernando, M.E. Telemedicine for diabetes care: The DIABTel approach towards diabetes telecare. *Med. Inform.* **1996**, *21*, 283–295. [CrossRef]
7. Martínez-García, M.A.; Fernández-Rosales, M.S.; López-Domínguez, E.; Hernández-Velázquez, Y.; Domínguez-Isidro, S. Telemonitoring system for patients with chronic kidney disease undergoing peritoneal dialysis: Usability assessment based on a case study. *PLoS ONE* **2018**, *13*, e0206600. [CrossRef] [PubMed]
8. Tan, J.; Mehrotra, A.; Nadkarni, G.N.; He, J.C.; Langhoff, E.; Post, J.; Galvao-Sobrinho, C.; Thode, H.C., Jr.; Rohatgi, R. Telenephrology: Providing Healthcare to Remotely Located Patients with Chronic Kidney Disease. *Am. J. Nephrol.* **2018**, *47*, 200–207. [CrossRef] [PubMed]
9. Jeffs, L.; Jain, A.K.; Man, R.H.; Onabajo, N.; Desveaux, L.; Shaw, J.; Hensel, J.; Agarwal, P.; Saragosa, M.; Jamieson, T.; et al. Exploring the utility and scalability of a telehomecare intervention for patients with chronic kidney disease undergoing peritoneal dialysis-a study protocol. *BMC Nephrol.* **2017**, *18*, 155. [CrossRef]
10. Manya, M.; Neal, S.; Teena, C.; Susie, Q.L. Satisfaction and improvements in peritoneal dialysis outcomes associated with telehealth. *Appl. Clin. Inform.* **2017**, *26*, 214–225. [CrossRef]
11. Fernandes, N.M.; Bastos, M.G.; Oliveira, N.A.; Vale-Costa, A.; Bernardino, H.S. Telemedicine: Development of a distance care system for pre-dialysis chronic kidney disease patients. *J. Bras. Nefrol.* **2015**, *37*, 349–358. [CrossRef]
12. Li, J.; Wang, H.; Xie, H.; Mei, G.; Cai, W.; Ye, J.; Zhang, J.; Ye, G.; Zhai, H. Effects of Post-Discharge Nurse-Led Telephone Supportive Care for Patients with Chronic Kidney Disease Undergoing Peritoneal Dialysis in China: A Randomized Controlled Trial. *Perit. Dial. Int. J. Int. Soc. Perit. Dial.* **2014**, *34*, 272–288. [CrossRef]
13. Schachter, M.E.; Bargman, J.M.; Copland, M.; Hladunewich, M.; Tennankore, K.K.; Levin, A.; Oliver, M.; Pauly, R.P.; Perl, J.; Zimmerman, D.; et al. Rationale for a home dialysis virtual ward: Design and implementation. *BMC Nephrol.* **2014**, *15*, 33. [CrossRef]

14. Harrington, D.M.; Myers, L.; Eisenman, K.; Bhise, V.; Nayak, K.S.; Rosner, M.H. The use of a tablet computer platform to optimize the care of patients receiving peritoneal dialysis: A pilot study. *Blood Purif.* **2014**, *37*, 311–315. [CrossRef] [PubMed]
15. Ariza, J.G.; Walton, S.M.; Sanabria, M.; Bunch, A.; Vesga, J.; Rivera, A. Evaluating a remote patient monitoring program for automated peritoneal dialysis. *Perit. Dial. Int.* **2020**, *40*, 377–383. [CrossRef] [PubMed]
16. Milan-Manani, S.; Baretta, M.; Giuliani, A.; Virzì, G.M.; Martino, F.; Crepaldi, C.; Ronco, C. Remote monitoring in peritoneal dialysis: Benefits on clinical outcomes and on quality of life. *J. Nephrol.* **2020**, *33*, 1301–1308. [CrossRef] [PubMed]
17. Bunch, A.; Ardila, F.; Castaño, R.; Quiñonez, S.; Corzo, L. Through the Storm: Automated Peritoneal Dialysis with Remote Patient Monitoring during COVID-19 Pandemic. *Blood Purif.* **2020**, *50*, 279–282. [CrossRef]
18. Sanabria, M.; Buitrago, G.; Lindholm, B.; Vesga, J.; Nilsson, L.-G.; Yang, D.; Bunch, A.; Rivera, A.S. Remote Patient Monitoring Program in Automated Peritoneal Dialysis: Impact on Hospitalizations. *Perit. Dial. Int.* **2019**, *39*, 472–478. [CrossRef] [PubMed]
19. Sanabria, M.; Rosner, M.; Vesga, J.; Molano, T.A.; Corzo, L.; Rodriguez, P.; Rios, C.; Rivera, A.; Bunch, A. A remote management program in automated peritoneal dialysis patients in Colombia. *Nefrol. Latinoam.* **2018**, *15*, 47–51. [CrossRef]
20. Drepper, V.J.; Martin, P.Y.; Chopard, C.S.; Sloand, J.A. Remote Patient Management in Automated Peritoneal Dialysis: A Promising New Tool. *Perit. Dial. Int.* **2018**, *38*, 76–78. [CrossRef]
21. Méndez-Durán, A.; Méndez-Bueno, J.F.; Tapia-Yánez, T.; Muñoz-Montes, A.; Aguilar-Sánchez, L. Epidemiology of chronic kidney failure in Mexico. *Dial. Trasp.* **2010**, *31*, 7–11. [CrossRef]
22. Bello, A.K.; Alrukhaimi, M.; Ashuntantang, G.E.; Basnet, S.; Rotter, R.C.; Douthat, W.G.; Kazancioglu, R.; Köttgen, A.; Nangaku, M.; Powe, N.R.; et al. Complications of chronic kidney disease: Current state, knowledge gaps, and strategy for action. *Kidney Int. Suppl.* **2017**, *7*, 122–129. [CrossRef]
23. Bello, A.K.; Okpechi, I.G.; Osman, M.A.; Cho, Y.; Cullis, B.; Htay, H.; Jha, V.; Makusidi, M.A.; McCulloch, M.; Shah, N.; et al. Epidemiology of peritoneal dialysis outcomes. *Nat. Rev. Nephrol.* **2022**, *18*, 779–793. [CrossRef] [PubMed]
24. Ekeland, A.G.; Bowes, A.; Flottorp, S. Effectiveness of telemedicine: A systematic review of reviews. *Int. J. Med. Inform.* **2010**, *79*, 736–771. [CrossRef] [PubMed]
25. Stevenson, J.K.; Campbell, Z.C.; Webster, A.C.; Chow, C.K.; Tong, A.; Craig, J.C.; Campbell, K.L.; Lee, V.W. eHealth interventions for people with chronic kidney disease. *Cochrane Database Syst. Rev.* **2019**, *8*, CD012379. [CrossRef] [PubMed]
26. Tzinis, I. Technology Readiness Level. NASA Web Article, NASA. Available online: https://www.nasa.gov/directorates/heo/scan/engineering/technology/technology_readiness_level (accessed on 28 April 2021).
27. Jansen-Kosterink, S.; Broekhuis, M.; van-Velsen, L. Time to act mature-Gearing eHealth evaluations towards technology readiness levels. *Digit. Health* **2022**, *8*, 20552076221113396. [CrossRef]
28. Lyng, K.M.; Jensen, S.; Bruun-Rasmussen, M. A Paradigm Shift: Sharing Patient Reported Outcome via a National Infrastructure. In *MEDINFO 2019: Health and Wellbeing e-Networks for All*; IOS Press: Amsterdam, The Netherlands, 2019; Volume 264, pp. 694–698. [CrossRef]
29. Kitchenham, B. *Procedures for Performing Systematic Reviews*; Technical Report TR/SE-0401; Software Engineering Group; Keele University: Keele, UK, 2004; Volume 33, pp. 1–26.
30. Ali, O.; Shrestha, A.; Soar, J.; Wamba, S.F. Cloud computing-enabled healthcare opportunities, issues, and applications: A systematic review. *Int. J. Inf. Manag.* **2018**, *43*, 146–158. [CrossRef]
31. Dickersin, K.; Scherer, R.; Lefebvre, C. Identifying relevant studies for systematic reviews. *BMJ* **1994**, *309*, 1286–1291. [CrossRef]
32. Light, R.; Pillemer, D. *Summing Up: The Science of Reviewing Research*; Harvard University Press: Cambridge, MA, USA, 1984.
33. Nayak, K.S.; Ronco, C.; Karopadi, A.N.; Rosner, M.H. Telemedicine and Remote Monitoring: Supporting the Patient on Peritoneal Dialysis. *Perit. Dial. Int.* **2016**, *36*, 362–366. [CrossRef] [PubMed]
34. Aldecoa, Q.J.; Ibáñez, J.M. Technology readiness levels: TRLS: Una introducción. *Ind. Econ.* **2014**, *393*, 165–171.
35. National Council of Humanities, Science and Technology (CONAHCyT). Available online: https://conacyt.mx/wp-content/uploads/sni/marco_legal/criterios/Anexo_Nivel_de_Madurez_Tecnologica.pdf (accessed on 9 October 2022).
36. Ikizler, T.A.; Cano, N.J.; Franch, H.; Fouque, D.; Himmelfarb, J.; Kalantar-Zadeh, K.; Kuhlmann, M.K.; Stenvinkel, P.; TerWee, P.; Teta, D.; et al. Prevention and treatment of protein energy wasting in chronic kidney disease patients: A consensus statement by the International Society of Renal Nutrition and Metabolism. *Kidney Int.* **2013**, *84*, 1096–1107. [CrossRef]
37. Olivares-Gandy, H.J.; Domínguez-Isidro, S.; López-Domínguez, E.; Hernández-Velázquez, Y.; Tapia-McClung, H.; Jorge De-la-Calleja, J. A telemonitoring system for nutritional intake in patients with chronic kidney disease receiving peritoneal dialysis therapy. *Comput. Biol. Med.* **2019**, *109*, 1–13. [CrossRef]

Disclaimer/Publisher's Note: The statements, opinions and data contained in all publications are solely those of the individual author(s) and contributor(s) and not of MDPI and/or the editor(s). MDPI and/or the editor(s) disclaim responsibility for any injury to people or property resulting from any ideas, methods, instructions or products referred to in the content.

Fuzzy Logic System for Classifying Multiple Sclerosis Patients as High, Medium, or Low Responders to Interferon-Beta

Edgar Rafael Ponce de Leon-Sanchez [1,*,†], Jorge Domingo Mendiola-Santibañez [2,*,†], Omar Arturo Dominguez-Ramirez [3], Ana Marcela Herrera-Navarro [1], Alberto Vazquez-Cervantes [4], Hugo Jimenez-Hernandez [1] and Horacio Senties-Madrid [5]

1. Facultad de Informática, Universidad Autónoma de Querétaro, Querétaro 76230, Mexico; mherrera@uaq.mx (A.M.H.-N.); hugo.jimenez@uaq.edu.mx (H.J.-H.)
2. Facultad de Ingeniería, Universidad Autónoma de Querétaro, Querétaro 76010, Mexico
3. Centro de Investigación en Tecnologías de Información y Sistemas, Universidad Autónoma del Estado de Hidalgo, Pachuca 42039, Mexico; omar@uaeh.edu.mx
4. Centro de Ingeniería y Desarrollo Industrial, Querétaro 76125, Mexico; alberto.vazquez@cidesi.edu.mx
5. Hospital HMG Coyoacán, Ciudad de Mexico 04380, Mexico; sentiesmadridh@gmail.com
* Correspondence: eponcedeleon13@alumnos.uaq.mx (E.R.P.d.L.-S.); mendijor@uaq.mx (J.D.M.-S.)
† These authors contributed equally to this work.

Abstract: Interferon-beta is one of the most widely prescribed disease-modifying therapies for multiple sclerosis patients. However, this treatment is only partially effective, and a significant proportion of patients do not respond to this drug. This paper proposes an alternative fuzzy logic system, based on the opinion of a neurology expert, to classify relapsing–remitting multiple sclerosis patients as high, medium, or low responders to interferon-beta. Also, a pipeline prediction model trained with biomarkers associated with interferon-beta responses is proposed, for predicting whether patients are potential candidates to be treated with this drug, in order to avoid ineffective therapies. The classification results showed that the fuzzy system presented 100% efficiency, compared to an unsupervised hierarchical clustering method (52%). So, the performance of the prediction model was evaluated, and 0.8 testing accuracy was achieved. Hence, a pipeline model, including data standardization, data compression, and a learning algorithm, could be a useful tool for getting reliable predictions about responses to interferon-beta.

Keywords: fuzzy logic system; pipeline prediction model; multiple sclerosis

1. Introduction

Multiple sclerosis (MS) is a chronic inflammatory disease of the central nervous system (CNS) [1]. Although MS can take several different forms, the most common type is relapsing–remitting MS (RRMS), characterized by alternating periods of remission and intensification of symptoms [2]. The etiology of MS can include several factors, such as genetic susceptibility and viral infections [3–5], which activate the immune system, generating immune dysregulation, and producing an immune attack against the myelin covering of the CNS [6]. Studies have shown that susceptibility to MS is genetically dependent, but the specific gene factors remain largely unknown. It is known that peripheral self-antigen-specific immune cells are activated during the antigen presentation process, and that they enter the CNS through the disrupted blood–brain barrier (BBB) [7]. The route of entry depends on the phenotype and activation state of the T cells. T cells play important roles in cellular immunity [8]. T cells are divided into helper T cells (Th) and regulatory T cells (Treg).

The autoimmune etiology of MS has been the target of the therapeutic approach to patients. Treatment of MS can be divided into treatment of MS symptoms, treatment of MS relapse, and treatment modifying disease progression. The main target of MS treatment is delaying the disease progression [9]. Interferon-beta (IFN-β) is one of the most widely prescribed disease-modifying therapies for RRMS patients. IFN-β has multiple pathways of action on the immune system. IFN-β inhibits the activated proliferation of T cells, and prevents the migration of activated immune cells through the BBB. Also, this drug inhibits the production of pro-inflammatory cytokines (e.g., IL-2, IL-12, IFN-γ), induces an increase in anti-inflammatory cytokines (e.g., IL-4, IL-5, IL-10 and TGF-β), and promotes re-myelination in CNS [10,11]. IFN-β can also prevent the differentiation of inflammatory Th1/Th17 cells, and it can change the phenotype of Th cells from inflammatory Th1 to anti-inflammatory Th2 cells. Studies have shown that IFN-β can significantly improve the clinical symptoms of patients, reduce the annual recurrence rate, and delay the progress of the disease [12]. However, IFN-β is only partially efficient, and a significant proportion of MS patients do not respond to this treatment, with the proportion of non-responders ranging from 20 to 50% [13]. Hence, in this paper, a pipeline model based on potential biomarkers associated with the response to IFN-β is proposed, to predict whether MS patients are potential candidates to be treated with this drug. Studies have researched the effect of gene polymorphisms on therapeutic responses to IFN-β, which can affect the efficacy of this therapy. Bustamante et al. [14] analyzed the relationship between single-nucleotide polymorphisms (SNPs) disposed in type I IFN-induced genes, genes becoming the toll-like receptor (TLR) pathway, and genes encoding neurotransmitter receptors, and the response to IFN-β treatment in MS patients. Martinez et al. [15] evaluated the effect of polymorphisms in some genes (CD46, CD58, FHIT, IRF5, GAPVD1, GPC5, GRBRB3, MxA, PELI3, and ZNF697) on responses to IFN-β treatment among RRMS patients. From seven selected SNPs, PELI3 and GABRR3 polymorphisms were exposed, to be related to IFN-β responses.

Genome-wide research is generated in large numbers of data, and there is a need for soft computing methods (SCMs)—such as artificial neural networks, fuzzy systems, evolutionary algorithms, or metaheuristic and swarm intelligence algorithms—that can deal with this amount of data [16]. Studies of fuzzy systems have only focused on MS diagnosis. Ayangbekun & Jimoh [17] designed a fuzzy inference system for diagnosing five brain diseases: Alzheimer's, Creutzfeldt–Jakob, Huntington's, MS, and Parkinson's. Hosseini et al. [18] developed a clinical decision support system (CDSS), to help specialists diagnose MS with a relapsing–remitting phenotype. Matinfar et al. [19] proposed an expert system for MS diagnosis, based on clinical symptoms and demographic characteristics. However, it is necessary to design new expert systems that can classify the possible responses to treatments in MS patients. Other studies have applied machine learning (ML) techniques to diagnose early MS. Goyal et al. [20] trained a random forest (RF) model with the serum level of eight cytokines (IL-1β, IL-2, IL-4, IL-8, IL-10, IL-13, IFN-γ, and TNF-α) in MS patients, to detect predictors for disease. Chen et al. [21] implemented a support vector machine (SVM) model, using gene expression profiles to identify potential biomarkers for MS diagnosis. CXCR4, ITGAM, ACTB, RHOA, RPS27A, UBA52, and RPL8 genes were detected. Among the studies that suggest genetics can predict the pharmacological response to a treatment, Fagone et al. [22] trained an uncorrelated reduced centroid (UCRC) algorithm to identify a subset of genes that could predict the responses to natalizumab in RRMS patients. A specific gene expression profile of CD4+ T cells could characterize the responsiveness.

Although the studies presented above have shown the efficiency of IFN-β at improving the clinical symptoms of MS patients, a proportion of patients did not respond to this treatment. Genome-wide analytical studies have been conducted, in order to identify genetic factors associated with the responses to IFN-β treatment. Gurevich et al. [23] identified a subgroup of secondary progressive MS (SPMS) patients presenting a gene expression signature similar to that of RRMS patients who are clinical responders to IFN-β treatment. SPMS patients were classified using unsupervised hierarchical clustering, according to IFN-inducible gene expression profiling identified in RRMS clinical responders to treatment. Although, the hierarchical clustering method is easy to implement, it rarely provides the best solution, due to lots of arbitrary decisions. Clarelli et al. [24] detected genetic factors that affect the long-term response to IFN-β. The found pathways associated with inflammatory processes and presynaptic membrane, i.e., the genes related to the glutamatergic system (GRM3 and GRIK2), play a potential role in the response to IFN-β. Jin et al. [25] implemented a feature selection method based on differentially correlated edges (DCE), to identify the most relevant genes associated with the response to IFN-β treatment in RRMS patients. Of the 23 identified genes, 7 had a confidence score > 2: CXCL9, IL2RA, CXCR3, AKT1, CSF2, IL2RB, and GCA. Because the analyzed data were unlabeled, the responder category was restricted to patients whose first relapse time was more than five years (60 months), resulting in nine responders and nine non-responders. So, seven patients were excluded from the analysis. Hence, we attempt to address some of the issues above in this research. The main contributions of this paper are as follows:

- An alternative fuzzy system based on expert knowledge, with linguistic rules to classify RRMS patients as high, medium, or low responders to IFN-β treatment.
- A pipeline prediction model, including a data preprocessing technique, a transformation technique for data compression, and a learning algorithm for making predictions on new data. The prediction model is trained with biomarkers associated with the IFN-β response for predicting whether MS patients are potential candidates to be treated with this drug, in order to avoid ineffective therapies.

2. Materials and Methods

The strategy followed in this research is described in the flowchart of Figure 1, which divides the proposal into four stages.

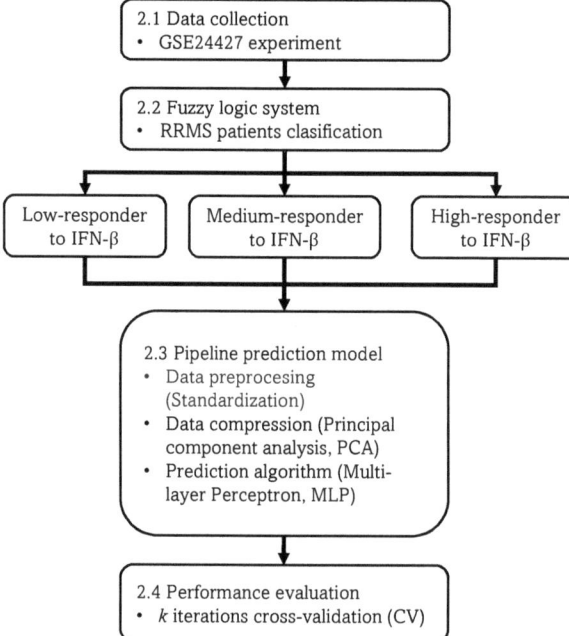

Figure 1. Proposed methodology. The gene data, demographic, and clinical characteristics are collected. Then, the RRMS patients are classified by the fuzzy logic system. A pipeline prediction model is implemented, including data standardization, PCA for data compression, and an MLP algorithm for making predictions. Finally, the k-iterations CV is implemented, for evaluating the model prediction performance.

2.1. Data Collection

The dataset was collected from the GSE24427 expression profiling by array experiment, available in the public repository of genomic data GEO [26]. Through the GPL96 [HG-U133A] platform (Affymetrix Human Genome U133A Array), the genome-wide expression profiles of peripheral blood mononuclear cells from 25 RRMS patients were obtained. Patients were treated with subcutaneous IFN-beta-1b (Betaferon, Bayer Healthcare) at the standard dose (250 µg every other day). Patient blood samples were drawn before first-dose, second-dose, 1st-month, 12th-month, and 24th-month IFN-β injection. The expression summary values were analyzed by GEO2R, an interactive web tool that allows viewing of a specific gene expression through the profile graph tab. On the one hand, the GPL96 platform enabled us to see demographic and clinical characteristics of RRMS patients, which were used as input variables for the proposed fuzzy system, and these are presented in Table 1.

Table 1. Demographic and clinical characteristics.

Sample	Gender	Age	EDSS [1] 1st Month	EDSS [1] 24th Month
1	Female	63	4	5.5
2	Male	45	1	1
3	Female	25	1	1
4	Female	27	4	3.5
5	Female	51	3	2.5
6	Female	41	2	4.5
7	Female	44	4	3
8	Male	30	1.5	2
9	Female	26	4	3.5
10	Male	42	1	1
11	Male	29	2	2.5
12	Female	28	1.5	2.5
13	Female	48	1	1
14	Female	47	3.5	3
15	Female	42	2	3
16	Female	50	3.5	3.5
17	Male	37	1.5	4.5
18	Female	43	2	2
19	Male	54	3	2
20	Male	40	1	1
21	Female	48	2	2
22	Female	38	2	3
23	Male	18	1.5	2
24	Female	24	1	1
25	Male	38	1	1

[1] Expanded disability status scale.

On the other hand, through the GPL96 platform, the expression values of 15 biomarkers associated with the response to IFN-β—IL-2, IL-12, IFN-γ, TNF-α, IL-4, IL-10, TGF-β, CD46, CD58, FHIT, IRF5, GAPVD1, GPC5, GRM3, and GRIK2—were collected and integrated into an Excel spreadsheet, for training the proposed pipeline prediction model. For example, the IL-2 and IL-4 cytokines expression values are displayed in Figures 2 and 3. The database is the same as the one used by Jin et al. [25]. However, the biomarkers are a little different.

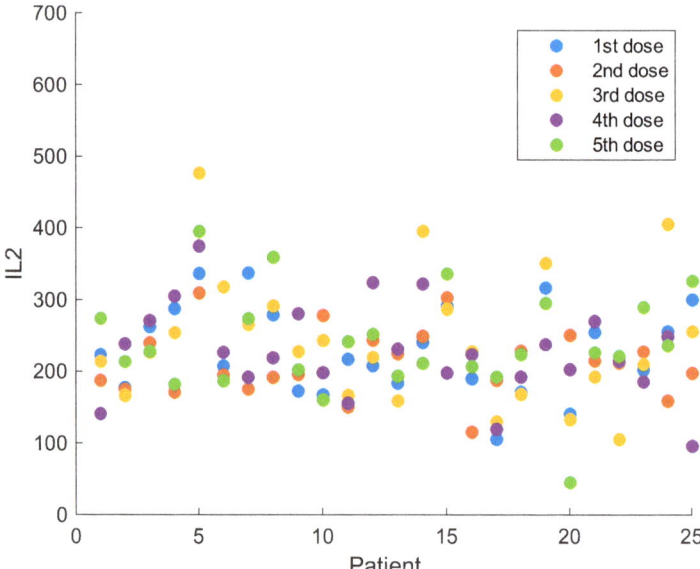

Figure 2. IL-2 cytokine. The expression values of 25 RRMS patients corresponding to five doses: before first-dose, second-dose, 1st-month, 12th-month, and 24th-month IFN-β injection.

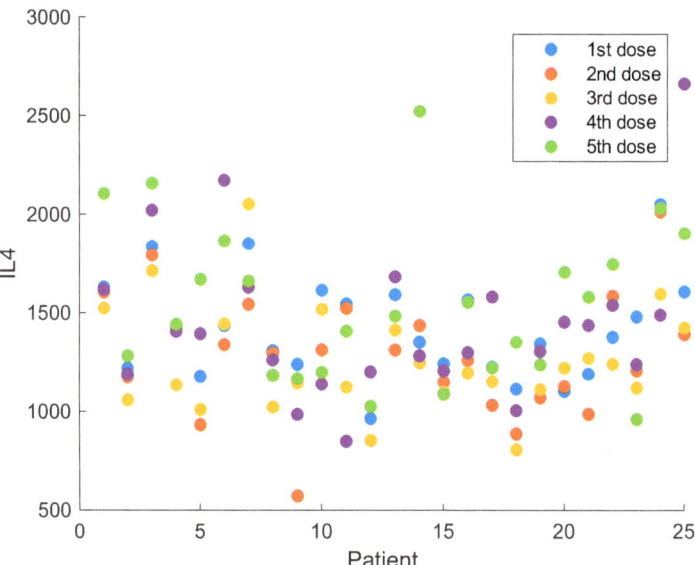

Figure 3. IL-4 cytokine. The expression values seem more scattered than IL-2 cytokine.

2.2. Fuzzy Logic System

Fuzzy systems are structures based on fuzzy sets and fuzzy logic theories for processing inaccurate information [27]. Their main property includes symbolic knowledge representation in a form of fuzzy conditional (if-then) rules. The typical structure of a fuzzy system is described in Figure 4.

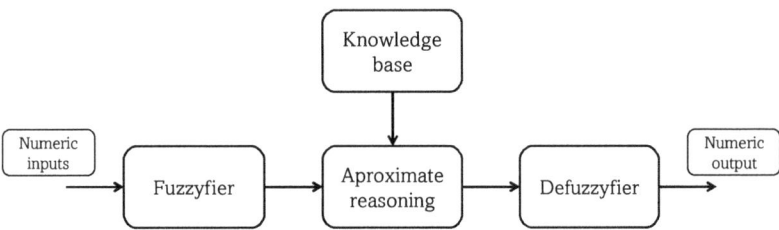

Figure 4. Fuzzy system structure. The fuzzyfier transforms the values of input variables into an N-dimensional fuzzy set A (linguistic values of the output variable) defined on the universe $\mathbb{X}$, by means of approximate reasoning (inference engine) using expert knowledge, which is represented as a set of fuzzy conditional rules (knowledge base). The result of the approximate reasoning is a fuzzy set $B(y)$. The defuzzyfier computes a representative numerical output y_0 from the result of fuzzy set $B(y)$ defined on the universe $\mathbb{Y}$.

The fuzzyfier can be defined as the membership function $\mu_A(x)$ of the fuzzy set A. Demographic and clinical characteristics of RRMS patients are used as input variables for the fuzzyfier. The numerical output y_0 is computed using the center of gravity (COG) method [28], as in Equation (1):

$$y_0 = \frac{\sum_{i=1}^n y_i \mu_B(y_i)}{\sum_{i=1}^n \mu_B(y_i)}, \tag{1}$$

where $\mu_B(y)$ represents the membership function of fuzzy set $B(y)$. The proposed fuzzy system is designed through the Fuzzy Logic Designer App of MATLAB R2023a software. The structure of the proposed fuzzy system is based on the Mamdani–Assilan fuzzy system (MAFS) [29], which includes a set of conditional fuzzy rules, in the form of Equation (2), that can be determined by a human expert:

$$\mathcal{R} = \{R^i\}_{i=1}^N = \{if\ and_{n=1}^N\ (X_n\ is\ L_{A_n}^{(i)}),\ then\ Y\ is\ L_B^{(i)}\}_{i=1}^I, \tag{2}$$

where I is the number of rules, X_n represents the input linguistic variables, Y is the output linguistic variable, and L_{A_n} and L_B are the linguistic values, defined by fuzzy sets A_N, and B on universes $\mathbb{X}_N$ and $\mathbb{Y}$, respectively. In this paper, the input linguistic variables describing the demographic characteristics—including gender and age—and the clinical characteristics—including expanded disability status scale (EDSS) 1st month and 24th month—are defined: $\mathcal{N}_1$ = "mean gender"; $\mathcal{N}_2$ = "mean age"; $\mathcal{N}_3$ = "mean EDSS 1st month"; $\mathcal{N}_4$ = "mean EDSS 24th month". The sets of possible linguistic values are collections of different labels describing the gender, age, EDSS 1st month, and EDSS 24th month: L_{A_1} = {"female", "male"}; L_{A_2} = {"pediatric", "adult", "elderly"}; L_{A_3} = {"low", "medium", "high"}; L_{A_4} = {"low", "medium", "high"}, and response to IFN-β: L_B = {"low", "medium", "high"}. To each one of the labels, the fuzzy sets $A_N^{(i)}$ are assigned, defined on the universe $\mathbb{X}_N$, which represents the range of possible values. The whole description of the defined linguistic variables is presented in Table 2.

For example, the graphics of the membership functions $\mu_{A_2^{(1)}}(age)$, $\mu_{A_4^{(1)}}$ (EDSS 24th month), and μ_B(Response to IFN $- \beta$) of the fuzzy sets $A_2^{(1)}$, $A_4^{(1)}$, and $B(y)$ are displayed in Figures 5–7, respectively.

Table 2. Linguistic variables description.

Membership Function	Fuzzy Set	Universe of Discourse	Parameters and Type
$\mu_{A_1^{(1)}}(gender)$	$A_1^{(1)}$	$\mathbb{X}_1$: [0 a 1]	Female: [−0.75; −0.083; 0.083; 0.75] Trapezoidal Male: [0.25; 0.916; 1.083; 1.75] Trapezoidal
$\mu_{A_2^{(1)}}(age)$	$A_2^{(1)}$	$\mathbb{X}_2$: [0 a 100] years	Pediatric: [−37.5; −4.167; 4.167; 37.5] Trapezoidal Adult: [8.333; 50; 91.666] Triangular Elderly: [62.5; 95.83; 104.2; 137.5] Trapezoidal
$\mu_{A_3^{(1)}}(EDSS\ ^1\ 1st\ month)$	$A_3^{(1)}$	$\mathbb{X}_3$: [0 a 10] units	Low: [−3.75; −0.416; 1.0; 5.0] Trapezoidal Medium: [1.0; 5.0; 9.0] Triangular High: [5.0; 9.0; 10.42; 13.75] Trapezoidal
$\mu_{A_4^{(1)}}(EDSS\ ^1\ 24th\ month)$	$A_4^{(1)}$	$\mathbb{X}_4$: [0 a 10] units	Low: [−3.75; −0.416; 1.0; 5.0] Trapezoidal Medium: [1.0; 5.0; 9.0] Triangular High: [5.0; 9.0; 10.42; 13.75] Trapezoidal
$\mu_B(Response\ to\ IFNb)$	$B(y)$	$\mathbb{Y}$: [0 a 1] units	Low: [−0.375; −0.04167; 0.1; 0.5] Trapezoidal Medium: [0.1; 0.5; 0.9] Triangular High: [0.5; 0.9; 1.042; 1.375] Trapezoidal

[1] Expanded disability status scale.

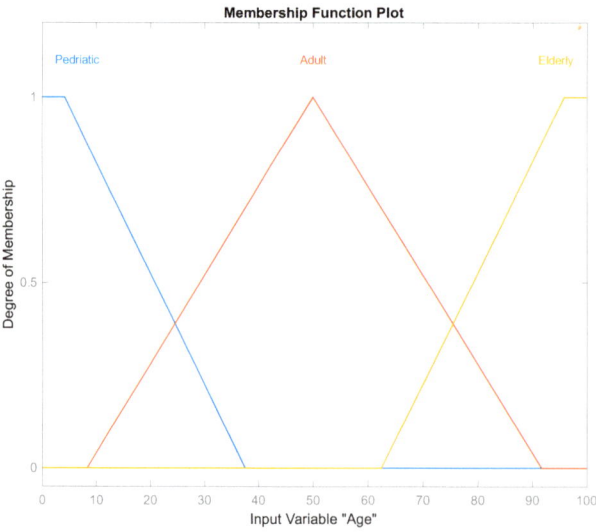

Figure 5. Set of linguistic values, which are three labels describing the "age" input variable, corresponding to fuzzy set $A_2^{(1)}$.

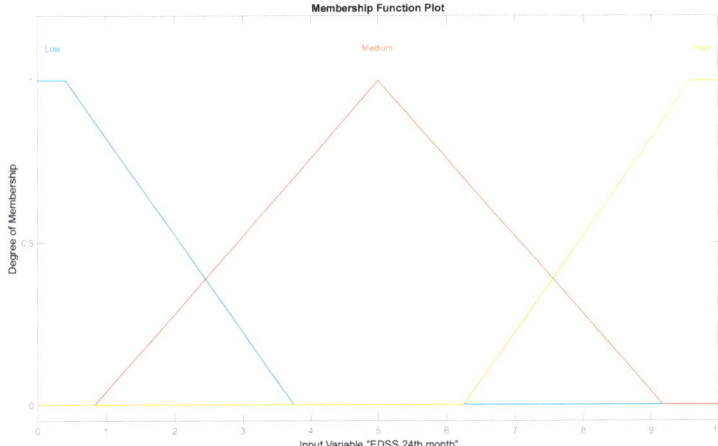

Figure 6. Set of linguistic values, which are three labels describing the "EDSS 24th month" input variable, corresponding to fuzzy set $A_4^{(1)}$.

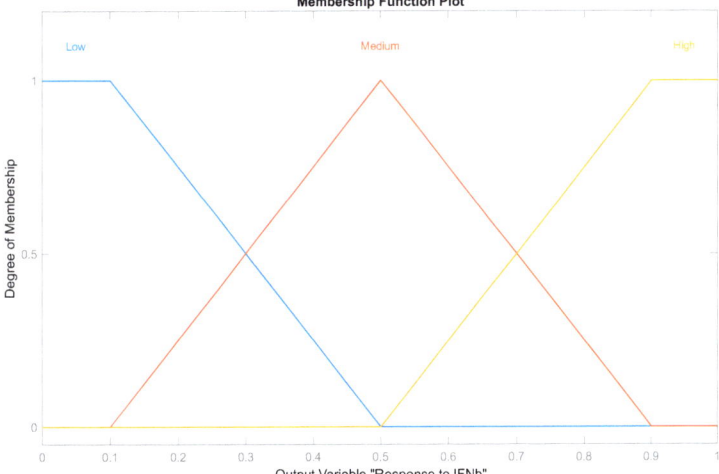

Figure 7. Set of linguistic values, which are three labels describing the "Response to IFN-β" output variable, corresponding to fuzzy set $B(y)$.

The fuzzy conditional rules (knowledge base) are meant to decide the influence of the input variables on responses to IFN-β treatment. Tables 3 and 4 display the 36 defined rules, according to the opinion of a neurology expert:

Table 3. Fuzzy rules definition (first part).

#	Rule
1	If gender is female and age is adult, and if EDSS 1st month is low and EDSS 24th month is low, then response to IFNb is medium.
2	If gender is female and age is adult, and if EDSS 1st month is medium and EDSS 24th month is medium, then response to IFNb is medium.

Table 3. Cont.

#	Rule
3	If gender is female and age is adult, and if EDSS 1st month is high and EDSS 24th month is high, then response to IFNb is medium.
4	If gender is female and age is elderly, and if EDSS 1st month is low and EDSS after 24th month is low, then response to IFNb is medium.
5	If gender is female and age is elderly, and if EDSS 1st month is medium and EDSS 24th month is medium, then response to IFNb is medium.
6	If gender is female and age is elderly, and if EDSS 1st month is high and EDSS 24th month is high, then response to IFNb is medium.
7	If gender is male and age is adult, and if EDSS 1st month is low and EDSS 24th month is low, then response to IFNb is medium.
8	If gender is male and age is adult, and if EDSS 1st month is medium and EDSS 24th month is medium, then response to IFNb is medium.
9	If gender is male and age is adult, and if EDSS 1st month is high and EDSS 24th month is high, then response to IFNb is medium.
10	If gender is male and age is elderly, and if EDSS 1st month is low and EDSS 24th month is low, then response to IFNb is medium.
11	If gender is male and age is elderly, and if EDSS 1st month is medium and EDSS 24th month is medium, then response to IFNb is medium.
12	If gender is male and age is elderly, and if EDSS 1st month is high and EDSS 24th month is high, then response to IFNb is medium.
13	If gender is female and age is adult, and if EDSS 1st month is low and EDSS 24th month is medium, then response to IFNb is low.
14	If gender is female and age is adult, and if EDSS 1st month is low and EDSS 24th month is high, then response to IFNb is low.
15	If gender is female and age is adult, and if EDSS 1st month is medium and EDSS 24th month is high, then response to IFNb is low.
16	If gender is female and age is elderly, and if EDSS 1st month is low and EDSS 24th month is medium, then response to IFNb is low.
17	If gender is female and age is elderly, and if EDSS 1st month is low and EDSS 24th month is high, then response to IFNb is low.
18	If gender is female and age is elderly, and if EDSS 1st month is medium and EDSS 24th month is high, then response to IFNb is low.

Table 4. Fuzzy rules definition (second part).

#	Rule
19	If gender is male and age is adult, and if EDSS 1st month is low and EDSS 24th month is medium, then response to IFNb is low.
20	If gender is male and age is adult, and if EDSS 1st month is low and EDSS 24th month is high, then response to IFNb is low.
21	If gender is male and age is adult, and if EDSS 1st month is medium and EDSS 24th month is high, then response to IFNb is low.
22	If gender is male and age is elderly, and if EDSS 1st month is low and EDSS 24th month is medium, then response to IFNb is low.
23	If gender is male and age is elderly, and if EDSS 1st month is low and EDSS 24th month is high, then response to IFNb is low.
24	If gender is male and age is elderly, and if EDSS 1st month is medium and EDSS 24th month is high, then response to IFNb is low.
25	If gender is female and age is adult, and if EDSS 1st month is high and EDSS 24th month is medium, then response to IFNb is high.
26	If gender is female and age is adult, and if EDSS 1st month is high and EDSS 24th month is low, then response to IFNb is high.
27	If gender is female and age is adult, and if EDSS 1st month is medium and EDSS 24th month is low, then response to IFNb is high.
28	If gender is female and age is elderly, and if EDSS 1st month is high and EDSS 24th month is medium, then response to IFNb is high.

Table 4. Cont.

#	Rule
29	If gender is female and age is elderly, and if EDSS 1st month is high and EDSS 24th month is low, then response to IFNb is high.
30	If gender is female and age is elderly, and if EDSS 1st month is medium and EDSS 24th month is low, then response to IFNb is high.
31	If gender is male and age is adult, and if EDSS 1st month is high and EDSS 24th month is medium, then response to IFNb is high.
32	If gender is male and age is adult, and if EDSS 1st month is high and EDSS 24th month is low, then response to IFNb is high.
33	If gender is male and age is adult, and if EDSS 1st month is medium and EDSS 24th month is low, then response to IFNb is high.
34	If gender is male and age is elderly, and if EDSS 1st month is high and EDSS 24th month is medium, then response to IFNb is high.
35	If gender is male and age is elderly, and if EDSS 1st month is high and EDSS 24th month is low, then response to IFNb is high.
36	If gender is male and age is elderly, and if EDSS 1st month is medium and EDSS 24th month is low, then response to IFNb is high.

2.3. Pipeline Prediction Model

A pipeline is a tool for setting a learning model, including a data preprocessing technique (for instance standardization for feature scaling), a transformation technique (such as PCA for data compression), and a learning algorithm (like MLP) for making predictions on new data. The structure of the proposed pipeline is shown in Figure 8.

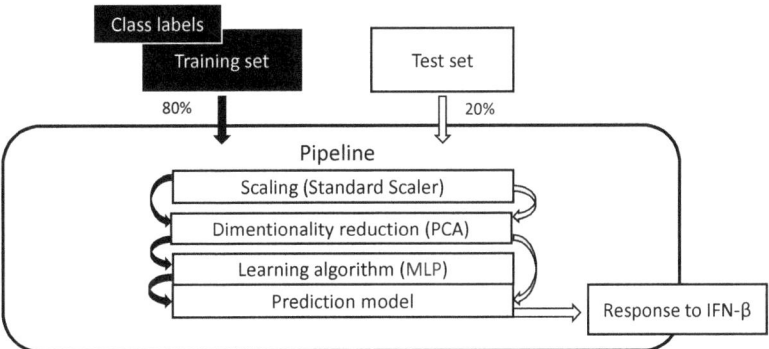

Figure 8. Structure of proposed pipeline model, including feature scaling, data compression, and prediction algorithm.

PCA is a technique of dimensionality reduction, which transforms data from a high-dimensional space to a space of lower dimensions. The dimension reduction is achieved by selecting the principal components (directions of maximum variance) as a basis set for the new space [30]. Applications of PCA include analysis of genome data and gene expression levels. For extracting the principal components, the data are standardized; then, the covariance matrix is built, to store the pairwise covariances between features. For example, the covariance between two features x_j and x_k can be computed by Equation (3):

$$\sigma_{jk} = \frac{1}{n} \sum_{i=1}^{n} (x_j^{(i)} - \mu_j)(x_k^{(i)} - \mu_k), \qquad (3)$$

where μ_j and μ_k are the representative samples of the j and k features, respectively (μ_k, $\mu_j = 0$, because of the standardization). The eigenvectors of the covariance matrix represent the principal components, and the eigenvalues define the magnitude of the eigenvectors, so the eigenvalues have to be ordered by decreasing the magnitude [31]. The ratio of an explained variance of an eigenvalue λ_j is the fraction of the eigenvalue and the total sum of the eigenvalues, as shown by Equation (4):

$$\frac{\lambda_j}{\sum_{j=1}^{d} \lambda_j} \qquad (4)$$

MLP is a supervised learning algorithm that uses the backpropagation technique for learning. The structure of MLP consists of an input layer of neurons that receive the $X = x_1, x_2, \ldots, x_m$ sample inputs, one or more hidden layers of neurons that convert the values from the previous layer to a weighted linear summation, $w_1 x_1 + w_2 x_2 + \ldots + w_m x_m$, followed by a non-linear activation function that is used to learn the weights, and then the output layer that predicts the class label of the samples [32]. During the learning stage, MLP compares the true class labels to the continuous output values of the non-linear activation function, to compute the prediction error and update the weights. The hyperparameters of MLP are arbitrarily set as follows: solver = 'sgd', activation = 'tanh', and learning_rate_init = 0.01.

2.4. Performance Evaluation

One of the key steps in building an ML or deep learning (DL) model is estimating its performance with new data. A model can suffer underfitting (high bias) if the model is too simple, or can suffer overfitting (high variance) if the model is too complex for the subjacent training data [31]. In order to get an acceptable bias–variance rate, the k-iterations cross-validation (CV) technique is implemented, which can obtain reliable estimates of the model's generalization performance.

In the k-iterations CV, the training dataset is randomly split into k iterations without replacement, where $k - 1$ iterations are used for model training, and 1 iteration is used for performance evaluation. This process is repeated k times, to obtain k models and performance estimates. Then, the average performance of the models is computed by Equation (5), based on the independent iterations, to obtain a performance estimate:

$$E = \frac{1}{k} \sum_{i=1}^{k} E_i \qquad (5)$$

Typically, the k-iterations CV is used for model fitting, to find the optimal values of the hyperparameters that produce satisfactory generalization performance. Also, the confusion matrix (CM) is computed, which reports the count of the predictions of a classifier [33]: true positives (TP), true negatives (TN), false positives (FP), and false negatives (FN).

3. Results

For this paper, a fuzzy logic system based on MAFS was implemented, to classify RRMS patients as high, medium, or low responders to IFN-β treatment. Also, for comparison purposes, a hierarchical clustering technique was implemented, to classify the same patients. After the dataset outputs were labeled, the gene features were used as training inputs for the proposed pipeline prediction model.

3.1. Fuzzy Logic System

At fuzzification stage, the membership values were computed for each one of the input variables. Tables 5 and 6 display the computed values of each membership function for all the samples.

Table 5. Fuzzification results (gender and age).

Sample	$\mu_{Female}(Gender)$	$\mu_{Male}(Gender)$	$\mu_{Pediatric}(Age)$	$\mu_{Adult}(Age)$	$\mu_{Elderly}(Age)$
1	1	0	0	0.687	0.015
2	0	1	0	0.88	0
3	1	0	0.375	0.4	0
4	1	0	0.315	0.448	0
5	1	0	0	0.975	0
6	1	0	0	0.784	0
7	1	0	0	0.856	0
8	0	1	0.225	0.479	0
9	1	0	0.345	0.424	0
10	0	1	0	0.808	0
11	0	1	0.255	0.496	0
12	1	0	0.258	0.472	0
13	1	0	0	0.952	0
14	1	0	0	0.928	0
15	1	0	0	0.808	0
16	1	0	0	1	0
17	0	1	0.015	0.688	0
18	1	0	0	0.832	0
19	0	1	0	0.903	0
20	0	1	0	0.76	0
21	1	0	0	0.952	0
22	1	0	0	0.712	0
23	0	1	0.585	0.232	0
24	1	0	0.405	0.376	0
25	0	1	0	0.712	0

Table 6. Fuzzification results (EDSS 1st month and EDSS 24th month).

Sample	$\mu_{Low}(EDSS^{1})$	$\mu_{Med}(EDSS^{1})$	$\mu_{High}(EDSS^{1})$	$\mu_{Low}(EDSS^{2})$	$\mu_{Med}(EDSS^{2})$	$\mu_{High}(EDSS^{2})$
1	0.25	0.75	0.0	0.0	0.875	0.125
2	1.0	0.0	0.0	1.0	0.0	0.0
3	1.0	0.0	0.0	1.0	0.0	0.0
4	0.25	0.75	0.0	0.375	0.625	0.0
5	0.5	0.5	0.0	0.625	0.375	0.0
6	0.75	0.25	0.0	0.125	0.875	0.0
7	0.25	0.75	0.0	0.5	0.5	0.0
8	0.875	0.125	0.0	0.75	0.25	0.0
9	0.25	0.75	0.0	0.25	0.75	0.0
10	1.0	0.0	0.0	1.0	0.0	0.0
11	0.75	0.25	0.0	0.625	0.375	0.0
12	0.875	0.125	0.0	0.625	0.375	0
13	1.0	0.0	0.0	1.0	0.0	0.0
14	0.375	0.625	0.0	0.5	0.5	0.0
15	0.75	0.25	0.0	0.5	0.5	0.0
16	0.375	0.625	0.0	0.375	0.625	0.0
17	0.875	0.125	0.0	0.125	0.875	0.0
18	0.75	0.25	0.0	0.75	0.25	0.0
19	0.5	0.5	0.0	0.75	0.25	0.0
20	1.0	0.0	0.0	1.0	0.0	0.0
21	0.75	0.25	0.0	0.75	0.25	0.0
22	0.75	0.25	0.0	0.5	0.5	0.0
23	0.875	0.125	0	0.75	0.25	0.0
24	1.0	0.0	0.0	1.0	0.0	0.0
25	1.0	0.0	0.0	1.0	0.0	0.0

[1] 1st month. [2] 24th month.

At the approximate reasoning stage, each one of the 36 inference rules from the knowledge base were evaluated with the obtained membership values from Tables 5 and 6. For example, considering the input values of #7 sample (gender: "female", age: "44", EDSS 1st month: "4", and EDSS 24th month: "3"), the inference engine results are shown in Table 7. In this case, only four rules—1, 2, 13, and 27—had an inference result different to zero. Figure 9 displays the evaluation graph of previous inference rules.

Table 7. Inference results for #7 sample.

#	Rule	Inference Engine
1	If gender is "female" and age is "adult", and if EDSS 1st month is "low" and EDSS 24th month is "low", then the response to IFNb is "medium".	min(1.0, 0.856, 0.25, 0.5) = 0.25
2	If gender is "female" and age is "adult", and if EDSS 1st month is "medium" and EDSS 24th month is "medium", then the response to IFNb is "medium".	min(1.0, 0.856, 0.75, 0.5) = 0.5
13	If gender is "female" and age is "adult", and if EDSS 1st month is "low" and EDSS 24th month is "medium", then the response to IFNb is "low".	min(1.0, 0.856, 0.25, 0.5) = 0.25
27	If gender is "female" and age is "adult", and if EDSS 1st month is "medium" and EDSS 24th month is "low", then the response to IFNb is "high".	min(1.0, 0.856, 0.75, 0.5) = 0.5

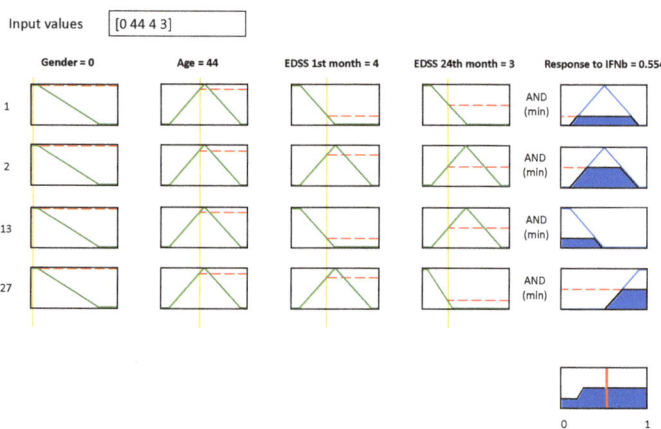

Figure 9. Evaluation graph of the 1, 2, 13, and 27 inference rules. The result graph consists of the combination of the four rules' inference values.

At defuzzification stage, the numerical outputs were computed, substituting the inference engine results into Equation (1), based on the inference result graphs. For example, for the inference results of #7 sample according to the result graph of Figure 10, the numerical output was computed as follows:

$$y_0 = \frac{0*0.25 + 0.1*0.25 + 0.2*0.25 + 0.3*0.5 + 0.4*0.5 + \cdots + 0.9*0.5 + 1*0.5}{0.25 + 0.25 + 0.25 + 0.5 + 0.5 + \cdots + 0.5 + 0.5} \quad (6)$$

$$y_0 = \frac{2.675}{4.75} = 0.563 \approx 0.554 \quad (7)$$

The small difference in calculation was due to the fuzzy system implementation in Matlab software providing more accurate results than by hand.

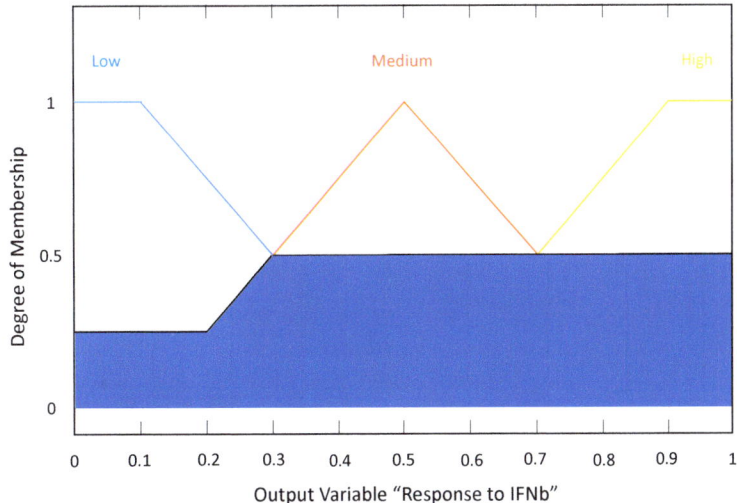

Figure 10. Inference result graph of #7 sample, which includes the values of the "Low", "Medium", and "High" linguistic labels.

Finally, a classification of high, medium, and low responders to the IFN-β drug was carried out, by three different methods: (1) opinion of a neurology expert, (2) proposed fuzzy system, and (3) agglomerative clustering model. The results are displayed in Table 8.

Table 8. Classification of response to IFN-β. The resulting numerical values of defuzzification less than 0.5 are considered as low responder (LR), those equal to 0.5 as medium responder (MR), and those greater than 0.5 as high responder (HR). For comparison purposes, the input data of Table 1 were preprocessed by the StandardScaler technique, and they were used to train a prediction model of agglomerative clustering (n_clusters = 3).

Sample	Expert Opinion	Fuzzy System (Deffuzification)	Agglomerative Clustering
1	LR	0.459 ⇒ LR	HR
2	MR	0.5 ⇒ MR	LR
3	MR	0.5 ⇒ MR	MR
4	HR	0.529 ⇒ HR	HR
5	HR	0.527 ⇒ HR	HR
6	LR	0.337 ⇒ LR	HR
7	HR	0.554 ⇒ HR	HR
8	LR	0.474 ⇒ LR	LR
9	HR	0.53 ⇒ HR	HR
10	MR	0.5 ⇒ MR	LR
11	LR	0.472 ⇒ LR	LR
12	LR	0.445 ⇒ LR	MR
13	MR	0.5 ⇒ MR	MR
14	HR	0.527 ⇒ HR	HR
15	LR	0.446 ⇒ LR	MR
16	MR	0.5 ⇒ MR	HR
17	LR	0.302 ⇒ LR	HR
18	MR	0.5 ⇒ MR	MR

Table 8. Cont.

Sample	Expert Opinion	Fuzzy System (Deffuzification)	Agglomerative Clustering
19	HR	0.554 ⇒ HR	LR
20	MR	0.5 ⇒ MR	LR
21	MR	0.5 ⇒ MR	MR
22	LR	0.446 ⇒ LR	MR
23	LR	0.463 ⇒ LR	LR
24	MR	0.5 ⇒ MR	MR
25	MR	0.5 ⇒ MR	LR

As Table 6 shows, 100% of the outputs were correctly labeled by the proposed fuzzy system, while 52% were correctly labeled by agglomerative clustering according to an expert opinion.

3.2. Pipeline Prediction Model

Once the dataset output labels had been classified, the pipeline prediction model was implemented, for making predictions on new data. First, the gene expression values were scaled by the StandardScaler technique. Then, the PCA technique was used, to reduce the dimensionality of the gene dataset by compressing it into a new subspace, so that only the subset of the eigenvectors (principal components) that contained more information (maximum variance) were selected. Figure 11 shows the results of the explained variance ratio of the eigenvalues.

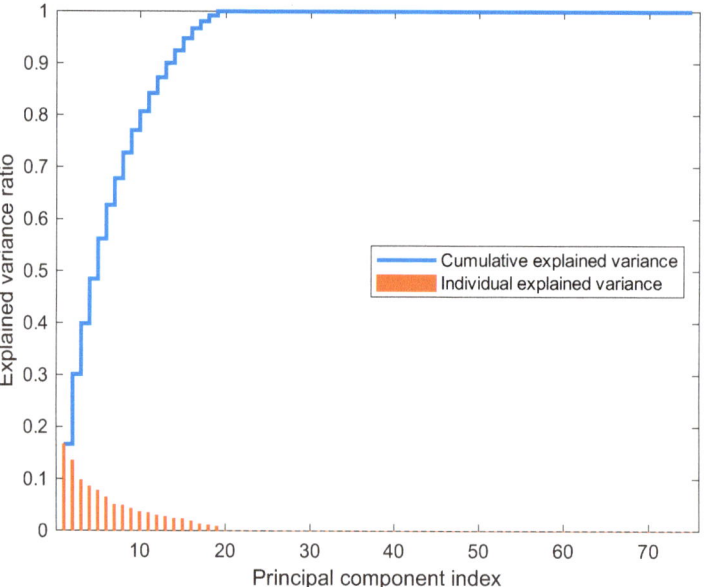

Figure 11. Explained variance ratio. The first principal component by itself accounts for almost 20% of the total variance. Furthermore, the first two combined principal components represent approximately 40% of the variance.

Figure 12 shows the graph used to determine the optimal value of the number of principal components (n_components) for the PCA technique to achieve the high testing accuracy of the MLP prediction algorithm.

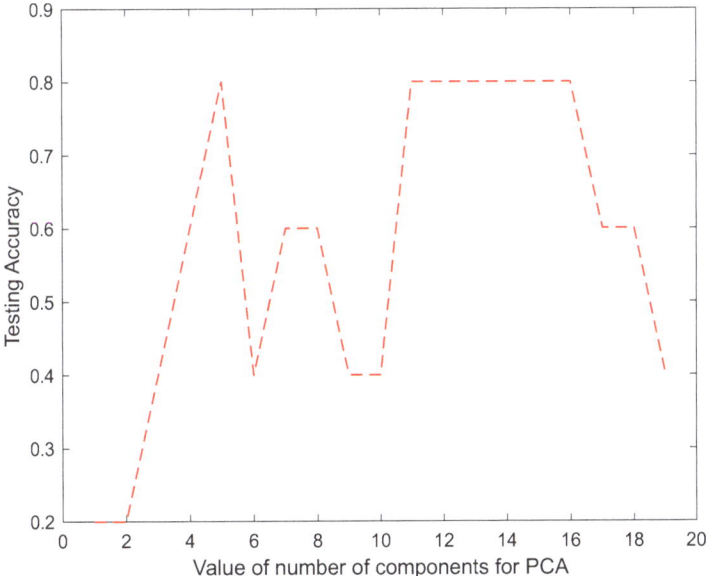

Figure 12. Optimal value of n_components. The value of the n_components is arbitrarily set to 13, for attaining a 0.8 average testing accuracy.

3.3. Performance Evaluation

In this paper, the $k = 8$-iterations CV technique was implemented for evaluating the prediction model performance. Table 9 presents the CV accuracy results for each fold. The maximun CV accuracy was achieved at the 7th and 8th folds, and the average estimate performance was 0.521 +/− 0.327.

Table 9. K-iterations cross-validation results.

Fold	CV Accuracy
1	0.333
2	0.667
3	0.333
4	0.333
5	0.500
6	0.000
7	1.000
8	1.000

The input data (1875 samples) were divided into 80% X_train (1500 samples) and 20% X_test (375 samples), according to Pareto analysis [34], in order to avoid overfitting. In addition, the output labels (25 samples) were divided into 80% y_train (20 samples) and 20% y_test (5 samples), for validation. The CM was computed with test and predicted data, and the results are shown in Figure 13.

The CM results represents one high-responder patient who was correctly predicted as a high responder, one low-responder patient who was correctly predicted as a low responder, one low-responder patient who was wrongly predicted as a medium responder, and two medium-responder patients who were correctly predicted as medium responders. Based on previous results, the prediction model achieved 0.8 testing accuracy.

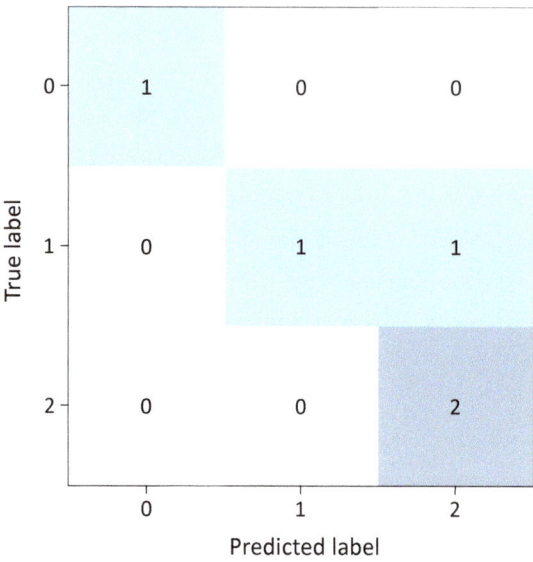

Figure 13. Confusion matrix results: (0) high responder to IFN-β, (1) low responder to IFN-β, and (2) medium responder to IFN-β.

4. Discussion

While binary logic generates only two output types—[0, 1]—fuzzy inference engines use approximate reasoning based on generalized rules of inference. Hence, fuzzy systems are convenient methods for decision support, due to their ability to process inaccurate information. For this paper, an alternative fuzzy system based on expert knowledge was implemented, for decision support in classification of the response to IFN-β treatment of RRMS patients. Demographic and clinical characteristics were used as input variables to the fuzzy system. As shown in Table 8, the classification of the proposed fuzzy system achieved better results than the agglomerative clustering, because the latter did not consider the intrinsic properties of the data, it simply used the distance between the data points to group them into clusters. A software issue in the fuzzy system design was to set a small number of input variables: the greater the number of variables, the greater the data processing time.

It is important to mention that at the beginning of the fuzzy system design, a proposal of fuzzy rules definition was reviewed by the expert neurologist, who considered only two output linguistic labels: "low" and "high" responder to IFN-β. Under these conditions, 88% efficiency was obtained in the results. After validating the results, the expert recommended adding an extra label—"medium"—to classify MS patients who had the same EDSS level at the beginning as at the end of treatment. After redefining the fuzzy rules, 100% efficiency was achieved.

Once the dataset output labels were classified by the fuzzy system, a pipeline prediction model was implemented, including data standardization, data compression through the PCA technique, and an MLP learning algorithm. The pipeline model was trained with 15 biomarkers associated with the response to IFN-β for predicting whether RRMS patients were potential candidates to be treated with this drug. As shown in Figure 12, by setting 13 principal components for PCA, 0.8 testing accuracy was achieved. The use of the PCA technique for data compression provides some advantages: (1) the reduced dimension has the property of keeping most of the useful information, while reducing noise and other undesirable data, (2) the time and memory used in the data processing are smaller, (3) it provides a way to understand and visualize the structure of complex datasets. The use of the k-iterations CV technique helps to obtain a good bias–variance rate. The highest CV accuracy was achieved at the 7th and 8th folds, as shown in Table 9. One disadvantage in evaluating the prediction model performance was that the test samples size was too small. Therefore, the number of iterations for the CV technique was limited to eight.

ML algorithms can find natural patterns in the data, and they are a useful alternative in the field of bio-informatics. These algorithms have been implemented to improve the MS diagnosis [20,21] and to help specialists to predict the response to drug treatments in MS patients [22,25]. Table 10 presents a comparison of the performance results of some ML applications in MS study.

Table 10. Performance results comparison of ML applications in MS study.

Author	Prediction	ML Technique	Accuracy
Fagone et al. [22]	Response to Natalizumab	UCRC	0.892
Goyal et al. [20]	MS diagnostic	RF	0.909
Jin et al. [25]	Response to IFN-β	SVM	0.809
Chen et al. [21]	MS diagnostic	SVM	0.930
Actual Paper	Response to IFN-β	MLP	0.521 +/− 0.327 [1]

[1] Average estimate performance achieved by k=8-iterations cross-validation.

The results obtained in this paper could be a reference for future works, using other genes related to the response to IFN-β treatment, as training data. Also, new prediction models, such as evolutionary or DL algorithms, could be designed, to improve model performance.

5. Conclusions

In general, IFN-β treatment effectively reduces the rate of relapse and delays the progression of neurological disability in MS patients. However, a percentage of patients do not respond, or partially respond to this drug. In this paper, the proposed fuzzy system, based on the opinion of an expert, demonstrated high efficiency in decision support, and it could be a useful tool in labeling classes, such as classification of the response to IFN-β therapy.

Although genome research is complex, there are ML methods—for instance, the proposed pipeline model—that can effectively deal the gene data for obtaining reliable predictions, to guide specialists in the selection of MS patients who may obtain the greatest benefit from IFN-β treatment. Biomarkers—in particular IL-2, IL-12, IFN-γ, TNF-α, IL-4, IL-10, TGF-β, CD46, CD58, FHIT, IRF5, GAPVD1, GPC5, GRM3, and GRIK2—can be convenient predictive variables for improving the comprehension of the influence of IFN-β therapy in MS patients.

Author Contributions: Conceptualization, E.R.P.d.L.-S. and A.M.H.-N.; methodology, E.R.P.d.L.-S. and A.M.H.-N.; software, E.R.P.d.L.-S.; validation, H.S.-M.; formal analysis, J.D.M.-S., O.A.D.-R., A.V.-C. and H.J.-H.; investigation, E.R.P.d.L.-S.; resources, J.D.M.-S., O.A.D.-R., A.M.H.-N., A.V.-C. and H.J.-H.; writing—original draft preparation, E.R.P.d.L.-S.; writing—review and editing, E.R.P.d.L.-S. and J.D.M.-S.; supervision, J.D.M.-S., O.A.D.-R., A.V.-C. and H.J.-H.; project administration, E.R.P.d.L.-S., J.D.M.-S. and O.A.D.-R. All authors have read and agreed to the published version of the manuscript.

Funding: This research received no external funding.

Data Availability Statement: The implemented pseudo-codes and the collected dataset are available at https://github.com/ponceraf2020/Pipeline-model.git (accessed on 30 May 2023).

Conflicts of Interest: The authors declare no conflict of interest.

References

1. Milo, R.; Miller, A. Revised diagnostic criteria of multiple sclerosis. *Autoimmun. Rev.* **2014**, *13*, 518–524. [CrossRef] [PubMed]
2. Martynova, E.; Khaibullin, T.; Salafutdinov, I.; Markelova, M.; Laikov, A.; Lopukhov, L.; Liu, R.; Sahay, K.; Goyal, M.; Baranwal, M.; et al. Seasonal Changes in Serum Metabolites in Multiple Sclerosis Relapse. *Int. J. Mol. Sci.* **2023**, *24*, 3542. [CrossRef] [PubMed]
3. Tarlinton, R.E.; Martynova, E.; Rizvanov, A.A.; Khaiboullina, S.; Verma, S. Role of viruses in the pathogenesis of multiple sclerosis. *Viruses* **2020**, *12*, 643. [CrossRef] [PubMed]
4. Zarghami, A.; Li, Y.; Claflin, S.B.; van der Mei, I.; Taylor, B.V. Role of environmental factors in multiple sclerosis. *Expert Rev. Neurother.* **2021**, *21*, 1389–1408. [CrossRef] [PubMed]
5. Dominguez-Mozo, M.I.; Perez-Perez, S.; Villarrubia, N.; Costa-Frossard, L.; Fernandez-Velasco, J.I.; Ortega-Madueño, I.; Garcia-Martinez, M.A.; Garcia-Calvo, E.; Estevez, H.; Luque Garcia, J.L.; et al. Herpesvirus antibodies, vitamin d and short-chain fatty acids: Their correlation with cell subsets in multiple sclerosis patients and healthy controls. *Cells* **2021**, *10*, 119. [CrossRef]
6. Rodríguez Murúa, S.; Farez, M.F.; Quintana, F.J. The immune response in multiple sclerosis. *Annu. Rev. Pathol. Mech. Dis.* **2022**, *17*, 121–139. [CrossRef]
7. Pinheiro, M.A.L.; Kooij, G.; Mizee, M.R.; Kamermans, A.; Enzmann, G.; Lyck, R.; Schwaninger, M.; Engelhardt, B.; de Vries, H.E. Immune cell trafficking across the barriers of the central nervous system in multiple sclerosis and stroke. *Biochim. Biophys. Acta (BBA)-Mol. Basis Dis.* **2016**, *1862*, 461–471. [CrossRef]
8. Liu, Y.; Zheng, M.; Ma, Z.; Zhou, Y.; Huo, J.; Zhang, W.; Liu, Y.; Guo, Y.; Zhou, X.; Li, H.; et al. Design, synthesis, and evaluation of PD-L1 degraders to enhance T cell killing activity against melanoma. *Chin. Chem. Lett.* **2022**, *34*, 107762. [CrossRef]
9. Szpakowski, P.; Ksiazek-Winiarek, D.; Glabinski, A. Targeting Antigen-Presenting Cells in Multiple Sclerosis Treatment. *Appl. Sci.* **2021**, *11*, 8557. [CrossRef]
10. Mirandola, S.R.; Hallal, D.E.; Farias, A.S.; Oliveira, E.C.; Brandão, C.O.; Ruocco, H.H.; Damasceno, B.P.; Santos, L.M. Interferon-beta modifies the peripheral blood cell cytokine secretion in patients with multiple sclerosis. *Int. Immunopharmacol.* **2009**, *9*, 824–830. [CrossRef]
11. Kay, M.; Hojati, Z.; Dehghanian, F. The molecular study of IFNβ pleiotropic roles in MS treatment. *Iran. J. Neurol.* **2013**, *12*, 149.
12. Cohan, S.L.; Hendin, B.A.; Reder, A.T.; Smoot, K.; Avila, R.; Mendoza, J.P.; Weinstock-Guttman, B. Interferons and multiple sclerosis: Lessons from 25 years of clinical and real-world experience with intramuscular interferon beta-1a (Avonex). *CNS Drugs* **2021**, *35*, 743–767. [CrossRef]
13. Río, J.; Nos, C.; Tintoré, M.; Borrás, C.; Galán, I.; Comabella, M.; Montalban, X. Assessment of different treatment failure criteria in a cohort of relapsing–remitting multiple sclerosis patients treated with interferon β: Implications for clinical trials. *Ann. Neurol. Off. J. Am. Neurol. Assoc. Child Neurol. Soc.* **2002**, *52*, 400–406. [CrossRef]
14. Bustamante, M.F.; Morcillo-Suárez, C.; Malhotra, S.; Rio, J.; Leyva, L.; Fernández, O.; Zettl, U.K.; Killestein, J.; Brassat, D.; García-Merino, J.A.; et al. Pharmacogenomic study in patients with multiple sclerosis: Responders and nonresponders to IFN-β. *Neurol. Neuroimmunol. Neuroinflamm.* **2015**, *2*. [CrossRef]
15. Martínez-Aguilar, L.; Pérez-Ramírez, C.; del Mar Maldonado-Montoro, M.; Carrasco-Campos, M.I.; Membrive-Jiménez, C.; Martínez-Martínez, F.; García-Collado, C.; Calleja-Hernández, M.Á.; Ramírez-Tortosa, M.C.; Jiménez-Morales, A. Effect of genetic polymorphisms on therapeutic response in multiple sclerosis relapsing-remitting patients treated with interferon-beta. *Mutat. Res. Mutat. Res.* **2020**, *785*, 108322. [CrossRef]
16. KARLIK, B. Soft computing methods in bioinformatics: A comprehensive review. *Math. Comput. Appl.* **2013**, *18*, 176–197. [CrossRef]
17. Ayangbekun, O.; Jimoh Ibrahim, A. Fuzzy logic application to brain diseases diagnosis. *J. Emerg. Trends Comput. Inf. Sci.* **2015**, *6*, 144–148.
18. Hosseini, A.; Asadi, F.; Arani, L.A. Development of a knowledge-based clinical decision support system for multiple sclerosis diagnosis. *J. Med. Life* **2020**, *13*, 612. [CrossRef]
19. Matinfar, F.; Golpaygani, A.T. A fuzzy expert system for early diagnosis of multiple sclerosis. *J. Biomed. Phys. Eng.* **2022**, *12*, 181. [CrossRef]
20. Goyal, M.; Khanna, D.; Rana, P.S.; Khaibullin, T.; Martynova, E.; Rizvanov, A.A.; Khaiboullina, S.F.; Baranwal, M. Computational Intelligence Technique for Prediction of Multiple Sclerosis Based on Serum Cytokines. *Front. Neurol.* **2019**, *10*, 781. [CrossRef]
21. Chen, X.; Hou, H.; Qiao, H.; Fan, H.; Zhao, T.; Dong, M. Identification of blood-derived candidate gene markers and a new 7-gene diagnostic model for multiple sclerosis. *Biol. Res.* **2021**, *54*, 1–12. [CrossRef] [PubMed]
22. Fagone, P.; Mazzon, E.; Mammana, S.; Di Marco, R.; Spinasanta, F.; Basile, M.S.; Petralia, M.C.; Bramanti, P.; Nicoletti, F.; Mangano, K. Identification of CD4+ T cell biomarkers for predicting the response of patients with relapsing-remitting multiple sclerosis to natalizumab treatment. *Mol. Med. Rep.* **2019**, *20*, 678–684. [CrossRef] [PubMed]
23. Gurevich, M.; Miron, G.; Falb, R.Z.; Magalashvili, D.; Dolev, M.; Stern, Y.; Achiron, A. Transcriptional response to interferon beta-1a treatment in patients with secondary progressive multiple sclerosis. *BMC Neurol.* **2015**, *15*, 1–8. [CrossRef]
24. Clarelli, F.; Liberatore, G.; Sorosina, M.; Osiceanu, A.; Esposito, F.; Mascia, E.; Santoro, S.; Pavan, G.; Colombo, B.; Moiola, L.; et al. Pharmacogenetic study of long-term response to interferon-β treatment in multiple sclerosis. *Pharmacogenom. J.* **2017**, *17*, 84–91. [CrossRef] [PubMed]

25. Jin, T.; Wang, C.; Tian, S. Feature selection based on differentially correlated gene pairs reveals the mechanism of IFN-therapy for multiple sclerosis. *Bioinform. Genom.* **2020**, *8*, 8812. [CrossRef]
26. National Center for Biotechnology Information (NCBI)—Gene Expression Omnibus (GEO) Database 2010. Available online: https://www.ncbi.nlm.nih.gov/geo/geo2r (accessed on 15 January 2023).
27. Rutkowska, D. *Neuro-Fuzzy Architectures and Hybrid Learning*; Springer Science & Business Media: Berlin/Heidelberg, Germany, 2001; Volume 85.
28. Van Leekwijck, W.; Kerre, E.E. Defuzzification: Criteria and classification. *Fuzzy Sets Syst.* **1999**, *108*, 159–178. [CrossRef]
29. Prokopowicz, P.; Czerniak, J.; Mikołajewski, D.; Apiecionek, Ł.; Ślezak, D. *Theory and Applications of Ordered Fuzzy Numbers: A Tribute to Professor Witold Kosiński*; Springer Open: Warsaw, Poland, 2017.
30. Sanguansat, P. *Principal Component Analysis: Engineering Applications*; IntechOpen: Rijeka, Croatia, 2012.
31. Mirjalili, V.; Raschka, S. *Python Machine Learning: Machine Learning and Deep Learning with Python, Scikit-Learn and TensorFlow*; Packt: Birmingham, UK, **2019**.
32. Casalino, G.; Castellano, G.; Consiglio, A.; Nuzziello, N.; Vessio, G. MicroRNA expression classification for pediatric multiple sclerosis identification. *J. Ambient. Intell. Humaniz. Comput.* **2021**, 1–10. [CrossRef]
33. Salamai, A.A.; El-kenawy, E.S.M.; Abdelhameed, I. Dynamic voting classifier for risk identification in supply chain 4.0. *CMC Comput. Mater. Contin.* **2021**, *69*, 3749–3766. [CrossRef]
34. Roccetti, M.; Delnevo, G.; Casini, L.; Mirri, S. An alternative approach to dimension reduction for pareto distributed data: A case study. *J. Big Data* **2021**, *8*, 1–23. [CrossRef]

Disclaimer/Publisher's Note: The statements, opinions and data contained in all publications are solely those of the individual author(s) and contributor(s) and not of MDPI and/or the editor(s). MDPI and/or the editor(s) disclaim responsibility for any injury to people or property resulting from any ideas, methods, instructions or products referred to in the content.

Article

Segmentation of Retinal Blood Vessels Using Focal Attention Convolution Blocks in a UNET

Rafael Ortiz-Feregrino, Saul Tovar-Arriaga *, Jesus Carlos Pedraza-Ortega and Juvenal Rodriguez-Resendiz

Faculty of Engineering, Universidad Autónoma de Querétaro, Santiago de Querétaro 76010, Mexico; rafaortizferegrino@gmail.com (R.O.-F.); carlos.pedraza@uaq.mx (J.C.P.-O.); juvenal@uaq.edu.mx (J.R.-R.)
* Correspondence: saul.tovar@uaq.mx

Abstract: Retinal vein segmentation is a crucial task that helps in the early detection of health problems, making it an essential area of research. With recent advancements in artificial intelligence, we can now develop highly reliable and efficient models for this task. CNN has been the traditional choice for image analysis tasks. However, the emergence of visual transformers with their unique attention mechanism has proved to be a game-changer. However, visual transformers require a large amount of data and computational power, making them unsuitable for tasks with limited data and resources. To deal with this constraint, we adapted the attention module of visual transformers and integrated it into a CNN-based UNET network, achieving superior performance compared to other models. The model achieved a 0.89 recall, 0.98 AUC, 0.97 accuracy, and 0.97 sensitivity on various datasets, including HRF, Drive, LES-AV, CHASE-DB1, Aria-A, Aria-D, Aria-C, IOSTAR, STARE and DRGAHIS. Moreover, the model can recognize blood vessels accurately, regardless of camera type or the original image resolution, ensuring that it generalizes well. This breakthrough in retinal vein segmentation could improve the early diagnosis of several health conditions.

Keywords: retinal blood vessels; artificial intelligence; convolutional neural networks; attention module; segmentation

Citation: Ortiz-Feregrino, R.; Tovar-Arriaga, S.; Pedraza-Ortega, J.C.; Rodriguez-Resendiz, J. Segmentation of Retinal Blood Vessels Using Focal Attention Convolution Blocks in a UNET. *Technologies* **2023**, *11*, 97. https://doi.org/10.3390/technologies11040097

Academic Editor: Pietro Zanuttigh

Received: 1 June 2023
Revised: 29 June 2023
Accepted: 5 July 2023
Published: 13 July 2023

Copyright: © 2023 by the authors. Licensee MDPI, Basel, Switzerland. This article is an open access article distributed under the terms and conditions of the Creative Commons Attribution (CC BY) license (https://creativecommons.org/licenses/by/4.0/).

1. Introduction

AI plays a prominent role in various fields, including programming-assistance tools such as OpenIA Copilot [1], protein prediction using Deep Mind and the model in [2], congenital disease prediction [3], and lesion segmentation in medical imaging, as demonstrated by recent research in X-ray imaging for COVID-19 [4]. These examples illustrate how AI can achieve metrics comparable to an expert's, making it a promising solution for automating daily tasks that intelligent algorithms can efficiently tackle.

In recent years, deep learning (DL), a subfield of machine learning (ML), has experienced significant growth within the AI domain. This expansion is well-founded because DL models do not require direct guidance from an expert nor the manual modification of complex hyperparameters to achieve suitable performance. Instead, many previously labeled examples are sufficient to initiate the learning process. This attribute of neural networks enables the exploration of complex domains, such as medicine, without requiring the presence of a domain expert at all times.

In our study, we employed CNN models [5] and attention modules based on those incorporated by visual transformers [6] to segment retinal blood vessels. The retina, a delicate layer at the back of the eye, plays a crucial role in our vision, as its connections go directly to the brain [7]. The segmentation of retinal blood vessels can aid in detecting degenerative diseases such as diabetic retinopathy [7], cardiovascular problems [8], and many other congenital conditions. Unfortunately, these diseases have a high prevalence worldwide [9].

The UNET model has emerged as a popular choice among developers and researchers for addressing segmentation challenges across diverse domains, for example, segmenting lesions caused by diabetic retinopathy [10], the segmentation of brain tumors using a modified model [11,12], and the segmentation of other skin lesions [13]. However, its versatility continues. The model can also segment images taken by UAVs and perform activities such as those described in [14]. These models are built upon the UNET architecture, which is highly regarded for its adaptability and straightforward customization capabilities, rendering it an ideal candidate for exploring novel concepts. Adaptations of the model include adding convolutions and supplementary connections, as demonstrated in [13], and incorporating comprehensive attention blocks, as showcased in the referenced article, to mentioned a few.

Most object prediction, segmentation, and classification models that use images as inputs are primarily based on CNNs [5]. Meanwhile, transformers [15], a type of architecture mainly used in NLP, are now improving the existing metrics by their distinct way of "paying attention". As a result, image-based transformer models [16] have emerged and are proving to achieve comparable, or at times better, results than traditional CNNs. However, these models come with two significant challenges: the requirement of big training data and the intensive computational power they consume during training. Some techniques, such as transfer learning (TL) [17], are used when the dataset is limited. The source and target datasets should have similar domains to maximize TL benefits.

Another commonly used approach is to take the attention mechanism of transformers as an independent module that provides some of the benefits of transformers without requiring a vast amount of training data. The approach that can be seen in [18] combines MHSA with convolutions to generate a bottleneck transformer (BT) that can be viewed as an attention module. This type of implementation using transformers and convolutions is the basis of this work, along with a focus inspired by the focal transformer presented by [19]; this allowed us to improve state-of-the-art metrics using the U-Net network as a backbone and adding modules that we call a "Focal Attention Convolution Block" (FACB).

This study's proposed FACB offers a distinctive characteristic of seamless integration into any CNN as an additional module. This integration does not require altering the backbone of the underlying models, making it similar to a plug-and-play component. This approach provides the advantages of the attention mechanisms seen in vision transformers without necessitating extensive modifications to the base model.

The FACB consists of two main parts. Firstly, the initial stage of the FACB captures information from the input data at various levels, which we refer to as windows. These windows can have different dimensions, such as small, medium, or large, and can be singular or multiple. Ultimately, the output of these windows is concatenated, providing information from different regions of the input data. This concatenated information is then passed to the second block of the FACB, which comprises several attention modules operating in parallel to process the input information. The versatility and compatibility of the FACB with matrix operations enable its implementation at any stage within a CNN architecture. This flexibility allows researchers to seamlessly incorporate the FACB module into existing CNNs, enhancing the network's capabilities without significant structural changes.

The remainder of this paper is organized as follows. Section 2 presents the proposed method's main idea, the databases used, the preprocessing, data augmentation, and details about the FACB and its use in the UNET model. Section 3 presents a comparative table with the state of the art and images of the inference of the proposed model. Finally, the discussion and conclusions are presented in Sections 4 and 5, respectively.

In recent years, there have been significant advancements in the field of medical imaging, particularly in the area of retinal segmentation. Retinal segmentation is the process of identifying and separating different structures within the retina, which can be critical for diagnosing and treating a variety of eye diseases.

The automated segmentation of blood vessels has long been a challenging task in computer vision and artificial intelligence research. Among various AI approaches, ML models have proven effective in segmenting and classifying these structures. There has been a growing adoption of DL models in recent years in this domain [20]. DL models offer a significant advantage in generalizing across diverse domains [21]. Talking specifically about DL, vision transformers [22] are becoming important in the area but the ideal conditions to apply these models are seldom present. Instead of using a complete transformer, we can take the part that pays attention. Attention modules in a convolutional network allow for capturing local and global spatial relationships in images more efficiently. Unlike full transformers, which require computing all interactions between pair elements, attention modules in a convolutional network can selectively focus on relevant regions and reduce computational complexity. This is especially beneficial for images, where visual information is highly structured and spatial relationships between pixels play a crucial role. Attention modules in a convolutional network also enable the better interpretation and visualization of results by providing attention maps highlighting the image's most important regions. In addition, attention modules can be easily incorporated into existing convolutional network architectures, facilitating their implementation and leveraging both approaches' benefits. Attention modules in a convolutional network offer an efficient and effective way to model spatial relationships in images, transcending computational limitations and taking advantage of the intrinsic visual structure of images.

Khanal, A. et al. [23] proposed a stochastic training scheme for deep neural networks that robustly balances precision and recall. Their method yielded a better balance of precision and recall relative to state-of-the-art techniques, resulting in higher F1 scores. However, their method can be misleading for unbalanced datasets. Gegundez-Arias, et al. [24] present a new method for vascular tree segmentation. The method outperforms other U-Net-based methodologies in terms of accuracy, requiring fewer hyperparameters and lower computational complexity. However, the major limitation of the practical integration is the limited number of examples available for network training. Galdran, A. et al. [25] reflect on the need to construct algorithmically complex methodologies for retinal vessel segmentation. It suggests that minimalistic models, adequately trained, can attain results that do not significantly differ from more complex approaches. The authors suggest that research should switch to modern high-resolution datasets rather than rely on old datasets.

In the work of Tang, P. et al. [26], the main contributions include a novel ensemble model based on multiproportional red and green channels that outperform other existing methods concerning two primary performance metrics (segmentation accuracy and AUC) and the first instance of the red channel providing performance gains. In the work of Ma, Y. et al. [27], WA-Net was developed to improve the segmentation accuracy of retinal blood vessels. Cross-training between datasets was performed to verify the model's generalization performance, and the results showed that WA-Net extracts more detailed blood vessels and has a superior performance. However, there are still some limitations, such as the need for more effective data augmentation and a long computational time due to the introduction of weight normalization.

Tuyet, V.T.H. et al. [28] proposed a method for retinal vessel segmentation using three periods: a salient edge map in the retinal vessel image, feature extraction using CNN in a salient map, and segmentation based on the pixel level of the Sobel operator in saliency. The Jaccard index value of the proposed method was found to be higher than other approaches. Therefore, the number of layers or operators for the salient region map can be improved in the future. Park, K.-B. et al. [29] proposed M-GAN, which outperformed previous studies with respect to accuracy, IoU, F1 score, and MCC. It derived balanced precision and recall together through the FN loss function. The proposed method with an adversarial discriminator showed better segmentation performance than a method without a discriminator.

Compared to state-of-the-art methods, the proposed method of Zhuo, Z. et al. [30] achieved extremely competitive performances on the DRIVE and STARE datasets. The results of cross-training show that the method has strong robustness and is faster than other CNN-based algorithms. However, the proposed method does not have a network structure that reduces the number of parameters while guaranteeing effective segmentation. Zhuo, Z. et al. [31] present a novel retinal vessel segmentation architecture that combines a U-Net with generative adversarial networks and a weighted feature matching loss. This architecture was evaluated on three retinal segmentation datasets (DRIVE, CHASE-DB1, and STARE). It showed improved performance compared to previous methods, with higher confidence scores, F1 score, sensitivity, specificity, accuracy, AUC-ROC, mean-IOU, and SSIM. However, the model suffers from a high false-positive rate.

2. Materials and Methods

Demonstrating the generalization of an artificial neural network model is a crucial aspect of its training. It requires diverse data to ensure the model recognizes and learns from patterns outside the training dataset. To achieve this, we utilized many publicly available datasets, each presenting unique characteristics in terms of their composition, such as variations in image resolution, centering, and color saturation, as depicted in Figure 1. These differences result from the use of various devices and the work of different individuals, resulting in images that differ significantly while all within the same domain (the retina).

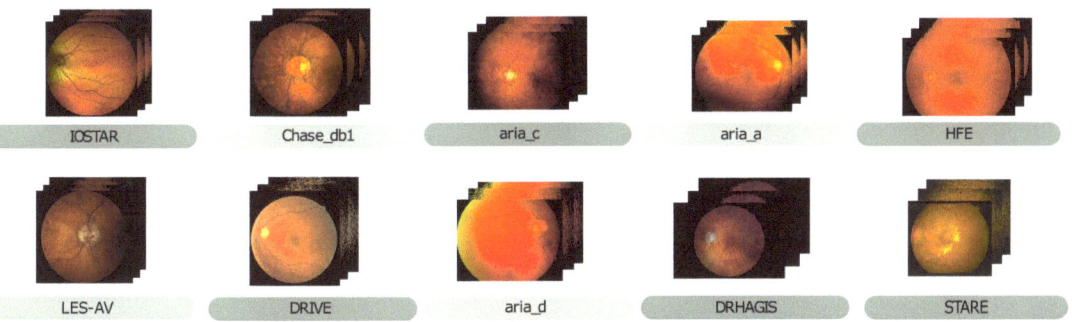

Figure 1. Although the images belong to the same retina domain, these datasets possess unique properties, as demonstrated by differences such as varying degrees of centering toward the optic nerve or the macula, mixed image resolutions, and differences in color saturation and brightness levels.

The blocks of a general methodology using DL models are very similar regardless of the task to be performed. In our case, as can be seen in Figure 2, the general development of our implementation, which contains the classic blocks of preprocessing and training/validation but inside each block, has specific processes.

2.1. Preprocessing

One of the main reasons data preprocessing is essential is the nature of deep neural networks. These networks are designed to learn functional patterns and representations from input data. However, the quality of the input data can vary widely and may contain noise, redundancy, outliers, or even irrelevant information. Preprocessing helps mitigate these problems by performing normalization, denoising, and the reduction of dimensionality. It is crucial to find a balance in data preprocessing and not to exaggerate when homogenizing the data since a model also benefits from the natural noise with which the data were taken.

2.1.1. Data Cleaning

The preprocessing block can be considered the kitchen where the food of our model is prepared. It is fundamental when handling various datasets because if garbage enters the model, we cannot expect good results no matter how robust the architecture is. Therefore, we must at least clean the data and try to normalize it as much as possible while maintaining the essence of each database.

The images contained in the datasets are very different in their composition, with some having a better resolution than others. In addition, the difference in saturation in the red channel is noticeable, so finding a balance when preprocessing the images is vital since we must consider the diversity of these images. RGB retinal images usually tend to have a reddish hue in their composition, which causes a considerable saturation in the red channel. The blue channel has the slightest presence in the image and can be considered the opposite of the red channel. Finally, the green channel is the one that best preserves the balance of all the channels, thus providing more precise and sharper information. Therefore, it was decided to only use the green channel. To keep the original information of the images the same, we chose to perform histogram equalization on the channel we used, i.e., the green one. After the equalization, we looked for an algorithm that would improve the sharpness and brightness without altering the images too much, so according to its performance, we decided to use CLAHE [32], which gives us a general enhancement of the image without altering too much of the original information of the images. The preprocessing applied to an image can be seen in the central block in Figure 2.

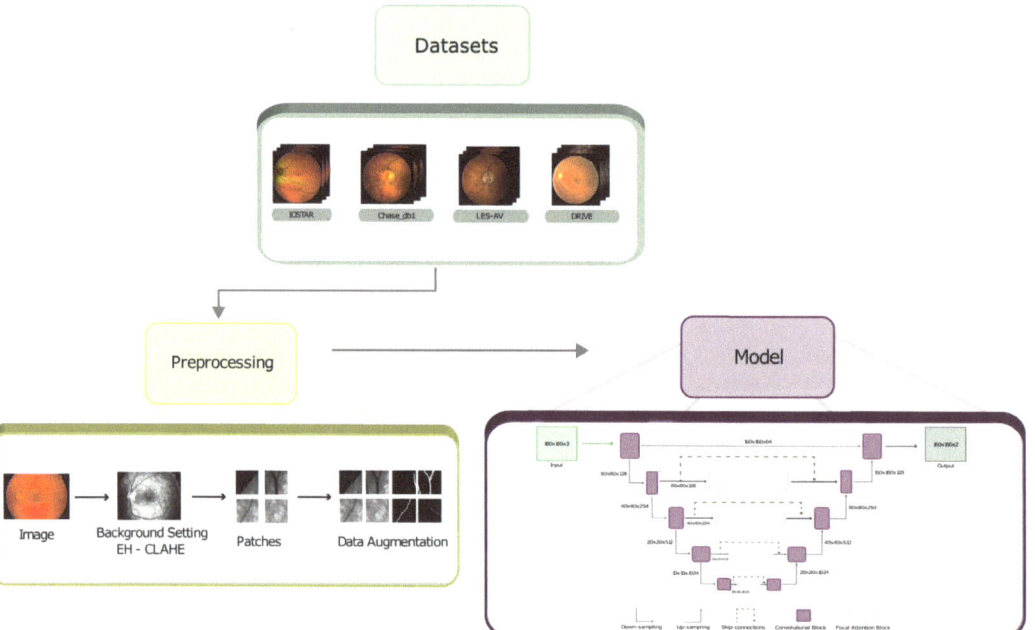

Figure 2. General methodology. The first block refers to the collection of the datasets. The central block represents the preprocessing applied to each of the datasets. The last block refers to the training and validation of the proposed model.

2.1.2. Data Augmentation

A disadvantage of public datasets is the small number of examples they contain, so an increase in data has to be implemented. For this reason, we divided each image into small segments called patches with a size of (160 × 160 × 3) and obtained 25,402 patches for training and 2900 for validation, trying to maintain a ratio of 90–10. An example of the patches can be seen in Figure 2.

In addition to the generation of patches as a method of data augmentation, it was decided to perform an automatic augmentation prior to model input. This means that before delivering an example to the model, it goes through a process that can rotate the image, displace it, increase its color saturation, add noise, and remove a part of the image, in order to force the model to generalize as much as possible and avoid overtraining. However, this process does not always happen for all the input examples since the aim is to generate the most significant difference between each data point. The decision as to which items undergo this process and which do not is randomly determined for each epoch.

2.2. The Model

When addressing specific tasks in the medical domain, such as disease prediction in medical imaging or lesion segmentation, choosing a good DL model becomes critical. A well-selected model has the potential to improve the accuracy of diagnosis and treatment, which, in turn, can lead to better clinical outcomes and more effective medical care.

2.2.1. Model Overview

As mentioned before, transformers are becoming very relevant in work with images, despite how inconvenient their training can be. Thus, using small modules that provide some of the benefits is a viable approach. In this work, the principal idea is the incorporation of a module consisting of main blocks. The first one is in charge of extracting as much information from the input as possible using different regions. These regions start from the center of the input and can extend to the edges; the idea is to collect as much information in the input directions, which contain valuable information from different points, and the second block pays attention with its different attention modules to the information the first block provides. The general idea of the focal attention convolution block (FACB) can be seen in Figure 3.

Figure 4 shows the block called CBAM bottleneck multihead self attention (CB-BMHSA), which incorporates an attention module inspired by the focal attention (FT) [19], which delivers fine and granular information to an MHSA. For our implementation, we took the FT idea of generating information in different regions of the image in parallel and used a CBAM [33] attention module to obtain spatial and channel information and then pass the information to a BT [18], which is a variant of the multihead self attention (MHSA); however, this was designed to work with convolutions, which means that there is no need to flatten the input features as a traditional MHSA would need.

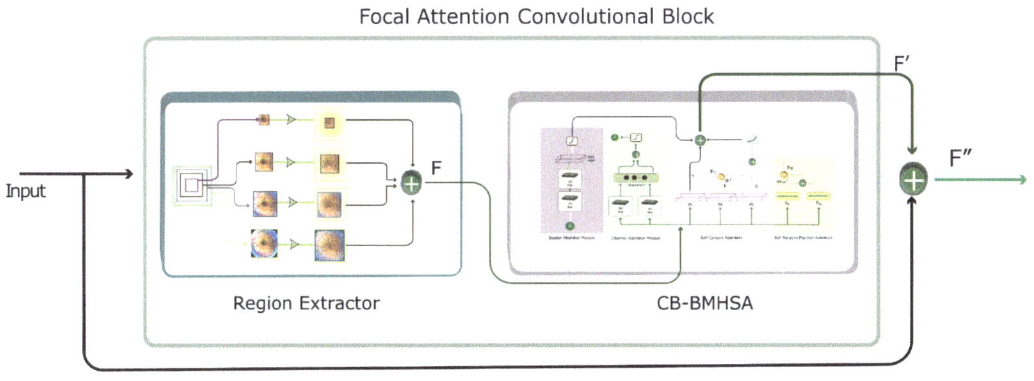

Figure 3. General FACB that contains two principal blocks.

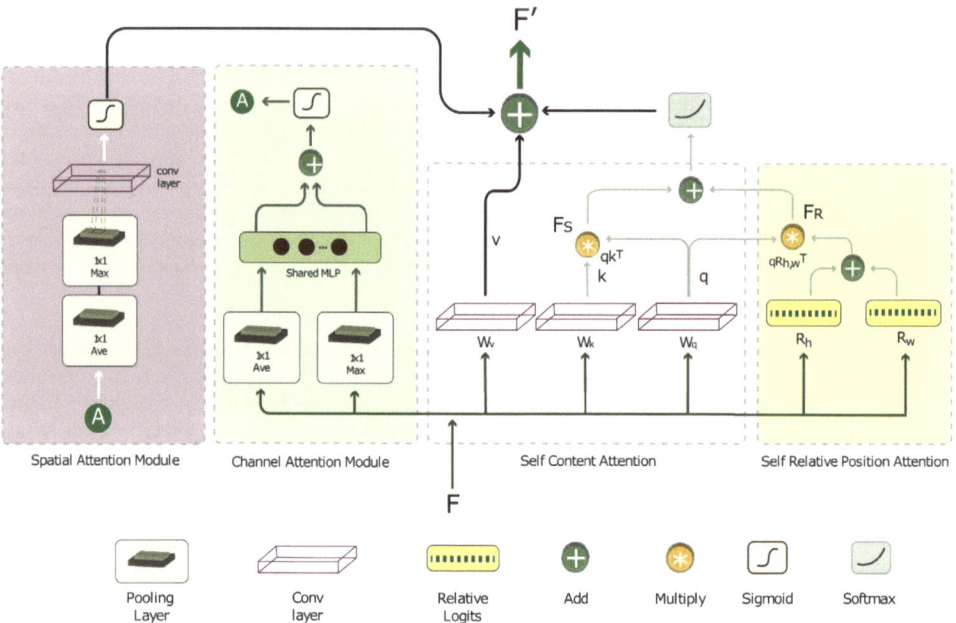

Figure 4. The CB-BMHSA attention block.

The input information of the CB-BMHSA comes from a block that we call the region extractor (RE). This block was inspired by the FAB used in the transformer in [19]. The block presented in the article, as mentioned earlier, has the disadvantage that it does not work with convolutions because it is 100% adapted to the use with transformers, so we decided to generate an alternative that can be coupled to the use of CNNs. In Figure 5, we can observe this block in more detail. The RE obtains the information from a window called Pr, which delimits the region of information to be extracted from the input features. Then, the window data are grouped into Pw sets using AvgPooling to generate subwindows with information from different regions of the input that will be summed to have general and granular details. Finally, when summing each subwindow, we use zero padding to avoid dimensionality problems. This procedure can be repeated for N number of windows as shown in Equation (1). It should be noted that FACB was designed for features with square dimensions. In addition, the FACB can be implemented on any CNN, extracting information from different regions of the input and paying attention to that information using different attention modules.

$$\sum_{n=1}^{N} ZeroPadding(AvgrPooling(Pr^{[n]})))\tag{1}$$

FACB can be implemented in any part of a traditional CNN, as the only requirement is that the input dimensions are square (m × m). Typically, classical CNN models always use square inputs and outputs, making the FACB an enhancer that allows the ability to pay attention to be added to models such as RESNET-50, RESNET-100 [34], VGG-16 [35], EfficientNet [36], Mobilnet [37], GoogleNet [38], and of course, the model that we use as a base in this article, the U-NET [39]. The U-NET was chosen because it is a well-studied model that has a wide variety of variants. These models have been applied to the segmentation of lesions, blood vessels, and other parts of the retina, which are components that can be used to compare and demonstrate that the FACB helps to enhance the performance of the model without modifying the main skeleton. The final model can be seen in Figure 6, to which FACB was added in the intermediate connections from the second downsampling.

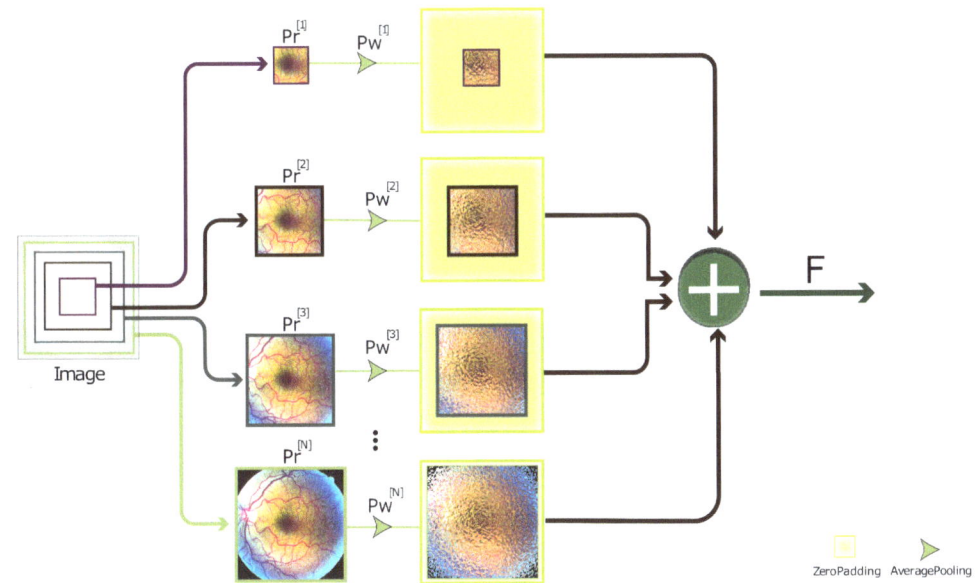

Figure 5. RE block that feeds the CB-BMHS block.

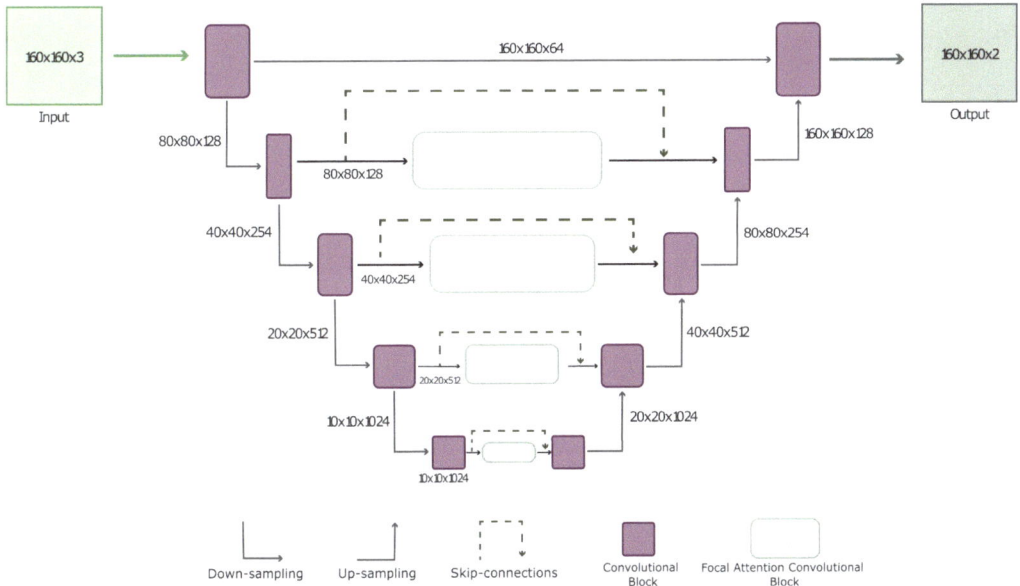

Figure 6. The complete U-Net architecture using FACB at the four points downstream.

2.2.2. UNET with FACB

When training a transformer either for natural language processing or a visual transformer for working with images, we know beforehand that this will require high computational power and a large amount of data. Similarly, adding attention modules increases the processing time of the model: not to the level of a transformer but more than that of a conventional CNN. Therefore, different configurations of the U-NET were tested with the FACB to determine the cost benefit. In this way, we noticed that the results of the metrics were similar when using the FACB in the whole model as to when it was applied from the second output of the convolution block; however, the computational cost changed a lot, i.e., when using the first option, the training time increased by 50%.

3. Results

The model was evaluated with different metrics that would give us important information about its performance. Several authors only consider metrics such as accuracy or AUC, i.e., metrics that, for this task, do not reflect the reality of the prediction since the number of pixels in the images that represent an absence of blood vessels is much higher than the number of pixels that contain blood vessels. This means that, although our model did not detect any blood vessels in the image, the accuracy rate and the sensitivity were high, since it focuses on the true negatives, which are more abundant than true positives. Thus, the accuracy, recall, and F1 score better reflect the model's performance speaking exclusively for this task. Table 1 shows a comparison of our model and the bests in the state of the art. As can be seen, not all the models were trained on several datasets, and some were only trained on a single dataset, leaving uncertainty as to whether the model would generalize or would only perform well on that dataset.

Table 1. The table shows a comparison of the best models and our proposal. As seen at least in one metric, the proposed model achieved better results than the state of the art. Moreover, the table indicates a good generalization over all datasets. The best results are in bold.

Dataset	Author	Accuracy	AUC	Precision	Recall	Specificity	F1
DRIVE	Park, K.-B. et al. [29]	97.06	98.68	83.02	83.46	98.36	83.24
	Galdran, A. et al. [25]	-	98.1	-	-	-	-
	Chen, D. et al. [40]	96.22	**98.78**	-	85.76	99.32	81.60
	UNET with FACB	**97.9**	93.6	**91.7**	**88.1**	99.0	**89.9**
CHASEDB1	Park, K.-B. et al. [29]	97.36	98.59	-	-	-	81.1
	Galdran, A. et al. [25]	-	98.47	-	-	-	-
	Chen, D. et al. [40]	98.12	**99.25**	-	84.93	**99.66**	82.73
	UNET with FACB	**99.1**	97.0	94.8	**94.4**	99.5	**94.6**
HRF	Park, K.-B. et al. [29]	97.61	**98.52**	79.72	-	-	79.72
	Galdran, A. et al. [25]	-	98.25	-	-	-	-
	Tang, P. et al. [26]	96.31	98.43	-	76.53	98.66	77
	UNET with FACB	**97.7**	91.3	**87.2**	**83.8**	**98.9**	**85.4**
STARE	Park, K.-B. et al. [29]	**98.76**	**99.73**	84.17	83.24	**99.38**	83.7
	Galdran, A. et al. [25]	-	98.28	-	-	-	-
	Chen, D. et al. [40]	97.96	99.53	-	87.93	99.37	88.36
	UNET with FACB	97.9	93.6	**91.7**	**88.1**	99	**89.9**
LES-AV	Galdran, A. et al. [25]	-	**97.34**	-	-	-	-
	UNET with FACB	99.3	97.1	94.7	94.6	99.6	94.6
IOSTAR	Guo, C. et al. [41]	97.13	**98.73**	-	80.82	98.54	-
	Li, X. et al. [42]	95.44	96.23	-	73.22	98.02	-
	Wu, H. et al. [43]	97.06	98.65	-	82.55	98.30	-
	UNET with FACB	**99.3**	97.1	94.7	**94.6**	**99.6**	94.6
ARIA (mean)	Tajbakhsh, N. et al. [44]	-	-	-	-	-	72
	UNET with FACB	97.3	89.9	86.4	81.4	96.1	**83.2**

The metrics that interest us, as mentioned above, are precision, recall, and F1 score. As can be seen, our model achieved superior results as compared to the state-of-the-art models in almost all the datasets for these metrics. The only datasets for which the model did not achieve similar numbers to the others were Arias A-C-D, but this reaffirms what we mentioned earlier: the accuracy and AUC for these datasets are competitive metrics but do not reflect the performance of the model since the rest of the metrics had much lower results.

In Figures 7 and 8, we can observe the inference on an image of each dataset. This inference shows us the outstanding performance of our model in the most challenging regions, such as the areas closest to the optic nerve, the bifurcations and intersections, the points where the path ends, the centralization towards the optic nerve or the macula, and the thinnest blood vessels and arteries. The comparison column provides valuable information on the differences between the prediction and the actual output. The yellow color represents the true positives of the segmentation, the red pixels are the false positives, and the green pixels are the false negatives. This final column allows us to directly compare the results in Table 1 and the model prediction. If we only look at the prediction and mask columns individually, it is difficult to infer whether the result is consistent with what is presented in the metrics output.

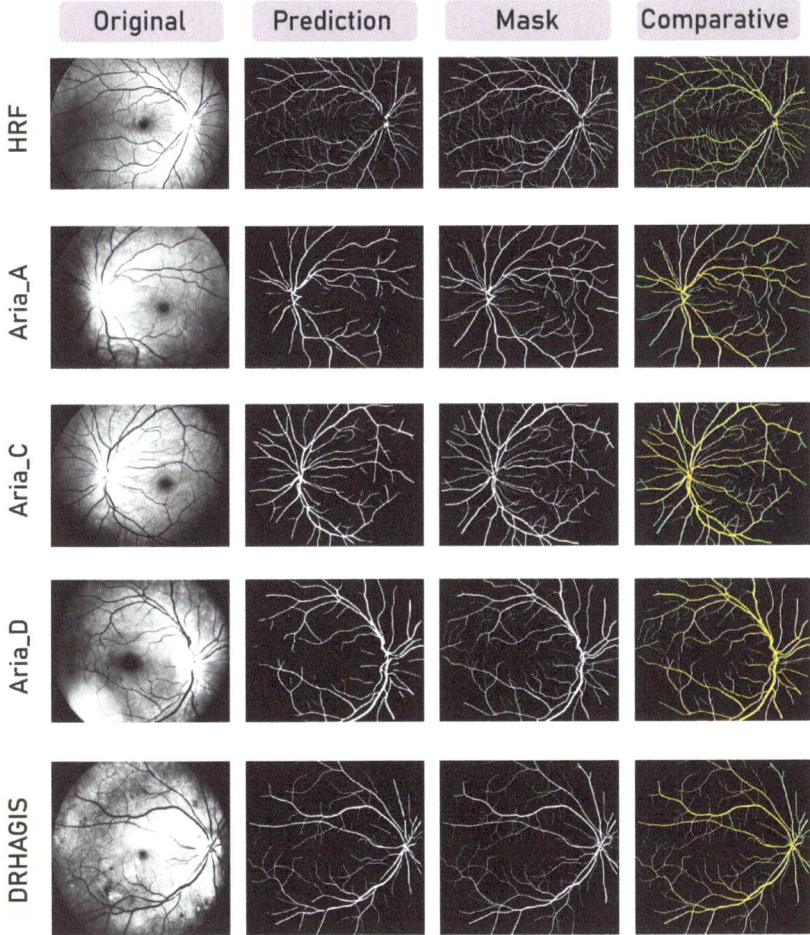

Figure 7. Model predictions part 1.

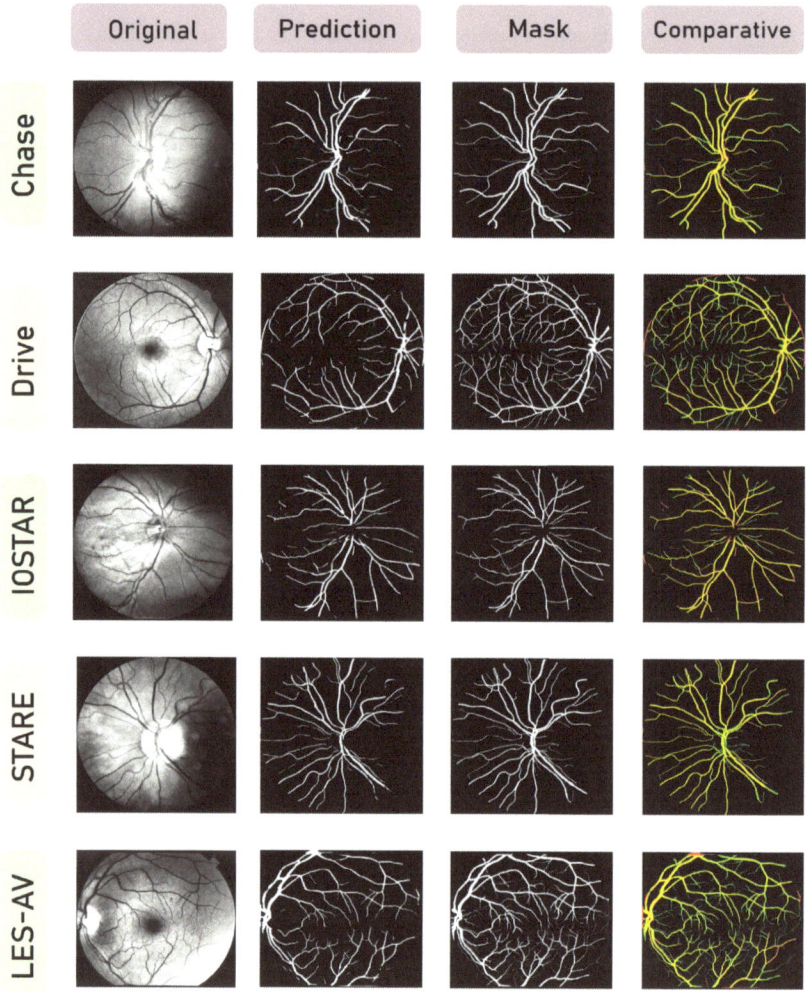

Figure 8. Model predictions part 2.

4. Discussion

The results of this paper indicate that the model has excellent performance in the most challenging regions of the datasets, such as the areas closest to the optic nerve, the bifurcations, and the intersections. This indicates that the FACB helps the UNET model pay attention to the segmentation's critical areas. However, the accuracy and AUC metrics for the Arias A-C-D datasets were lower than in the other datasets, reaffirming that these metrics only sometimes reflect the model's performance. Nevertheless, the inference on the images also showed outstanding results, indicating that the model can accurately detect blood vessels and arteries even in the thinnest areas. Further research should focus on exploring the implications of these results and investigating potential future research directions. There are two crucial considerations to bear in mind when utilizing the FACB. Firstly, incorporating multiple FACBs into neural networks results in a substantial increase in memory requirements. Consequently, integrating such models becomes feasible with access to robust hardware resources. Secondly, training the model on high-resolution images exceeding 200 pixels poses a formidable challenge due to the corresponding surge in memory demands. To tackle this issue, it is vital to explore strategies, such as the one proposed in this article, which involve partitioning images into smaller patches. This approach enables the successful implementation of FACBs by mitigating the excessive memory demands associated with larger image resolutions.

5. Conclusions

This research shows that our model performs in the most challenging regions for retinal vessel segmentation. This study has implications for previous retinal vessel segmentation studies and provides potential future research directions. This approach in retinal vein segmentation could improve the early diagnosis of several health conditions. Moreover, the model's ability to recognize blood vessels accurately regardless of camera type or the original image resolution suggests that it generalizes well and could be used in various applications. This was achieved by combining the existing attention modules and the regional extractor, and combining the best of the two classic convolutional networks and care delivery approaches. Another significant consideration is that some datasets are more challenging than others. They demand the use of all the possible datasets available to train and evaluate the model in order to achieve generalization. It is possible to enhance segmentation performance by conducting fine-tuning on each dataset. However, the ultimate objective is to achieve generalizability regardless of the image type presented to the model.

The overarching goal for future research is to achieve robustness and generalizability regardless of the image type presented in the model. Continued efforts in fine-tuning techniques, dataset augmentation, and advancements in model architectures and attention mechanisms can contribute to developing highly accurate and versatile DL models for medical image analysis.

Combining the strengths of attention mechanisms and CNNs, we can develop models that capture relevant features, focus on informative regions, improve interpretability, and achieve more accurate segmentation results. These advancements can contribute to early disease diagnosis, precision medicine, and improved patient care in medical imaging.

Future research efforts can focus on optimizing the architecture and training strategies for attention-guided CNN models. This may involve exploring different attention mechanisms, designing novel network architectures that effectively fuse attention and convolutional operations, and developing specialized loss functions that encourage accurate vessel segmentation. Additionally, investigating the transferability and generalizability of attention-guided CNN models across different datasets and imaging modalities can further improve their practical applicability in various medical image analysis tasks.

Author Contributions: Conceptualization, R.O.-F.; Methodology, R.O.-F.; Software, R.O.-F.; Validation, S.T.-A. and J.C.P.-O.; Formal analysis, R.O.-F.; Investigation, R.O.-F.; Resources, S.T.-A., J.C.P.-O. and J.R.-R.; Data curation, R.O.-F.; Writing original draft preparation, R.O.-F.; Writing review and editing, S.T.-A., J.C.P.-O. and J.R.-R.; Visualization, R.O.-F. and J.R.-R.; Supervision, S.T.-A. All authors have read and agreed to the published version of the manuscript.

Funding: This research received no external funding.

Institutional Review Board Statement: Not applicable.

Informed Consent Statement: Not applicable.

Data Availability Statement: The datasets used in this work are publicly accessible. For more information, see the respective articles of each dataset. In addition, the complete source code can be obtained from the following repository: https://github.com/FereBell/Focal-Attention-Convolution-Blocks (accessed on 15 June 2023).

Conflicts of Interest: The authors declare no conflict of interest.

References

1. Dakhel, A.M.; Majdinasab, V.; Nikanjam, A.; Khomh, F.; Desmarais, M.C.; Jiang, Z.M. GitHub Copilot AI Pair Programmer: Asset or Liability? *J. Syst. Softw.* **2023**, *203*, 111734. [CrossRef]
2. Jumper, J.; Evans, R.; Pritzel, A.; Green, T.; Figurnov, M.; Ronneberger, O.; Tunyasuvunakool, K.; Bates, R.; Žídek, A.; Potapenko, A.; et al. Highly Accurate Protein Structure Prediction with AlphaFold. *Nature* **2021**, *596*, 583–589. [CrossRef] [PubMed]
3. Yu, Z.; Wang, K.; Wan, Z.; Xie, S.; Lv, Z. Popular Deep Learning Algorithms for Disease Prediction: A Review. *Clust. Comput.* **2023**, *26*, 1231–1251. [CrossRef] [PubMed]
4. Arias-Garzón, D.; Alzate-Grisales, J.A.; Orozco-Arias, S.; Arteaga-Arteaga, H.B.; Bravo-Ortiz, M.A.; Mora-Rubio, A.; Saborit-Torres, J.M.; Serrano, J.Á.M.; De La Iglesia Vayá, M.; Cardona-Morales, O.; et al. COVID-19 Detection in X-ray Images Using Convolutional Neural Networks. *Mach. Learn. Appl.* **2021**, *6*, 100138. [CrossRef] [PubMed]
5. Alzubaidi, L.; Zhang, J.; Humaidi, A.J.; Al-Dujaili, A.; Duan, Y.; Al-Shamma, O.; Santamaría, J.; Fadhel, M.A.; Al-Amidie, M.; Farhan, L. Review of Deep Learning: Concepts, CNN Architectures, Challenges, Applications, Future Directions. *J. Big Data* **2021**, *8*, 53. [CrossRef]
6. Dosovitskiy, A.; Beyer, L.; Kolesnikov, A.; Weissenborn, D.; Zhai, X.; Unterthiner, T.; Dehghani, M.; Minderer, M.; Heigold, G.; Gelly, S.; et al. An Image Is Worth 16 × 16 Words: Transformers for Image Recognition at Scale 2021. *arXiv* **2020**, arXiv2010.11929.
7. Front Matter. In *Computational Retinal Image Analysis*; Elsevier: Amsterdam, The Netherlands, 2019; pp. i–iii, ISBN 978-0-08-102816-2.
8. Poplin, R.; Varadarajan, A.V.; Blumer, K.; Liu, Y.; McConnell, M.V.; Corrado, G.S.; Peng, L.; Webster, D.R. Prediction of Cardiovascular Risk Factors from Retinal Fundus Photographs via Deep Learning. *Nat. Biomed. Eng.* **2018**, *2*, 158–164. [CrossRef] [PubMed]
9. Cheloni, R.; Gandolfi, S.A.; Signorelli, C.; Odone, A. Global Prevalence of Diabetic Retinopathy: Protocol for a Systematic Review and Meta-Analysis. *BMJ Open* **2019**, *9*, e022188. [CrossRef] [PubMed]
10. Sambyal, N.; Saini, P.; Syal, R.; Gupta, V. Modified U-Net Architecture for Semantic Segmentation of Diabetic Retinopathy Images. *Biocybern. Biomed. Eng.* **2020**, *40*, 1094–1109. [CrossRef]
11. Rehman, M.U.; Cho, S.; Kim, J.H.; Chong, K.T. BU-Net: Brain Tumor Segmentation Using Modified U-Net Architecture. *Electronics* **2020**, *9*, 2203. [CrossRef]
12. Rehman, M.U.; Cho, S.; Kim, J.; Chong, K.T. BrainSeg-Net: Brain Tumor MR Image Segmentation via Enhanced Encoder–Decoder Network. *Diagnostics* **2021**, *11*, 169. [CrossRef]
13. Anand, V.; Gupta, S.; Koundal, D.; Nayak, S.R.; Barsocchi, P.; Bhoi, A.K. Modified U-NET Architecture for Segmentation of Skin Lesion. *Sensors* **2022**, *22*, 867. [CrossRef] [PubMed]
14. Zou, K.; Chen, X.; Zhang, F.; Zhou, H.; Zhang, C. A Field Weed Density Evaluation Method Based on UAV Imaging and Modified U-Net. *Remote. Sens.* **2021**, *13*, 310. [CrossRef]
15. Vaswani, A.; Shazeer, N.; Parmar, N.; Uszkoreit, J.; Jones, L.; Gomez, A.N.; Kaiser, L.; Polosukhin, I. Attention Is All You Need. In Proceeding of the Advances in Neural Information Processing Systems 30 (NIPS 2017), Long Beach, CA, USA, 4–9 December 2017.
16. Liu, Z.; Hu, H.; Lin, Y.; Yao, Z.; Xie, Z.; Wei, Y.; Ning, J.; Cao, Y.; Zhang, Z.; Dong, L.; et al. Swin Transformer V2: Scaling Up Capacity and Resolution. In Proceedings of the IEEE/CVF Conference on Computer Vision and Pattern Recognition (CVPR), New Orleans, LA, USA, 24 June 2022.
17. Zhuang, F.; Qi, Z.; Duan, K.; Xi, D.; Zhu, Y.; Zhu, H.; Xiong, H.; He, Q. A Comprehensive Survey on Transfer Learning. *Proc. IEEE* **2020**, *109*, 43–76. [CrossRef]
18. Srinivas, A.; Lin, T.-Y.; Parmar, N.; Shlens, J.; Abbeel, P.; Vaswani, A. Bottleneck Transformers for Visual Recognition. In Proceedings of the 2021 IEEE/CVF Conference on Computer Vision and Pattern Recognition (CVPR), Nashville, TN, USA, 25 June 2021; IEEE: Piscatway, NJ, USA, 2021; pp. 16514–16524.
19. Yang, J.; Li, C.; Zhang, P.; Dai, X.; Xiao, B.; Yuan, L.; Gao, J. Focal Self-Attention for Local-Global Interactions in Vision Transformers. *arXiv* **2021**, arXiv2107.00641.
20. Moccia, S.; De Momi, E.; El Hadji, S.; Mattos, L.S. Blood Vessel Segmentation Algorithms—Review of Methods, Datasets and Evaluation Metrics. *Comput. Methods Programs Biomed.* **2018**, *158*, 71–91. [CrossRef]
21. Ciecholewski, M.; Kassjański, M. Computational Methods for Liver Vessel Segmentation in Medical Imaging: A Review. *Sensors* **2021**, *21*, 2027. [CrossRef]
22. Maurício, J.; Domingues, I.; Bernardino, J. Comparing Vision Transformers and Convolutional Neural Networks for Image Classification: A Literature Review. *Appl. Sci.* **2023**, *13*, 5521. [CrossRef]
23. Khanal, A.; Estrada, R. Dynamic Deep Networks for Retinal Vessel Segmentation. *Front. Comput. Sci.* **2020**, *2*, 35. [CrossRef]
24. Gegundez-Arias, M.E.; Marin-Santos, D.; Perez-Borrero, I.; Vasallo-Vazquez, M.J. A New Deep Learning Method for Blood Vessel Segmentation in Retinal Images Based on Convolutional Kernels and Modified U-Net Model. *Comput. Methods Programs Biomed.* **2021**, *205*, 106081. [CrossRef]
25. Galdran, A.; Anjos, A. State-of-the-art retinal vessel segmentation with minimalistic models. *Front. Nat.* **2022**, *12*, 6174 [CrossRef]
26. Tang, P.; Liang, Q.; Yan, X.; Zhang, D.; Coppola, G.; Sun, W. Multi-Proportion Channel Ensemble Model for Retinal Vessel Segmentation. *Comput. Biol. Med.* **2019**, *111*, 103352. [CrossRef]
27. Ma, Y.; Li, X.; Duan, X.; Peng, Y.; Zhang, Y. Retinal Vessel Segmentation by Deep Residual Learning with Wide Activation. *Comput. Intell. Neurosci.* **2020**, *2020*, 1–11. [CrossRef] [PubMed]

28. Tuyet, V.T.H.; Binh, N.T. Improving Retinal blood vessels Segmentation via Deep Learning in Salient Region. *SN Comput. Sci.* **2020**, *1*, 248. [CrossRef]
29. Park, K.-B.; Choi, S.H.; Lee, J.Y. M-GAN: Retinal Blood Vessel Segmentation by Balancing Losses Through Stacked Deep Fully Convolutional Networks. *IEEE Access* **2020**, *8*, 146308–146322. [CrossRef]
30. Zhuo, Z.; Huang, J.; Lu, K.; Pan, D.; Feng, S. A Size-Invariant Convolutional Network with Dense Connectivity Applied to Retinal Vessel Segmentation Measured by a Unique Index. *Comput. Methods Programs Biomed.* **2020**, *196*, 105508. [CrossRef]
31. Kamran, S.A.; Hossain, K.F.; Tavakkoli, A.; Zuckerbrod, S.L.; Sanders, K.M.; Baker, S.A. RV-GAN: Segmenting Retinal Vascular Structure in Fundus Photographs Using a Novel Multi-Scale Generative Adversarial Network. In Proceedings of the International Conference on Medical Image Computing and Computer-Assisted Intervention, Strasbourg, France, 27 September 27–1 October 2021; Volume 12908, pp. 34–44.
32. Yadav, G.; Maheshwari, S.; Agarwal, A. Contrast Limited Adaptive Histogram Equalization Based Enhancement for Real Time Video System. In Proceedings of the 2014 International Conference on Advances in Computing, Communications and Informatics (ICACCI), Delhi, India, 24–27 September 2014; IEEE: Piscatway, NJ, USA, 2014; pp. 2392–2397.
33. Woo, S.; Park, J.; Lee, J.-Y.; Kweon, I.S. CBAM: Convolutional Block Attention Module. In Proceedings of the European Conference on Computer Vision (ECCV), Munich, Germany, 8–14 September 2018.
34. He, K.; Zhang, X.; Ren, S.; Sun, J. Deep Residual Learning for Image Recognition. In Proceedings of the IEEE Conference on Computer Vision and Pattern Recognition (CVPR), Boston, MA, USA, 7–12 June 2015.
35. Simonyan, K.; Zisserman, A. Very Deep Convolutional Networks for Large-Scale Image Recognition. *arXiv* **2015**, arXiv:1409.1556.
36. Tan, M.; Le, Q.V. EfficientNet: Rethinking Model Scaling for Convolutional Neural Networks. *arXiv* **2020**, arXiv:1905.11946.
37. Howard, A.G.; Zhu, M.; Chen, B.; Kalenichenko, D.; Wang, W.; Weyand, T.; Andreetto, M.; Adam, H. MobileNets: Efficient Convolutional Neural Networks for Mobile Vision Applications. *arXiv* **2017**, arXiv:1704.04861.
38. Szegedy, C.; Liu, W.; Jia, Y.; Sermanet, P.; Reed, S.; Anguelov, D.; Erhan, D.; Vanhoucke, V.; Rabinovich, A. Going Deeper with Convolutions. *arXiv* **2014**, arXiv:1409.4842.
39. Ronneberger, O.; Fischer, P.; Brox, T. U-Net: Convolutional Networks for Biomedical Image Segmentation. In Proceedings of the Medical Image Computing and Computer-Assisted Intervention–MICCAI 2015: 18th International Conference, Munich, Germany, 5–9 October 2015.
40. Chen, D.; Yang, W.; Wang, L.; Tan, S.; Lin, J.; Bu, W. PCAT-UNet: UNet-like Network Fused Convolution and Transformer for Retinal Vessel Segmentation. *PLoS ONE* **2022**, *17*, e0262689. [CrossRef]
41. Guo, C.; Szemenyei, M.; Yi, Y.; Xue, Y.; Zhou, W.; Li, Y. Dense Residual Network for Retinal Vessel Segmentation. In Proceedings of the ICASSP 2020–2020 IEEE International Conference on Acoustics, Speech and Signal Processing (ICASSP), Barcelona, Spain, 4–8 May 2020; IEEE: Piscataway, NJ, USA, 2020; pp. 1374–1378.
42. Li, X.; Jiang, Y.; Li, M.; Yin, S. Lightweight Attention Convolutional Neural Network for Retinal Vessel Image Segmentation. *IEEE Trans. Ind. Inf.* **2021**, *17*, 1958–1967. [CrossRef]
43. Wu, H.; Wang, W.; Zhong, J.; Lei, B.; Wen, Z.; Qin, J. SCS-Net: A Scale and Context Sensitive Network for Retinal Vessel Segmentation. *Med. Image Anal.* **2021**, *70*, 102025. [CrossRef]
44. Tajbakhsh, N.; Lai, B.; Ananth, S.; Ding, X. ErrorNet: Learning Error Representations from Limited Data to Improve Vascular Segmentation. In Proceedings of the 2020 IEEE 17th International Symposium on Biomedical Imaging (ISBI), Iowa City, IA, USA, 3–7 April 2020.

Disclaimer/Publisher's Note: The statements, opinions and data contained in all publications are solely those of the individual author(s) and contributor(s) and not of MDPI and/or the editor(s). MDPI and/or the editor(s) disclaim responsibility for any injury to people or property resulting from any ideas, methods, instructions or products referred to in the content.

Review

Digital Technologies to Provide Humanization in the Education of the Healthcare Workforce: A Systematic Review

María Gonzalez-Moreno [1,2], Carlos Monfort-Vinuesa [1,3,4], Antonio Piñas-Mesa [5] and Esther Rincon [1,3,*]

1. Psycho-Technology Lab., Universidad San Pablo-CEU, CEU Universities, Urbanización Montepríncipe, 28660 Boadilla del Monte, Spain; mgmoreno@ceu.es (M.G.-M.); carlos.monfortvinuesa@ceu.es (C.M.-V.)
2. Departamento de Ciencias Médicas Básicas, Facultad de Medicina, Universidad San Pablo-CEU, CEU Universities, Campus de Montepríncipe, Urbanización Montepríncipe, 28660 Boadilla del Monte, Spain
3. Departamento de Psicología y Pedagogía, Facultad de Medicina, Universidad San Pablo-CEU, CEU Universities, Urbanización Montepríncipe, 28660 Boadilla del Monte, Spain
4. Departamento de Medicina Interna, HM Hospitales, Universidad San Pablo-CEU, CEU Universities, Urbanización Montepríncipe, 28660 Boadilla del Monte, Spain
5. Departamento de Humanidades, Facultad Humanidades y CC Comunicación, Universidad San Pablo-CEU, CEU Universities, Paseo Juan XXIII 8, 28040 Madrid, Spain; anpime@ceu.es
* Correspondence: maria.rinconfernande@ceu.es; Tel.: +34-913-724-700 (ext. 15076)

Citation: Gonzalez-Moreno, M.; Monfort-Vinuesa, C.; Piñas-Mesa, A.; Rincon, E. Digital Technologies to Provide Humanization in the Education of the Healthcare Workforce: A Systematic Review. *Technologies* 2023, 11, 88. https://doi.org/10.3390/technologies11040088

Academic Editors: Juvenal Rodriguez-Resendiz, Gerardo I. Pérez-Soto, Karla Anhel Camarillo-Gómez and Saul Tovar-Arriaga

Received: 21 May 2023
Revised: 27 June 2023
Accepted: 28 June 2023
Published: 5 July 2023

Copyright: © 2023 by the authors. Licensee MDPI, Basel, Switzerland. This article is an open access article distributed under the terms and conditions of the Creative Commons Attribution (CC BY) license (https://creativecommons.org/licenses/by/4.0/).

Abstract: Objectives: The need to incentivize the humanization of healthcare providers coincides with the development of a more technological approach to medicine, which gives rise to depersonalization when treating patients. Currently, there is a culture of humanization that reflects the awareness of health professionals, patients, and policy makers, although it is unknown if there are university curricula incorporating specific skills in humanization, or what these may include. Therefore, the objectives of this study are as follows: (1) to identify what type of education in humanization is provided to university students of Health Sciences using digital technologies; and (2) determine the strengths and weaknesses of this education. The authors propose a curriculum focusing on undergraduate students to strengthen the humanization skills of future health professionals, including digital health strategies. Methods: A systematic review, based on the scientific literature published in EBSCO, Ovid, PubMed, Scopus, and Web of Science, over the last decade (2012–2022), was carried out in November 2022. The keywords used were "humanization of care" and "humanization of healthcare" combined both with and without "students". Results: A total of 475 articles were retrieved, of which 6 met the inclusion criteria and were subsequently analyzed, involving a total of 295 students. Three of them (50%) were qualitative studies, while the other three (50%) involved mixed methods. Only one of the studies (16.7%) included digital health strategies to train humanization. Meanwhile, another study (16.7%) measured the level of humanization after training. Conclusions: There is a clear lack of empirically tested university curricula that combine education in humanization and digital technology for future health professionals. Greater focus on the training of future health professionals is needed, in order to guarantee that they begin their professional careers with the precept of medical humanities as a basis.

Keywords: humanization of care; humanization of healthcare; medical humanities; undergraduate education; digital technology

1. Introduction

Digital health is defined as the use of digital technologies for health [1]. It has gradually become a relevant topic of healthcare practice [2], including tools such as mobile health technology (mHealth), virtual reality (VR), or artificial intelligence (AI). Despite digital health having enormous potential [3], it is still poorly incorporated into healthcare workforce [4]. Therefore, several authors have claimed that digital health training should be included in the curriculum for healthcare professionals [4–6], so that they can reach their full potential [3,7,8]. One way to do this could be to combine digital health technology and training in humanization for the healthcare workforce, paving the way for future challenges and opportunities.

In this respect, the humanistic training of health professionals has given rise to quite different interpretations, including the understanding of "medical humanities" (MH). These range from positions that present MH as a discipline falling between objective technique and compassionate ethical attitude [9], to interdisciplinary approaches that help to integrate and interpret human experiences of illness [10], understanding the human condition of health and illness [11], and medical practice. Classically, among the tools used, mention is made of creative training processes such as the use of literature, art, creative writing, cinema forums, the narrative of patients and healthcare workers, ethical decision making, anthropology, and history in pursuit of the goals of humanized medical education [12].

The term humanization has an ethical connotation, as it refers to the evaluation of human actions according to values and, specifically, to the treatment of others in different human relationships. In the healthcare setting, a recently published and tested model [13–15] defined humanization as a "set of personal competencies that allow for the development of professional practice within the healthcare environment, respecting and ensuring dignity and respect for human beings. It is, therefore, an activity focused on improving physical, mental, and emotional healthcare, from the perspective of both patients and health professionals themselves". For this reason, supported by the MH disciplines, educational tools are developed to improve personal skills in the humanization of future health professionals, which are aimed at improving the personal dimension of care.

Different studies suggest that this integration of MH in education can help students develop essential qualities such as professionalism, self-awareness, social and communication skills, and reflective practice [16,17]. It also encourages a more holistic approach to patient care [18]. A US multi-institutional survey [19] showed that medical students' exposure to the humanities correlates with positive personal qualities and reduces stress, mitigates burnout, fosters resilience, and promotes well-being [18,20,21]. Finally, it helps them to think like doctors [22].

The inclusion of the humanities in medical education may offer significant potential benefits to individual future physicians and to the medical community. There should be no debate about the definition and precise role of the humanities in medical education, since this training must be present from the beginning of studies in health sciences, providing tools to health workers, which need to be continuously developed and recycled during their years of professional practice [23].

Wald et al. [24]. assert that it is essential for medical students to be taught, from the early stages of their training and throughout their careers, that the practice of medicine can never be black and white, and that—in line with Schon's view [25]—dealing with greyness, uncertainty, and doubt will always be present in professional practice and human complexity. The inclusion of MH could thus support the need for medical education to respond to this complexity and help provide the necessary framework to cultivate competent and compassionate physicians [24].

Quantifying the long-term impact of humanities training is a commendable goal. In Ousager and Johannessen's (2010) systematic review of 245 articles concerning the humanities in medical education [26], only nine papers provided evidence of attempts to document their long-term impact Humanities courses in medical schools should set specific, measurable goals for their curricula and determine methods to evaluate whether those goals are being met [27].

However, all this would be meaningless without a broader vision that leads from the disciplines of MH, through the different educational tools, to the personal competencies that enable "humanized medical care" (HMC). Therefore, several studies have tried to determine what this term (HMC) means concretely, and how best to implement it. In this regard, when analyzing the different actors involved in the delivery of HMC, the results show a clear gap between the expectations cited by stakeholders and the practices implemented in daily clinical practice [28].

When this concept is applied to the field of healthcare practice, it is noticed that two different expressions are used in the current literature, which, although related, refer to different specific realities: one is "humanization of healthcare"(HH); the other, "humanization of care" (HOC). The former, which is more inclusive, refers to actions aimed at the humanization of both healthcare management and practice, and focuses on all the agents involved in the clinical relationship. On the other hand, publications on the "humanization of care" focus on the role of nursing and specific actions such as the adoption of healthcare environments to make them more friendly.

The term "humanization of healthcare" (HH) originated in the scientific literature [29], yet there is currently no clear consensus on its definition [28,30]. The term HH implies consideration of the stakeholders involved in healthcare, such as patients themselves, patients' caregivers, health professionals and policy makers, as well the interaction between them all [28]. However, future care providers, i.e., current healthcare students, are not included among them.

This absence is noteworthy. As early as the nineteenth century, William Osler coined the phrase "The good physician treats the disease; the great physician treats the patient who has the disease" [31]. Lately Ronnie Mac Keith in his essay "The tyranny of the idea of cure" cautioned that "Patients are not uninterested vehicles of interesting diseases" [27]. Therefore, this concern is at the root of the training of future health professionals, highlighting the importance of embracing uncertainty and restoring the integral balance between the sciences and the humanities.

Other authors [32] considered three illustrative areas related to learning that could serve as teaching tools for the medical humanities. These three areas would aim to improve the understanding of patients' experiences of disease.

The HH has two dimensions that are the object of this humanization: firstly, the "structural", relating to the management of the means for health activity; and, secondly, the "personal", relating to the aptitudes and attitudes of healthcare workers. From the structural point of view, the automation and standardization of medical care, as well as the lack of time, have been highlighted as causes of dehumanizing treatment by health professionals [33]. It has even been suggested that patients are sometimes not treated as individuals, but as a "symptom cluster" [34]. Moreover, excessive bureaucracy, deficiencies in hospital structure, overcrowding, excessive workload, lack of material resources (as experienced during the COVID-19 pandemic), poor coordination between departments, and lack of assessments of dehumanizing behavior have been noted [35].

Regarding the skills and attitudes of healthcare personnel, these are linked to the ethcal codes that regulate the healthcare relationship with patients and their relatives. However, competence in the HOC, in addition to ethical competence, would include psychological competences such as empathic competence (closely linked to ethical competence). The concept of empathy towards the patient, respect for the patient's dignity, and consideration for the patient as an individual have been cited in several studies as fundamental aspects of this "humanized care" [28]. On the other hand, the concept of moral sensitivity, defined as the ability to be aware that one's actions may affect other people, has also been related to the concept of humanization [36].

Therefore, the MH are transformed into the HH when they are grounded in human action and values, both incorporated into specific academic curricula. This will promote the psychological, relational, and ethical attitudes of future health professionals that favor "humanized care" for the patient. Although it has not yet been empirically proven, it is postulated that humanistic training applied to the education of healthcare professionals can have a positive influence on improving the treatment of patients, as well as the management of healthcare environments to make them friendlier or more appropriate for patients and their vulnerabilities.

It is the responsibility of the scientific and teaching community to provide future health professionals with training in humanities for the exercise of their profession. Pursuing this goal and measuring the impact of humanization of healthcare through digital health strategies represents the challenge itself.

Given that there is no international consensus on the concept of humanization of healthcare, nor, therefore, on the best way to teach it, and that its effective implementation has been linked to students' prior training, as have digital health strategies, this research study aims to systematically review the existing scientific literature regarding the explicit training that undergraduate students receive regarding the concept of humanized healthcare, using digital technology. Specifically, the following objectives are sought: (1) to identify what type of education in humanization is provided to university students of health sciences, developed through digital health strategies; and (2) to determine its strengths and weaknesses.

The following sections will further explain the methodology developed for data collection, as well as the main outcomes and conclusions.

2. Methods

To develop the present study, the following methodology was adopted.

2.1. Overview

The current systematic review was performed and reported using the Preferred Reporting Items for Systematic Reviews and Meta-Analyses (PRISMA) Statement (see study protocol in Supplementary Material S1) [37–40]. The protocol was registered with the PROSPERO International Prospective Register of Systematic Reviews (CRD42022382146).

2.2. Selection Criteria

Articles were considered potentially relevant if they were published in English or Spanish between the years 2012 and 2022 inclusive and included a training protocol or education in the "humanization of healthcare" for undergraduate students of any of the branches of health sciences, provided as part of their university education. We excluded abstracts or conference papers, study protocols, narrative reviews, and articles that were published in a language other than English or Spanish, as well as those that did not include training or results in "humanization of healthcare" of university students (for example, training in "humanization of healthcare" for active professionals in hospital settings).

2.3. Outcomes

The primary outcomes were the types of humanization training provided to students at university and if digital technologies were used for that purpose; whether the humanized skills were assessed before or after training; and the efficacy of the training provided (in terms of increasing humanization skills in students). The secondary outcomes were the level of satisfaction of the students involved and the strengths and weaknesses of the forms of training examined.

2.4. Search Methodology

A comprehensive search was carried out in EBSCO (Academic Search Complete, CINAHL Plus with Full Text, Communication Source, eBook Collection, E-Journals, ERIC, Fuente Academica Premier, Humanities International Complete, MEDLINE, MLA Directory of Periodicals, MLA International Bibliography, OpenDissertations, PSICODOC, Psychology and Behavioral Sciences Collection, PsycInfo), PubMed, Scopus, Ovid, and WOS (Web of Science Core Collection) from its inception until November 2022. The detailed search strategies used in all the databases are provided in Supplementary Material S1. The original versions of all the research articles were retrieved for examination, and a search library was created using RefWorks©, a bibliography management program.

2.5. Data Collection and Analysis

For the sake of completeness, two reviewers (M.G.-M. and E.R.) independently evaluated and reviewed all the titles and abstracts of identified references to determine their eligibility for inclusion in the study. In case of discrepancies, a third author was consulted (C.M.-V.). After that, Cohen's kappa coefficient for inter-observer agreement [41] was calculated in order to determine the degree of agreement between the data of the two investigators (E.R. and M.G.-M.). The interpretation of the data obtained from Cohen's kappa was calculated using SPSS version 27 (IBM Corp., Armonk, NY, USA), based on the categories established by Douglas Altman [42] as 0.00–0.20 (poor), 0.21–0.40 (fair), 0.41–0.60 (moderate), 0.61–0.80 (good), and 0.81–1.00 (very good). One author (M.G.-M.) independently extracted data on outcomes from all the studies. All extracted data were reviewed for completeness by two reviewers (E.R. and C.M.-V.).

A data extraction tool was developed in Microsoft Excel, which was used to retrieve relevant information. Cross-checking was undertaken to identify any inaccuracies or oversights. Discrepancies were resolved amongst the core team with the involvement of the broader research team when necessary.

2.6. Data Extraction and Management

We extracted data on (1) publication year, (2) country, (3) study design, (4) study aim, (5) sample size, (6) mean participant age, (7) university course, (8) type of training provided (using digital technologies—yes/no—), (9) assessment of prior/subsequent level of humanization, (10) outcomes, and (11) student satisfaction, (12) strengths, and (13) weaknesses.

2.7. Quality of Studies Included

The study designs of the articles included varied widely; therefore, the quality of the studies included was appraised using the Mixed Methods Appraisal Tool (MMAT), developed in 2006 [43] and revised in 2018 [44]. Total scores with higher values indicated a lower risk of methodological bias (see Multimedia Supplementary Material S2). The critical evaluation of designing and developing educational interventions was made through a checklist for critically appraising reports of educational interventions [45] (see Supplementary Material S3). One author (C.M.-V.) independently extracted data on outcomes from all the studies. For completeness, all extracted data were reviewed by two reviewers (E.R. and A.P.-M.).

2.8. Statistical Analysis

Data were pooled using SPSS version 27 (IBM Corp., Armonk, NY, USA), allowing for frequency analysis (percentages).

3. Results

The following outcomes were obtained.

3.1. Search Results/Characteristics of Included Studies

The total number of articles retrieved when searching with the chosen keywords "humanization of care" and "humanization of healthcare" was 475. After discarding duplicates (288), 187 studies remained (39.3%) and these were evaluated on the basis of title and abstract. Of these, 174 (93%) were rejected as they clearly did not meet the inclusion criteria. Based on titles and abstracts, 13 (7%) articles were selected for full text screening; 7 (53.8%) out of these 13 [46–52] were discarded for various reasons (see Supplementary Material S4). A total of six publications (46.2%) were included in the end [53–58]. A PRISMA flow diagram is shown in Figure 1 [39]. Cohen's kappa was good ($\kappa = 0.72$) based on the categories developed by Altman [42]. All the chosen studies were deemed to be of sufficient quality to contribute equally to the thematic synthesis.

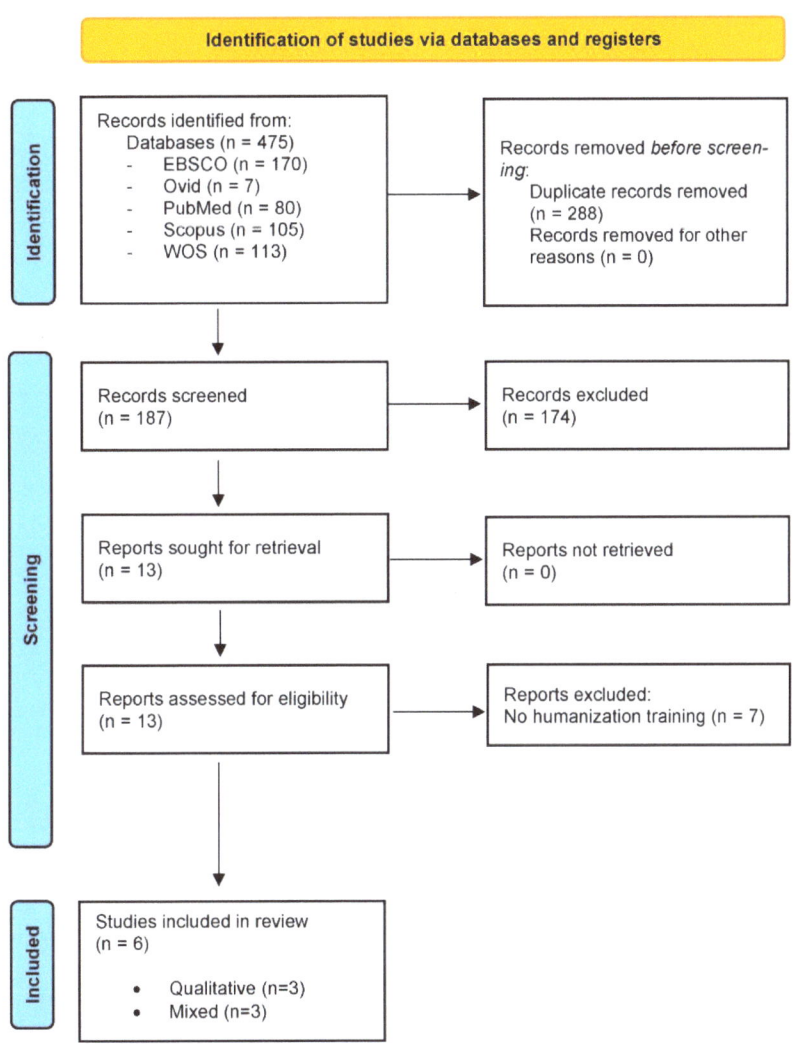

Figure 1. Flowchart of the systematic literature review.

3.2. General Characteristics of Included Studies

A review of the general features of the six papers evaluated shows (Table 1) that three of the studies (50%) were carried out in Spain [53,54,58], one in Brazil (16.7%) [56] one in Chile (16.7%) [57], and one in Canada (16.7%) [58], and that were published—except for a paper published in 2017 [53] (Feijoo-Cid et al., 2016) almost all (83.3%)—between 2019 and 2022.

Table 1. General characteristics of included studies (n = 6).

Authors	Country	Study Design	Study Aim	Sample Size	Mean Age of Participants	University Course	Type of Training Provided
Feijoo-Cid et al. [53]	Spain	Mixed	Evaluate nursing students' satisfaction with EPIN * as a teaching and learning methodology.	64	N/A	4th-year Degree in Nursing	No DT. Three expert patients (a woman living with HIV, a man caring for his dying parents, and a man living with cancer) shared their illness narratives with the students.
Jiménez-Rodriguez et al. [54]	Spain	Mixed	Evaluate the effects of virtual simulation-based training on developing and cultivating humanization competencies in undergraduate nursing students.	60	23.83	3rd-year Degree in Nursing	DT involved. A virtual platform of online video conferences provided by the university (Blackboard Collaborate LauncherTM) was used to develop six simulated scenarios related to basic healthcare at patients homes.
Létourneau et al. [55]	Canada	Qualitative	Provide a description of nursing students and nurses recommendations aimed at improving the development of humanistic caring.	26	24	1st-, 2nd-, 3rd-year Degree in Nursing	No DT. The selected methodological approach was Benner's (1994) interpretive phenomenology. Aims to understand human experiences in the particular worlds of research participants. In this case, the focus was on nursing students' and nurses' recommendations.
Mega et al. [56]	Brazil	Qualitative	This study aims to understand how literature can influence the humanistic training of medical students, creating a representative model based on the experience.	12	N/A	2nd-year Degree in Nursing	No DT. This research was guided by the methodological framework of Grounded Theory (GT), which aims to understand reality from the perception or meaning that a certain context or object has for the person, generating knowledge, increasing understanding, and providing a significant guide for action.
Moya et al. [57]	Chile	Mixed	The aim of the project consisted of establishing a systematic observation of the subject, taught in the second year of the degree, using the focus group methodology, which aims to foster growth and the realization of human potential.	112	N/A	2nd-year Degree in Nursing	No DT. To establish a systematic observation of second year students using the focus group methodology. About 15 students, seated in a circle together with a facilitator, met once a week (two continuous modules) in order to stimulate self-knowledge and communication with one another.
Sierras-Davó et al. [58]	Spain	Qualitative	To explore the meaning of the experience and knowledge acquired by nursing students from different European countries, trained through previous learning experiences in Healthcare Improvement Science.	21	25.5	Degree in Nursing	No DT. A phenomenological approach based on the Giorgi method was carried out through a group discussion of 21 European students. The analysis was also triangulated with three experienced researchers who broke down the data into eight units of meaning

* EPIN: "Expert Patient Illness Narratives". DT: Digital Technologies used.

The total number of participants in the various trials was 295 students, with the work of Moya et al. [57] in Chile comprising the largest number of participants, concretely 112 students in the second year of their nursing degree (Table 1).

Only in the study by Mega et al. [56] were the participants medical students, while the rest (83.3%) [53–55,57,58] involved students from different nursing courses. One study [53] specifies that the participants in the study were enrolled in the Module in Medical Anthropology. It is important to note that studies were developed in different countries (Table 1).

Finally, regarding the general features of the studies, it should be noted that only three of the six studies (50%) mention the mean age of the participants, which was 23 years old [53], 24 years old [55], and 25 years old [58]. The papers can be organized into several subgroups according to the number of participants and the type of study conducted. On the one hand, the studies involving a semi-structured interview, which is 50% of them [55,56,58], worked with groups of 12-26 participants, while those that worked with a training methodology (33%)—"Expert Patient Illness Narratives"(EPIN) [53] or virtual simulation-based training [54] —using surveys and quasi-experimental studies, evaluated between 60 and 64 participants. The work of [57], whose objective was to develop learning experience through guided group encounters and the use of surveys, involved 112 students. Only one study [54] included digital health strategies to improve humanization (Table 1).

3.3. Assessment of Methodological Quality of Included Studies

Of the six studies analyzed using the Mixed Methods Appraisal Tool (MMTA) [44], three of them (50%) are qualitative studies, while the other three (50%) correspond to mixed methods (see Supplementary Material S2). A detailed analysis of the six studies yields results that are consistent with the question posed and the tools chosen to answer it. The analysis of the results obtained from the corrected data and their interpretation shows consistency among the different studies.

3.4. Outcomes

With regard to the analysis of the results (Table 2), only one of the studies (16.7%) [54] evaluated the level of humanization before/after training, using digital technologies to develop it. Meanwhile, another study (16.7%) [53] evaluated the level of humanization after training.

In one study [53], after training, a self-administered written questionnaire was distributed, which had been developed by the authors specifically for this study, taking into account the learning objectives, skills, and abilities to be acquired by the students. Meanwhile, in another study [54], a validated questionnaire—the Healthcare Professional Humanization Scale (HUMAS)—was used to evaluate the acquisition of humanization competencies by comparing the levels obtained in these competencies at baseline (pre-test) and after the virtual simulation experience (post-test).

The results of the study by Feijoo-Cid et al. [53] show that students valued the use of EPIN in their nursing training, both in terms of expanding their current knowledge and acquiring new nursing skills. The nursing students were satisfied with EPIN as a learning and teaching methodology. On the one hand, they reported an improvement in various aspects of their training, as well as the integration of new knowledge, meaning, applicability of theory and critical reflection. On the other hand, EPIN also offered a new humanized perspective of care.

In this paper, however, it was emphasized that women more frequently found the new learning methodology helpful for expanding their competency "to rationalize the presence of the Health–Illness–Care triad in all groups, societies and historical moments"; therefore, it was established as a learning outcome. Men, however, found that this methodology facilitated the development of critical thinking, as well as the ability to identify situations of normalized or deviant care.

Table 2. Main outcomes of included studies (n = 6).

Authors	Assessment of Prior/Subsequent Level of Humanization	Outcomes	Student Satisfaction	Strengths	Weaknesses
Feijoo-Cid et al. [53]	After training, a self-administered written questionnaire developed by the authors specifically for this study was distributed, taking into account the learning objectives, skills, and abilities to be acquired by the students.	Students valued the use of EPIN in their nursing training, both in terms of expanding their current knowledge and acquiring new nursing skills. Women more frequently found the new learning methodology helpful for expanding their competency.	The results of this study show that nursing students found EPIN satisfactory as a learning and teaching methodology. They reported an improvement to various aspects of their training, as well as the integration of new knowledge, meaning, applicability of theory and critical reflection. On the other hand, EPIN also offered a new humanized perspective of care.	EPIN is presented as a practical, real, truthful, and complete way of applying theoretical concepts learned in the classroom, facilitating the acquisition of a certain level of narrative competence.	This study has some limitations that should be taken into account: - Little evidence of patients being included in the teaching process. - Results must be treated with caution as the questionnaire used was not validated. - The research conducted was based on a local experience, with data from an elective course, which may have had a positive impact on the results due to a high level of student motivation. - The greater proportion of female students (78.1%) may have obscured the perspective of male nurses.
Jiménez-Rodríguez et al. [54]	A validated questionnaire—the Healthcare Professional Humanization Scale (HUMAS)—was used to evaluate the acquisition of humanization competencies by comparing the levels obtained in these competencies at baseline (pre-test) and after the virtual simulation experience (post-test).	Following the virtual simulation sessions, students total scores for levels of humanization improved, as did their competencies in emotional understanding and self-efficacy.	N/A	The use of validated scales included self-efficacy or empathy as humanization competencies.	The main limitation of this study lies in the specific disadvantage of both simulated and real-life nursing video consultation, namely, technical issues. Ensuring adequate network access and the correct functioning of virtual platforms could mitigate these potential problems.
Létourneau et al. [55]	N/A	The phenomenological analysis of participants' recommendations revealed five key themes: (1) pedagogical strategies, (2) educators' approach, (3) considerations in teaching humanistic caring, (4) work overload, and (5) volunteerism and externship.	N/A	A key strength highlighted was the effect of clinical externships on humanistic caring, empathy and compassion.	Participants voiced a wide diversity of recommendations.
Mega et al. [56]	N/A	The model incorporates the idea that literature enhances the humanization of care and is capable of breaking away from the biomedical model.	N/A	Humanization of care was enhanced due to the representative model based on the categories created, of the experience that relays the students' satisfaction with literature in medical education.	There is a need for uniformity in the curriculum, with the aim of organizing the activity and learning opportunities for other students.

164

Table 2. Cont.

Authors	Assessment of Prior/Subsequent Level of Humanization	Outcomes	Student Satisfaction	Strengths	Weaknesses
Moya et al. [57]	N/A	A high proportion viewed the content and experiences in a positive light, claiming that the subject adds value to their life projects, and further reinforces the vocational dimension.	N/A	N/A	Lack of systematization of scientific quality for the evaluation of internal experience processes.
Sierras-Davó et al. [58]	N/A	Nursing empowerment and horizontal health organizations were two of the most recurrent units of meaning, together with professional values such as teamwork and humanization of care.	N/A	The theoretical-practical approach of the sessions and cultural diversity.	Due to the timing of the intervention, there is a lack of long-term results.

Turning to the work of Jiménez-Rodríguez et al. [54], their results show that following the virtual simulation sessions, students' total scores for levels of humanization improved, as did their competencies in emotional understanding and self-efficacy, with large effect sizes in all of them (rB = 0.508, rB = 0.713, and rB = 0.505, respectively). In other words, there was a significant improvement in the acquisition of humanization competencies according to the "Healthcare Professional Humanization Scale" model (HUMAS) [15].

As for the rest of the studies (66.7%) [55–58], none of them included assessments of pre- or post-training knowledge by means of any objective tools. In one study [55], the phenomenological analysis of participants' recommendations revealed five key themes: (1) pedagogical strategies, (2) educators' approach, (3) considerations in teaching humanistic caring, (4) work overload, and (5) volunteerism and externship. In other study [56], the researchers' assessment of the students accounts led them to conclude that the model incorporates the idea that literature enhances the humanization of care and is capable of departing from the biomedical model (Table 2).

In turn, a high percentage of the students who had participated in the research with groups guided [57], viewed the content and experiences in a positive light, claiming that the subject adds value to their life projects, and further reinforces the vocational dimension.

Finally, the training based on learning experiences in "Healthcare Improvement Science" [58] was valued positively, with empowerment and horizontal health organizations being two of the most recurrent units of meaning, together with professional values such as teamwork and humanization of care (Table 2).

3.5. Strengths and Weaknesses

With regard to the strengths and weaknesses (Table 2), one of the main strengths of the work of Feijoo-Cid et al. [53] refers that the training involved is shown to be a practical, real, truthful, and complete way of applying theoretical concepts learned in the classroom, and therefore facilitates the acquisition of a certain level of narrative competence. Furthermore, Sierras-Davó et al. [58] highlight the theoretical–practical approach of the sessions, in addition to the cultural diversity, as strengths.

Jiménez-Rodríguez et al. [54] see strength in using validated scales for measuring competence in humanization, while using digital technologies for student training. Other study [55] showed a key strength by highlighting the effect of clinical externships on humanistic caring, empathy, and compassion, which has rarely been reported.

The work of Mega et al. [56] addressed humanization of care as being enhanced due to a representative model of the experience—based on their created categories—that relays the students' satisfaction with literature in medical education.

Regarding the weaknesses, one study [53] indicate certain drawbacks that need to be considered: there was little evidence of patients being included in the teaching process; results must be treated with caution as the questionnaire used was not validated; the research conducted was based on a local experience, with data from an elective course, which may have had a positive impact on the results due to a high level of student motivation; and the greater proportion of female students (78.1%) may have obscured the perspective of male nurses. Other limitations would include technical aspects, as in the work of Jiménez-Rodríguez et al. [54], who underline the specific disadvantage of the lack of adequate network access and correct functioning of virtual platforms as the main limitation of their study.

Finally, the qualitative studies [55,57,58] speak of limitations associated with the qualitative study design itself, ranging from the diversity of recommendations given by the participants [55], to the need for uniformity in the curriculum [57], or the timing of the intervention [57]. In this regard, Mega et al. [56] show another limitation to be the lack of systematization in the evaluation of internal experience processes.

4. Discussion

Although previous systematic reviews have discussed the assessment of medical humanities in undergraduate students [59,60], to the best of our knowledge, this is the first systematic review that addresses how education in humanization is delivered to undergraduate health science students using digital technologies.

The main objectives were to determine the educational characteristics of the humanization programs studied, as well as to identify their advantages and limitations, in order to propose a university training model in "the humanization of healthcare" for future health professionals.

In recent decades, the intensified use of technology has improved the prevention, diagnosis, and treatment of diseases, which has generally enhanced medical care, increasing the quality, efficiency, and safety of patient treatment [61].

However, this technification in clinical practice implies, in many cases, a fragmentation in medical specialization, which results in great experts treating illness in a complex and specific manner, under highly specific conditions. Although this should favor and give greater security to the patient, in many cases, the patient feels as though they are seen as a set of symptoms, rather than as an individual with particular needs [33,62].

Automation and standardization of care, together with the fragmentation of work often-limited time and personnel, can lead to depersonalized and, for some patients and their families, dehumanized healthcare [61].

Some studies suggest that the general perception of patients, i.e., their satisfaction with their care, seems to be linked not only to the technical skills of the professionals treating them, the conditions of care in terms of waiting time, the total time in which they receive care, and the amount of information received, but also to a certain "humanistic" attitude exhibited by healthcare personnel [61].

In view of this demand, it is worth asking whether this "certain humanistic attitude" [61] can be taught and, if so, how it could be included in the curriculum of health sciences degrees, using digital health strategies. In other words, how should be taught the treatment of patients as people, in an increasingly digitalized healthcare environment?

Subjects in "humanization of healthcare"—defined as a mental, emotional, and moral attitude that forces the professional to continually rethink they own mental framework, and to reshape intervention habits so that they are oriented towards the good of the patient (a vulnerable person in need of care) [52] —allow for the identification of parameters such as the protection of patient values (autonomy, confidentiality, dignity), personalized treatment, and active listening, to be the focus of students' training.

The studies discussed, which all fall under the umbrella concept of "humanization of healthcare", involved different training strategies. However, they share common ground in focusing on an experience that aims to change the student's mental and emotional patterns so that their practice is oriented towards the patient's values as a result of their own personal and professional development due to these experiences.

The development of critical thinking about what the areas of professional practice are, the concepts on which it pivots (e.g., what is health, what is disease), the social implications of clinical–medical proceedings, and the mutual interdependence between social normativity and medical practice all imply the deliberate shaping of the student's thinking about how they will be as a health professional and, on a deeper level, what kind of health professional they want to be.

In the studies included, there is discussion of narrative training strategies [53] and phenomenological training strategies [55,58] as tools for cognitive learning of the humanistic competence of caregiving. Virtual simulation may be seen in a similar light [54]. The objective of these studies is to evaluate the effects of this kind of training on the development and honing of humanization competencies. In other cases [56,57], the research evaluates the competencies acquired in subjects specific to the degrees taken by students who have already developed their own curricular strategies, either through literature [56] or through guided encounter groups, as a method of enabling them to recognize themselves and others in their health–illness process [57].

This evidences the strong link between the idea of "humanization of healthcare" and the idea of care, namely, the vulnerability of the patient. This is reinforced by the health professional's own attitude towards training in this context. As an example, nursing students, i.e., those closest to care-based services who used the Expert Patient Illness Narratives (EPIN) methodology [53], were proactive in terms of improving different aspects of their training and integrating new knowledge, meaning, and applicability of theory, as well as in terms of strong critical reflection.

From the perspective of medical students accustomed to intrinsically practical competencies, humanities do not provide useful skills for clinical practice; they may understand and value the content but do not recognize its significance for professionalization [63]. As opposed to scientific, empirical evidence, which is built on medical–epidemiological knowledge, and which becomes a legitimizing factor for intervention on patients [64], the concept of "medical humanities" refers to a conscious habit of thought, a more humane and at the same time less tangible type of knowledge that is directly undervalued as "evidence" [65].

The belief that some professors of humanistic disciplines do not understand the reali-ty of medical practice, and that reflection on values is intrusive, unnecessary, and has little real effect on their empathy as professionals, has been raised [66]. In this regard, the differences according to gender as revealed by the answers to the training questionnaires from the methodology "Expert Patient Illness Narratives"(EPIN) are considered to be of particular importance [53].

Thus, women, who comprised the majority in the trial (88%), more frequently found that the new learning methodology helps them to "develop the competence to rationalize the presence of the Health–Illness–Care triad in all groups, societies and historical moments", thus being solidified as a learning outcome; while the men who participated in the study—only 22%—stated that the methodology helped them to "develop critical thinking and reasoning"; therefore, the ability to "identify normalized or deviant care situations" was established as a learning outcome. Very few students considered the methodology to be helpful in "incorporating therapeutics as a unit of analysis in health–illness–care processes". This remains a representation of the more rational and technological position of males (in a similar way to medical students), coherent with the positions presented by Shapiro [66].

The impact of all these strategies is evidenced by the level of satisfaction shown by the participants in the training sessions. Létourneau et al. [55] highlight the importance that nursing students placed on enhancing the development of humanistic caring as a core competency in education programs. Mega et al. [56] conclude that literature enhances the humanization of care and is able to establish a break from the biomedical model. In turn, a high percentage of the students who participated in the research with groups guided by facilitators valued the content and experiences positively [57], claiming that the subject that formed part of the study contributed to their life projects, and also reinforced the vocational dimension.

Training based on learning experiences in "Healthcare Improvement Science" was positively valued by the students in their training in humanization, with nursing empowerment and horizontal healthcare organizations being two of the most recurrent units of meaning, together with professional values such as teamwork and humanization in care [58].

The results of the questionnaire evaluations are even more significant. The study by Feijoo-Cid et al. [53] shows that nursing students found the Expert Patient Illness Narratives (EPIN) methodology satisfactory as a teaching and learning method. On the one hand, they described improvements in different areas of their training and the integration of new knowledge, meaning, applicability of theory, and critical reflection; on the other hand, EPIN also provided them with a new humanized perspective on care.

Finally, Jiménez-Rodríguez et al. [54] reported an increase in total humanization scores in post-testing after the sessions, quantified as statistically significant differences in the dimensions of emotional understanding and self-efficacy, as well as in the total score for the humanization scale (self-efficacy, sociability, affection, emotional understanding, and optimism). Surprisingly, it was the only study that used digital technologies to develop humanization training. In this study, there was a significant improvement in the acquisition of humanization competencies according to the model of the "Healthcare Professional Humanization Scale" (HUMAS) [13,15].

This outcome reveals the potential of the different strategies used, and on this basis, it allowed the authors to offer a training program in humanization for healthcare students. Given that much of medical education is currently framed in terms of competencies [67], it is argued that the humanization curriculum should be designed as an overarching competency, i.e., to be taught through the acquisition of other types of skills. As such, competency in humanization may be measured in a similar way to other outcomes in the medical and health sciences curriculum.

This program, which uses the HUMAS model [13,15] as its backbone, is innovative because it has been adapted to the university environment, especially considering that HUMAS was initially conceived as a humanization protocol for health professionals, instead of university students. Furthermore, through its scale, it becomes a tool that will allow for the evaluation of the competencies acquired by university students.

Thus, according to the dimensions contemplated by HUMAS [13,15], this training proposal would be aimed at the acquisition of the following skills, all of which are considered to be powerful recontextualization tools.

Optimistic disposition: This would be trained through "Expert Patient Illness Narratives" [53], by looking at accounts from patients who have witnessed positive aspects in the course of their illness, as well as accounts given by health professionals who have struggled in adverse situations, such as those that occurred during COVID-19.

Sociability: This would be trained through mindfulness techniques, self-reflection, and social skills, since they have proven useful for health professionals to recognize, regulate and demonstrate empathy [68,69].

Emotional understanding: Students would be trained through reading literary texts, which is an idea taken from one of the studies discussed [56] wherein literary texts were used to train humanization. Thus, Knight's text [70] would be used because she analyzes the model of the humanization of healthcare previously proposed by another author [71]. Furthermore, real patients would be included, who would observe the students' performance in various real cases in order to evaluate them in terms of their competency in "emotional understanding". As such, "involving patients in teaching and assessment" has previously been identified as important for improving person-centeredness [72].

Self-efficacy: Training would be provided through workshops that foster emotional intelligence, understood as the capacity for successful achievement and well-being.

Affection: This would be trained through problem-solving therapy [73], as well as through teaching different effective coping techniques and self-management of negative emotions through digital health technologies such as virtual or mixed reality.

The scarcity of standardized protocols which involve digital health strategies to train humanization in healthcare professionals highlights that it could be an innovative and promising research topic. Digital health strategies could be useful to widely disseminate humanization training to expert groups worldwide. Perhaps this kind of training, developed through digital health tools, could represent a fast way of disseminating successful humanization protocols that could become a productive and efficient practice to include in the undergraduate healthcare students' curriculum, enhancing their digital health literacy at the same time.

Limitations

There were even fewer studies that combine humanization and digital health strategies. Because only six studies were finally included in this systematic review, and they were developed in different countries, the conclusions must therefore be interpreted with caution.

Similar to previous studies [28], the number of participants in the included studies varied widely, ranging from 12 [56] to 112 [57]. There is a lack of studies involving students as future healthcare agents, who will have to provide care, based on the humanization of healthcare. This is one of the most important limitations found, and it is further exacerbated by the lack of studies conducted with non-nursing students.

In the present analysis, except for the case of Mega et al. [56], which involved medical students, all the studies involved nursing students [53–55,57,58].

5. Conclusions

There is a clear lack of university curricula that incorporate education in humanization for future health professionals involving digital technology, at least that are subject to empirical validation and therefore published in a journal paper. Greater focus on the training of future health professionals is needed in order to guarantee that they start their professional careers based on the precept of medical humanities.

As a second conclusion, this scarcity of university curricula incorporating humanization education is sustained by methodological and substantive problems.

From a methodological point of view, it is necessary to design curricula that include tools that allow for the development of skills culminating in the acquisition of competencies by university students that can be evaluated by the system. For this purpose, a training program has been provided as a training strategy to improve skills and competencies, based on HUMAS [13,15].

Finally, there is a need for more studies on medical graduates, and also studies involving more balanced groups of female and male participants, in order to analyze and gain a deeper understanding of the reason for this perception of "soft skills" as a prior step to the development of strategies for the revaluation of "health humanities". In this regard, it is important to highlight the need to develop awareness programs for undergraduate students so that they understand the influence of "soft skills" in their future professional practice.

Supplementary Materials: The following supporting information can be downloaded at: https://www.mdpi.com/article/10.3390/technologies11040088/s1, Supplementary Material S1: Study protocol; Supplementary Material S2: Quality assessment of included studies; Supplementary Material S3: Appraisal of included educational interventions; Supplementary Material S4: Reasons for studies' exclusion.

Author Contributions: Conceptualization, M.G.-M. and E.R.; methodology, E.R.; formal analysis, M.G.-M.; investigation, M.G.-M.; resources, M.G.-M.; data curation, M.G.-M. and E.R.; writing—original draft preparation, E.R. and M.G.-M.; writing—review and editing, M.G.-M., C.M.-V., A.P.-M. and E.R.; visualization, M.G.-M., C.M.-V., A.P.-M. and E.R.; supervision, E.R.; project administration, E.R.; funding acquisition, A.P.-M. All authors have read and agreed to the published version of the manuscript.

Funding: This work was supported by the grant "MPFI20AP" from the Universidad San Pablo CEU, CEU Universities (Madrid, Spain).

Informed Consent Statement: Not applicable.

Data Availability Statement: Not applicable.

Conflicts of Interest: The authors declare no conflict of interest.

References

1. World Health Organization. *WHO Guideline: Recommendations on Digital Interventions for Health System Strengthening. Executive Summary*; World Health Organization: Geneva, Switzerland, 2019; Available online: https://apps.who.int/iris/bitstream/handle/10665/311977/WHORHR-19.8-eng.pdf?ua=1 (accessed on 14 April 2023).
2. Khurana, M.P.; Raaschou-Pedersen, D.E.; Kurtzhals, J.; Bardram, J.E.; Ostrowski, S.R.; Bundgaard, J.S. Digital health competencies in medical school education: A scoping review and delphi method study. *BMC Med. Educ.* **2022**, *22*, 129. [CrossRef] [PubMed]
3. Makri, A. Bridging the digital divide in health care. *Lancet Digit. Health* **2019**, *1*, 204–205. [CrossRef]
4. Machleid, F.; Kaczmarczyk, R.; Johann, D.; Balčiūnas, J.; Atienza-Carbonell, B.; von Maltzahn, F.; Mosch, L. Perceptions of digital health education among european medical students: Mixed methods survey. *J. Med. Internet Res.* **2020**, *22*, e19827. [CrossRef] [PubMed]
5. Rampton, V.; Mittelman, M.; Goldhahn, J. Implications of artificial intelligence for medical education. *Lancet Digit. Health* **2020**, *2*, 111–112. [CrossRef]
6. Topol, E. The Topol Review. Preparing the Healthcare Workforce to Deliver the Digital Future. 2019. Available online: https://topol.hee.nhs.uk/wp-content/uploads/HEE-Topol-Review-2019.pdf (accessed on 14 April 2023).
7. Chen, Y.; Banerjee, A. Improving the digital health of the workforce in the COVID-19 context: An opportunity to future-proof medical training. *Futur. Healthc. J.* **2020**, *7*, 189–192. [CrossRef]
8. Wong, B.L.H.; Khurana, M.P.; Smith, R.D.; El-Omrani, O.; Pold, A.; Lotfi, A.; O'leary, C.A.; Saminarsih, D.S. Harnessing the digital potential of the next generation of health professionals. *Hum. Resour. Health* **2021**, *19*, 50. [CrossRef] [PubMed]
9. Pellegrino, E. The humanities in medical education: Entering the post-evangelical era. *Theor Med.* **1984**, *5*, 253–266. [CrossRef]
10. Evans, M. Reflections on the humanities in medical education. *Med. Educ.* **2002**, *36*, 508–551. [CrossRef] [PubMed]
11. Klugman, C.M. Medical humanities teaching in North American Allopathic and Osteopathic Medical Schools. *J. Med. Humanit.* **2017**, *39*, 473–481. [CrossRef]
12. Kirklin, D. The centre for medical humanities, royal free and University College Medical School, London, England. *Acad. Med.* **2003**, *78*, 1048–1053. [CrossRef] [PubMed]
13. Pérez-Fuentes, M.C.; Jurado, M.D.M.M.; Peco, I.H.; Linares, J.J.G. Propuesta de un Modelo de Humanización basado en las Competencias Personales: Modelo HUMAS [Proposal of a Humanization Model based on Personal Competencies: HUMAS Model]. *Eur. J. Health Res. EJHR* **2019**, *5*, 63–77. [CrossRef]
14. Pérez-Fuentes, M.D.C.; Herrera-Peco, I.; Molero Jurado, M.D.M.; Oropesa Ruiz, N.F.; Ayuso-Murillo, D.; Gázquez Linares, J.J. A cross-sectional study of empathy and emotion management: Key to a work environment for humanized care in nursing. *Front. Psychol.* **2020**, *11*, 1–10. [CrossRef] [PubMed]
15. Perez-Fuentes, M.D.C.; Herera-Peco, I.; Molero Jurado, M.D.M.; Oropesa Ruiz, N.F.; Ayuso-Murillo, D.; Gázquez Linares, J.J. The development and validation of the healthcare professional humanization scale (HUMAS) for nursing. *Int. J. Environ. Res. Public Health* **2019**, *16*, 3999. [CrossRef] [PubMed]
16. Mann, S. Focusing on Arts, Humanities to Develop Well-Rounded Physicians. Washington (DC). AAMC News. 2017. Available online: https://news.aamc.org/medical-education/article/focusing-artshumanities-well-rounded-physicians/ (accessed on 28 May 2018).
17. Wald, H.; Anthony, D.; Hutchinson, T.; Liben, S.; Smilovitch, M.; Donato, A. Professional identity formation in medical education for humanistic, resilient physicians: Pedagogic strategies for bridging theory to practice. *Acad. Med.* **2015**, *90*, 753–760. [CrossRef]
18. Gordon, J. Medical humanities: To cure sometimes, to relieve often, to comfort always. *Med. J. Aust.* **2005**, *182*, 5–8. [CrossRef]
19. Mangione, S.; Chakraborti, C.; Staltari, G.; Harrison, R.; Tunkel, A.R.; Liou, K.T.; Cerceo, E.; Voeller, M.; Bedwell, W.L.; Fletcher, K.; et al. Medical students' exposure to the humanities correlates with positive personal qualities and reduced burnout: A multi-institutional U.S. survey. *J. Gen. Intern. Med.* **2018**, *33*, 628–634. [CrossRef]
20. Wald, H.S.; Haramati, A.; Bachner, Y.; Urkin, J. Promoting resiliency for interprofessional faculty and senior medical students: Outcomes of a workshop using mind-body medicine and interactive reflective writing. *Med. Teach.* **2016**, *38*, 525–528. [CrossRef]
21. Wald, H.S.; White, J.; Reis, S.P.; Esquibel, A.Y.; Anthony, D. Grappling with complexity: Medical students' reflective writings about challenging patient encounters as a window into professional identity formation. *Med. Teach.* **2018**, *41*, 152–160. [CrossRef] [PubMed]
22. Chiavaroli, N. Knowing how we know: An epistemological rationale for the medical humanities. *Med. Educ.* **2017**, *51*, 13–21. [CrossRef]
23. Wershof Schwartz, A.; Abramson, J.S.; Wojnowich, I.; Accordino, R.; Ronan, E.J.; Rifkin, M.R. Evaluating the impact of the humanities in medical education. *Mt. Sinai J. Med. N. Y.* **2009**, *76*, 372–380. [CrossRef]
24. Wald, H.S.; McFarland, J.; Markovina, I. Medical humanities in medical education and practice. *Med. Teach.* **2019**, *41*, 492–496. [CrossRef] [PubMed]
25. Schon, D.A. *The Reflective Practitioner: How Professionals Think in Action*; Basic Books: New York, NY, USA, 1983.
26. Ousager, J.; Johannessen, H. Humanities in undergraduate medical education: A literature review. *Acad. Med. J. Assoc. Am. Med. Coll.* **2010**, *85*, 988–998. [CrossRef]
27. Ronen, G.M.; Kraus de Camargo, O.; Rosenbaum, P.L. How Can We Create Osler's "Great Physician"? Fundamentals for Physicians' Competency in the Twenty-first Century. *Med. Sci. Educ.* **2020**, *30*, 1279–1284. [CrossRef] [PubMed]

28. Busch, I.M.; Moretti, F.; Travaini, G.; Wu, A.W.; Rimondini, M. Humanization of care: Key elements identified by patients, caregivers, and healthcare providers. A Systematic Review. *Patient Patient-Cent. Outcomes Res.* **2019**, *12*, 461–474. [CrossRef] [PubMed]
29. Borbasi, S.; Galvin, K.T.; Adams, T.; Todres, L.; Farrelly, B. Demonstration of the usefulness of a theoretical framework for humanizing care with reference to a residential aged care service Australia. *J. Clin. Nurs.* **2012**, *22*, 881–889.
30. Sanz-Osorio, M.T.; Sastre-Rus, M.; Monistrol, O.; Pérez Criado, M.; Vallès, V.; Escobar-Bravo, M.A. Humanization of care in acute psychiatric hospitalization units: A scoping review. *J. Psychiatr. Ment. Health Nurs.* **2023**, *30*, 162–181. [CrossRef]
31. Centor, R.M. To be a great physician, you must understand the whole story. *Med. Gen. Med.* **2007**, *26*, 59.
32. Kemp, S.J.; Day, G. Teaching medical humanities in the digital world: Affordances of technology-enhanced learning. *Med. Humanit.* **2014**, *40*, 125–130. [CrossRef]
33. Behruzi, R.; Hatem, M.; Goulet, L.; Fraser, W.D. Perception of humanization of birth in a highly specialized hospital let's think differently. *Health Care Women Int.* **2014**, *35*, 127–148. [CrossRef]
34. Kienzle, H.F. Fragmentation of the doctor-patient relationship as a result of standardization and economization. *Z. Arztl Qual.* **2004**, *98*, 193–199. (In German)
35. Plumed, C. Una Aportación para la Humanización. *Arch. Hosp.* **2013**, *11*, 261–352. Available online: https://n9.cl/bi2mg (accessed on 14 April 2023).
36. Lutzen, K.; Nordstro, G.; Evertzon, M. Moral Sensitivity in Nursing Practice. *Scand. J. Caring Sci.* **1995**, *4*, 131–138. [CrossRef]
37. Booth, A.; Clarke, M.; Dooley, G.; Ghersi, D.; Moher, D.; Petticrew, M.; Stewart, L. The nuts and bolts of PROSPERO: An international prospective register of systematic reviews. *Syst. Rev.* **2012**, *1*, 2. [CrossRef] [PubMed]
38. Moher, D.; Shamseer, L.; Clarke, M.; Ghersi, D.; Liberati, A.; Petticrew, M.; Shekelle, P.; Stewart, L.A.; PRISMA-P Group. Preferred reporting items for systematic review and meta-analysis protocols (PRISMA-P) 2015 statement. *Syst. Rev.* **2015**, *4*, 1. [CrossRef] [PubMed]
39. Page, M.J.; McKenzie, J.E.; Bossuyt, P.M.; Boutron, I.; Hoffmann, T.C.; Mulrow, C.D.; Shamseer, L.; Tetzlaff, J.M.; Akl, E.A.; Brennan, S.E.; et al. The PRISMA 2020 statement: An updated guideline for reporting systematic reviews. *Br. Med. J.* **2021**, *372*, n71. [CrossRef]
40. Shamseer, L.; Moher, D.; Clarke, M.; Ghersi, D.; Liberati, A.; Petticrew, M.; Shekelle, P.; Stewart, L.A.; PRISMA-P Group. Preferred reporting items for systematic review and meta-analysis protocols (PRISMA-P) 2015: Elaboration and explanation. *BMJ* **2015**, *350*, g7647. [CrossRef]
41. Cohen, J. A coefficient of agreement for nominal scales. *Educ. Psychol. Meas.* **2016**, *20*, 37–46. [CrossRef]
42. Altman, D.G. *Practical Statistics for Medical Research*; Chapman and Hall: London, UK, 1990.
43. Pluye, P.; Gagnon, M.P.; Griffiths, F.; Johnson-Lafleur, J. A scoring system for appraising mixed methods research, and concomitantly appraising qualitative, quantitative and mixed methods primary studies in mixed studies reviews. *Int. J. Nurs. Stud.* **2009**, *46*, 529–546. [CrossRef]
44. Hong, Q.N.; Pluye, P.; Fàbregues, S.; Bartlett, G.; Boardman, F.; Cargo, M.; Dagenais, P.; Gagnon, M.-P.; Griffiths, F.; Nicolau, B.; et al. *Mixed Methods Appraisal Tool (MMAT). Registration of Copyright (#1148552)*; Canadian Intellectual Property Office, Industry Canada: Gatineau, QC, Canada, 2018.
45. Morrison, J.M.; Sullivan, F.; Murray, E.; Jolly, B. Evidence-based education: Development of an instrument to critically appraise reports of educational interventions. *Med. Educ.* **1999**, *33*, 890–893. [CrossRef]
46. Ávila, L.I.; Silveira, R.S.D.; Figueiredo, P.P.D.; Mancia, J.R.; Gonçalves, N.G.D.C.; Barlem, J.G.T. Moral construction of undergraduate nursing students to promote care humanization. *Texto Contexto-Enferm.* **2018**, *27*.
47. Casate, J.C.; Corrêa, A.K. A humanização do cuidado na formação dos profissionais de saúde nos cursos de graduação [The humanization of care in the education of health professionals in undergraduate courses]. *Rev. Esc. Enferm. USP* **2012**, *46*, 219–226. [CrossRef] [PubMed]
48. González-García, M.; Lana, A.; Zurrón-Madera, P.; Valcárcel-Álvarez, Y.; Fernández-Feito, A. Nursing Students' Experiences of Clinical Practices in Emergency and Intensive Care Units. *Int. J. Environ. Res. Public Health* **2020**, *17*, 5686. [CrossRef] [PubMed]
49. Jimenez-Herrera, M.F.; Font-Jimenez, I.; Bazo-Hernández, L.; Roldán-Merino, J.; Biurrun-Garrido, A.; Hurtado Pardos, B. Moral sensitivity of nursing students. Adaptation and validation of the moral sensitivity questionnaire in Spain. *PLoS ONE* **2022**, *17*, e0270049. [CrossRef]
50. Moreira, M.A.; Lustosa, A.M.; Dutra, F.; Barros Ede, O.; Batista, J.B.; Duarte, M.C. Public humanization policies: Integrative literature review. *Cien Saude Colet.* **2015**, *20*, 3231–3242. [CrossRef] [PubMed]
51. Muñoz-Angel, Y.M. Patrón de conocimiento personal identificado en narrativas de profesores de enfermería [Pattern of personal knowledge identified in nursing faculty narratives]. *Rev. Cuid.* **2019**, *10*, 1–19. [CrossRef]
52. Veras, R.M.; Passos, V.B.C.; Feitosa, C.C.M.; Fernandes, S.C.S. Different training models in health and student conceptions of humanized medical care. *Cien Saude Colet.* **2022**, *27*, 1781–1792, (In Portuguese and English). [CrossRef]
53. Feijoo-Cid, M.; Moriña, D.; Gómez-Ibáñez, R.; Leyva-Moral, J.M. Expert patient illness narratives as a teaching methodology: A mixed method study of student nurses satisfaction. *Nurse Educ. Today* **2017**, *50*, 1–7. [CrossRef]
54. Jiménez-Rodríguez, D.; Pérez-Heredia, M.; Molero Jurado, M.D.M.; Pérez-Fuentes, M.D.C.; Arrogante, O. Improving Humanization Skills through Simulation-Based Computers Using Simulated Nursing Video Consultations. *Healthcare* **2021**, *10*, 37. [CrossRef]

55. Létourneau, D.; Goudreau, J.; Cara, C. Nursing Students and Nurses' Recommendations Aiming at Improving the Development of the Humanistic Caring Competency. *Can. J. Nurs. Res.* **2022**, *54*, 292–303. [CrossRef]
56. Mega, M.N.; Bueno, B.C.; Menegaço, E.C.; Guilhen, M.P.; Pio, D.A.M.; Vernasque, J.R.D.S. Students' experience with literature in medical education. *Rev. Bras. Educ. Méd.* **2021**, *45*, e059. [CrossRef]
57. Moya, C.; Pulgar, C.; Trajtmann, A. Grupos de encuentro: Desarrollo de habilidades para la humanización del cuidado, a partir de la experiencia de aprendizaje de estudiantes de enfermería [Encounter groups: Development of skills for the humanization of care, based on the learning experience of nursing students]. *Foro Educ.* **2019**, *33*, 53–77.
58. Sierras-Davó, M.C.; Lillo-Crespo, M.; Verdú Rodríguez, P. Qualitative evaluation of an educational intervention on health improvement in nursing students. *Aquichan* **2021**, *21*, e2112. [CrossRef]
59. Carr, S.E.; Noya, F.; Phillips, B.; Harris, A.; Scott, K.; Hooker, C.; Mavaddat, N.; Ani-Amponsah, M.; Vuillermin, D.M.; Reid, S.; et al. Health Humanities curriculum and evaluation in health professions education: A scoping review. *BMC Med Educ.* **2021**, *21*, 568. [CrossRef]
60. Hoang, B.L.; Monrouxe, L.V.; Chen, K.S.; Chang, S.C.; Chiavaroli, N.; Mauludina, Y.S.; Huang, C.D. Medical Humanities Education and Its Influence on Students' Outcomes in Taiwan: A Systematic Review. *Front. Med.* **2022**, *9*, 857488. [CrossRef]
61. Lovato, E.; Minniti, D.; Giacometti, M.; Sacco, R.; Piolatto, A.; Barberis, B.; Papalia, R.; Bert, F.; Siliquini, R. Humanisation in the emergency department of an Italian hospital: New features and patient satisfaction. *Emerg. Med. J.* **2013**, *30*, 487–491. [CrossRef]
62. Langdon, L.O.; Toskes, P.P.; Kimball, H.R.M. The American Board of Internal Medicine Task Force on subspecialty internal medicine. Future roles and training of internal medicine subspecialists. *Ann. Intern. Med.* **1996**, *124*, 686–691. [CrossRef]
63. Quadrelli, A. La enseñanza de Humanidades en Medicina: Reflexiones a partir de una mirada antropológica [Teaching Humanities in Medicine: Reflections from an anthropological perspective]. *Páginas Educ.* **2013**, *6*, 127–137.
64. Castiel, L.D.; Álvarez-Dardet, C. *La Salud Persecutoria: Los Límites de la Responsabilidad [Persecutory Health: Limitations of Liability]*; Lugar Editorial: Buenos Aires, Argentina, 2010.
65. Van der Geest, S. Overcoming ethnocentrism: How social science and medicine relate and should relate to one another. *Soc. Sci. Med.* **1995**, *40*, 869–872. [CrossRef] [PubMed]
66. Shapiro, J.; Coulehan, J.; Wear, D.; Montello, M. Medical humanities and their discontents: Definitions, critiques, and implications. *Acad. Med.* **2009**, *84*, 192–198. [CrossRef] [PubMed]
67. Charon, R. Narrative medicine. A model for empathy, reflection, profession, and trust. *JAMA* **2001**, *286*, 1897–1902. [CrossRef]
68. Martín-Asuero, A.; Rodríguez Blanco, T.; Pujol-Ribera, E.; Berenguera, A.; Moix Queraltó, J. Evaluación de la efectividad de un programa de mindfulness en profesionales de atención primaria [Evaluation of the effectiveness of a mindfulness program in primary care professionals]. *Gac. Sanit.* **2013**, *27*, 521–528. [CrossRef] [PubMed]
69. Oro, P.; Esquerda, M.; Viñas-Salas, J.; Soler-González, J.; Pifarré, J. Mindfulness en estudiantes de medicina [Mindfulness in medical students]. *FEM Rev. Fund. Educ. Médica* **2015**, *18*, 305–312.
70. Knight, A. From nurse to service user: A personal cancer narrative. *Brit. J. Nurs.* **2018**, *27*, S18–S21. [CrossRef] [PubMed]
71. Todres, L.; Galvin, K.T.; Holloway, I. The humanization of healthcare: A value framework for qualitative research. *Int. J. Qual. Stud. Health Well-Being* **2009**, *4*, 68–77. [CrossRef]
72. Koopman, W.J.; LaDonna, K.A. Is person-centred medical education an aim or an empty promise? *Med. Educ.* **2022**, *56*, 472–474. [CrossRef] [PubMed]
73. Nezu, A.M.; Nezu, C.; D'Zurilla, T. Terapia de solución de problemas [Problem-solving therapy]. *Psicooncología* **2013**, *10*, 217–231.

Disclaimer/Publisher's Note: The statements, opinions and data contained in all publications are solely those of the individual author(s) and contributor(s) and not of MDPI and/or the editor(s). MDPI and/or the editor(s) disclaim responsibility for any injury to people or property resulting from any ideas, methods, instructions or products referred to in the content.

Article

Optimizing EMG Classification through Metaheuristic Algorithms

Marcos Aviles [1,*,†], Juvenal Rodríguez-Reséndiz [1,*,†] and Danjela Ibrahimi [2,3,†]

1. Facultad de Ingeniería, Universidad Autónoma de Querétaro, Querétaro 76010, Mexico
2. Facultad de Medicina, Universidad Autónoma de Querétaro, Querétaro 76176, Mexico; danjela.ibrahimi@uaq.mx
3. Brain Vision & Learning Center, Querétaro 76230, Mexico
* Correspondence: marcosaviles@ieee.org (M.A.); juvenal@uaq.edu.mx (J.R.-R.)
† These authors contributed equally to this work.

Abstract: This work proposes a metaheuristic-based approach to hyperparameter selection in a multilayer perceptron to classify EMG signals. The main goal of the study is to improve the performance of the model by optimizing four important hyperparameters: the number of neurons, the learning rate, the epochs, and the training batches. The approach proposed in this work shows that hyperparameter optimization using particle swarm optimization and the gray wolf optimizer significantly improves the performance of a multilayer perceptron in classifying EMG motion signals. The final model achieves an average classification rate of 93% for the validation phase. The results obtained are promising and suggest that the proposed approach may be helpful for the optimization of deep learning models in other signal processing applications.

Keywords: PSO; GWO; metaheuristic; multilayer perceptron; hyperparameters; EMG signals; optimization; deep learning

Citation: Aviles, M.; Rodríguez-Reséndiz, J.; Ibrahimi, D. Optimizing EMG Classification through Metaheuristic Algorithms. *Technologies* 2023, 11, 87. https://doi.org/10.3390/technologies11040087

Academic Editor: Mario Munoz-Organero

Received: 30 May 2023
Revised: 24 June 2023
Accepted: 28 June 2023
Published: 2 July 2023

Copyright: © 2023 by the authors. Licensee MDPI, Basel, Switzerland. This article is an open access article distributed under the terms and conditions of the Creative Commons Attribution (CC BY) license (https://creativecommons.org/licenses/by/4.0/).

1. Introduction

The classification of electromyographic (EMG) signals corresponding to movement is a fundamental task in biomedical engineering and has been widely studied in recent years. EMG signals are electrical records of muscle activity that contain valuable information about muscle contraction and relaxation patterns. The accurate classification of these signals is essential for various applications, such as EMG-controlled prosthetics, rehabilitation, and the monitoring of muscle activity [1].

One recently used method to classify EMG signals is the multilayer perceptron (MLP). This artificial neural network architecture has proven effective in signal processing and pattern classification. An MLP consists of several layers of interconnected neurons, each activated by a non-linear function. These layers include an input layer, one or more hidden layers, and an output layer. Although MLPs are suitable for the classification of EMG signals, their performance is strongly affected by the choice of hyperparameters. Hyperparameters are configurable values that are not learned directly from the dataset but do define the behavior and performance of the model. Some examples of hyperparameters in the MLP context are as follows [2–4]:

- Number of neurons in hidden layers: This hyperparameter determines the generalization power of the model. Too few neurons leads to underfitting, while too many leads to overfitting.
- Learning rate: This factor determines how much the network weights are adjusted during the learning process. A high learning rate prevents the model from converging, while a low learning rate slows the training process.
- Training periods: This indicates the number of times that the network weights were updated during training using the complete dataset. An insufficient number of epochs leads to the undertraining of the model, while too many epochs leads to overtraining.

- Training batch size: The number of training samples to use each time that the weights are updated. The batch size affects the stability of the training process and the speed of convergence of the model.

Traditionally, hyperparameter selection has involved a trial-and-error process of exploring different combinations of values to determine the best performance. However, this approach is time-consuming and computationally intensive, especially with a large search space. Automated hyperparameter search methods have been developed to address this problem [5]. In this context, it is proposed to use the particle swarm optimization (PSO) and gray wolf optimization (GWO) algorithms to select the hyperparameters of the MLP model automatically. These metaheuristic optimization algorithms effectively find the optimal solution in a given search space.

PSO and GWO work similarly, generating an initial set of possible solutions and iteratively updating them based on their performance. Each solution is a combination of MLP hyperparameters. The objective of these algorithms is to find the combination of hyperparameters that maximizes the performance of the MLP model in the classification of EMG signals [6].

The performed experiments show that hyperparameter optimization significantly improves the performance of MLP models in classifying EMG signals. The optimized MLP model achieved a classification accuracy of 93% in the validation phase, which is promising. The main motivations of this work are the following.

- Comparison of algorithms: The main objective of this study is to compare and analyze the selection of hyperparameters using metaheuristic algorithms. The PSO algorithm, one of the most popular, was implemented and compared with the GWO algorithm, which is relatively new. This comparison allows us to evaluate both algorithms' performance and efficiency in selecting hyperparameters in the context of the classification of EMG signals.
- Exploration of new possibilities: Although the PSO and GWO optimization algorithms have been widely used for feature selection in EMG signals, their application to optimize classifiers has yet to be fully explored. This study seeks to address this gap and examine the effectiveness of metaheuristic algorithms in improving rankings.

The current work is structured as follows. Section 2 provides a comprehensive literature review, offering insights into the proposed work. In Section 3, the methods and definitions essential for the development of the project are outlined. Section 4 presents the sequential steps to be followed in order to implement the proposed algorithm. The results and discoveries obtained are presented in Section 5. Section 6 presents the interpretation of the results from the perspective of previous studies and working hypotheses. Lastly, the areas covered by the scope of this work are presented in Section 7.

2. Related Works

In signal processing, particularly electromyography, various approaches have been proposed to enhance the accuracy of pattern recognition models. In 2018, Purushothaman et al. [7] introduced an efficient pattern recognition scheme for the control of prosthetic hands using EMG signals. The study utilized eight EMG channels from eight able-bodied subjects to classify 15 finger movements, aiming for optimal performance with minimal features. The EMG signals were preprocessed using a dual-tree complex wavelet transform. Subsequently, several time-domain features were extracted, including zero crossing, slope sign change, mean absolute value, and waveform length. These features were chosen to capture relevant information from the EMG signals.

The results demonstrated that the naive Bayes classifier and ant colony optimization achieved average precision of 88.89% in recognizing the 15 different finger movements using only 16 characteristics. This outcome highlights the effectiveness of the proposed approach in accurately classifying and controlling prosthetic hands based on EMG signals.

On the other hand, in 2019, Too et al. [8] proposed the use of Pbest-guide binary particle swarm optimization to select relevant features from EMG signals decomposed by a discrete wavelet transform, managing to reduce the features by more than 90% while maintaining average classification accuracy of 88%. Moreover, Sui et al. [9] proposed the use of the wavelet package to decompose the EMG signal and extract the energy and variance of the coefficients as feature vectors. They combined PSO with an enhanced support vector machine (SVM) to build a new model, achieving an average recognition rate of 90.66% and reducing the training time by 0.042 s.

In 2020, Kan et al. [10] proposed an EMG pattern recognition method based on a recurrent neural network optimized by the PSO algorithm, obtaining classification accuracy of 95.7%.

One year later, in 2021, Bittibssi et al. [11] implemented a recurrent neural network model based on long short-term memory, Convolution Peephole LSTM, and a gated recurrent unit to predict movements from sEMG signals. Various techniques were evaluated and applied to six reference datasets, obtaining prediction accuracy of almost 99.6%. In the same year, Li et al. [12] developed a scheme to classify 11 movements using three feature selection methods and four classification methods. They found that the TrAdaBoost-based incremental SVM method achieved the highest classification accuracy. The PSO method achieved classification accuracy of 93%.

Moreover, Cao et al. [13] proposed an sEMG gesture recognition model that combines feature extraction, genetic algorithm, and a support vector machine model with a new adaptive mutation particle swarm optimization algorithm to optimize the SVM parameters, achieving a recognition rate of 97.5%.

In 2022, Aviles et al. [14] proposed a methodology to classify upper and lower extremity electromyography (EMG) signals using feature selection GA. Their approach yielded average classification efficiency exceeding 91% using an SVM model. The study aimed to identify the most informative features for accurate classification by employing GA in feature selection.

Subsequently, Dhindsa et al. [15] utilized a feature selection technique based on binary particle swarm optimization to predict knee angle classes from surface EMG signals. The EMG signals were segmented, and twenty features were extracted from each muscle. These features were input into a support vector machine classifier for the classification task. The classification accuracy was evaluated using a reduced feature set comprising only 30% of the total features, to reduce the computational complexity and enhance efficiency. Remarkably, this reduced feature set achieved accuracy of 90.92%, demonstrating the effectiveness of the feature selection technique in optimizing the classification performance.

Finally, in 2022, Li et al. [16] proposed a lower extremity movement pattern recognition algorithm based on the Improved Whale Algorithm Optimized SVM model. They used surface EMG signals as input to the movement pattern recognition system, and movement pattern recognition was performed by combining the IWOA-SVM model. The results showed that the recognition accuracy was 94.12%.

3. Materials and Methods

This section shows the essential concepts applied in this work.

3.1. EMG Signals

An EMG signal is a bioelectric signal produced by muscle activity. When a muscle contracts, the muscle fibers are activated, generating an electrical current measured with surface electrodes. The recorded EMG signal contains information about muscle activity, such as force, movement, and fatigue. The EMG signal has a low amplitude, typically ranging from 0.1 mV to 10 mV. It is important to pre-process the signal to remove noise and amplify it before performing any analysis. Furthermore, the location of the electrodes on the muscle surface is crucial to obtain accurate and consistent EMG signals [17,18].

In the context of movement classification using EMG signals, movements made by a subject are recorded by surface electrodes placed on the skin over the muscles involved. The resulting EMG signals are processed to extract relevant features and train a classification model. Artifacts, such as unintentional electrode movements or electromagnetic interference, affect the quality of the EMG signals and reduce the accuracy of the classification model. Therefore, steps must be taken to ensure that the EMG signals are as clean and accurate as possible [17,19].

3.2. Multilayer Perceptron

The MLP is an artificial neural network for supervised learning tasks such as classification and regression. It is a feedforward network composed of several layers of interconnected neurons. Each neuron receives weighted inputs and applies a nonlinear activation function to produce an output. The backpropagation algorithm is commonly used to adjust the weights of the connections between neurons. This iterative process minimizes the error between the output of the network and the expected output based on a given training dataset [4,20].

The MLP consists of an input layer, a hidden layer, and an output layer. The input layer receives input features and forwards them to the hidden layer, and the hidden layer processes the features and passes them to the output layer. The output layer produces the final output, a classification result. The specific architecture of the MLP, including the number of neurons in each layer and the number of hidden layers, depends on the task and the input data [4,20]. Below, in the pseudocode in Algorithm 1, the MLP algorithm is presented.

Note that the following pseudocode assumes that the weight matrices and bias vectors have already been initialized and altered by a suitable algorithm and that the activation function σ has been chosen. The algorithm then takes an input vector x and passes it through the MLP to produce an output vector y. The intermediate variables a_l and h_l are the input and output of each hidden layer, respectively. The activation function σ is usually a non-linear function that allows the MLP to learn complex mappings between inputs and outputs.

Algorithm 1 Multilayer Perceptron

1: **Input:** Input vector x, weight matrices $W_{i,j}$ and bias vectors b_i, number of hidden layers L, activation function σ
2: **Output:** Output vector y
3: **for** $l = 1$ to L **do**
4:    **if** $l = 1$ **then**
5:       $a_l = W_{l-1,l} x + b_l$
6:    **else**
7:       $a_l = W_{l-1,l} \sigma(a_{l-1}) + b_l$
8:    $h_l = \sigma(a_l)$
9: $y = h_L$

3.3. Particle Swarm Optimization and Gray Wolf Optimizer

The PSO algorithm is an optimization method inspired by observing the collective behavior of a swarm of particles. Each particle represents a solution in the search space and moves based on its own experience and the experience of the swarm in general. The goal is to find the best possible solution to an optimization problem [21,22].

The PSO algorithm has proven effective in optimizing complex problems in various areas, including machine learning. This work uses PSO to optimize the hyperparameters of a multilayer perceptron in the classification of EMG signals. The pseudocode in Algorithm 2 shows the PSO algorithm [21].

Algorithm 2 Particle Swarm Optimization

1: **Input:** Number of particles N, maximum number of iterations T_{max}, parameters ω, ϕ_p, ϕ_g, initial positions x_i and velocities v_i
2: **Output:** Global best position p_{best} and its corresponding fitness value f_{best}
3: Initialize positions and velocities of particles: $x_i \leftarrow$ random, $v_i \leftarrow 0$
4: **for** $t = 1$ to T_{max} **do**
5:     **for** each particle $i = 1, \ldots, N$ **do**
6:         Evaluate fitness of current position: $f_i \leftarrow$ fitness function(x_i)
7:         **if** $f_i < f_{pbest_i}$ **then**
8:             Update personal best position: $p_{best_i} \leftarrow x_i$, $f_{pbest_i} \leftarrow f_i$
9:     Find global best position: $p_{best} \leftarrow \mathrm{argmin}_{p_{best_j}} f_{pbest_j}$
10:     **for** each particle $i = 1, \ldots, N$ **do**
11:         Update velocity: $v_i \leftarrow \omega v_i + \phi_p r_p (p_{best_i} - x_i) + \phi_g r_g (p_{best} - x_i)$
12:         Update position: $x_i \leftarrow x_i + v_i$
13: **Return:** p_{best} and f_{best}

In the algorithm, a set of parameters that regulate the speed and direction of movement of each particle is used. These parameters are the inertial weight ω, the cognitive learning coefficient ϕ_p, and the social learning coefficient ϕ_g. The current positions and velocities of the particles are also used, as well as the personal and global best positions found by the entire swarm [22].

On the other hand, the gray wolf optimizer is an algorithm inspired by the social behavior of gray wolves. This algorithm is based on the social hierarchy and the collaboration between wolves in a pack to find optimal solutions to complex problems. The algorithm starts with an initial population of wolves (candidate solutions) and uses an iterative process to improve these solutions. The positions of wolves are updated during each iteration based on their results, simulating a hunt and pack search. As the algorithm progresses, the wolves adjust their positions based on the quality of their solutions and feedback from the pack leaders. Lead wolves represent the best solutions found so far, and their influence ripples through the pack, helping to converge toward more promising solutions. The GWO has proven to be effective in optimizing complex problems in various areas, such as mathematical function optimization, pattern classification, parameter optimization, and engineering. The pseudocode in Algorithm 3 shows the GWO algorithm [6].

Algorithm 3 Gray Wolf Optimizer

1: Initialize the wolf population (initial solutions)
2: Initialize the position vector of the group leader ($\mathbf{X}^*$)
3: Initialize the position vector of the previous group leader ($\mathbf{X}^{**}$)
4: Initialize the iteration counter (t)
5: Define the maximum number of iterations (T_{max})
6: **while** $t < T_{max}$ **do**
7:     **for** each wolf in the population **do**
8:         Update the fitness value of the wolf
9:     Sort the wolves based on their fitness values (from lowest to highest)
10:     **for** each wolf in the population **do**
11:         **for** each dimension of the position vector **do**
12:             Generate random values (r_1, r_2)
13:             Calculate the update coefficient (A)
14:             Calculate the scale factor (C)
15:             Update the position of the wolfs
16:     Increment the iteration counter (t)
17: Obtain the wolf with the best fitness value ($\mathbf{X}^*$)

3.4. Hyperparameters

A hyperparameter is a parameter that is not learned from the data but is set before training the model. Hyperparameters dictate how the neural network learns and how the model is optimized. Ensuring the appropriate selection of hyperparameters is crucial in achieving the optimal performance of the model (Nematzadeh, 2022) [23].

When working with MLPs, several critical hyperparameters significantly impact the performance of the model. These include the number of hidden layers, the number of neurons within each layer, the chosen activation function, the learning rate, and the number of training epochs. The numbers of hidden layers and neurons per layer play a crucial role in the capacity of the network to capture intricate functions. Increasing these aspects enables the network to learn complex relationships within the data. However, it may also result in overfitting issues [3,24].

The activation function determines the nonlinearity of the network and, therefore, its ability to represent nonlinear functions. The most common activation function is the sigmoid function, but others, such as the ReLU function and the hyperbolic tangent function, are also frequently used [25].

The learning rate determines how much the network weights are adjusted in each training iteration. If the learning rate is too high, the network starts to oscillate and not converge, while a low learning rate causes the network to converge slowly and become stuck in local minima. The number of training epochs determines how often the entire dataset is processed during training. Too many epochs leads to overfitting, while too few epochs leads to the suboptimality of the model. In this work, the PSO and GWO algorithms are used to find the best values of the hyperparameters of the MLP network [3,25].

3.5. Sensitivity Analysis

In order to verify the impact that each of the characteristics selected by genetic algorithm (GA) has on the classification of the EMG signal, a sensitivity analysis is performed. This technique consists of removing one of the predictors during the classification process and recording the accuracy percentage. This is to observe how the output of the model is altered. If the classification percentage decreases, it indicates that the removed feature significantly impacts the prediction [14]. This procedure is performed once the features have been selected, to assess the importance of the chosen predictors through GA.

The procedure of calculating the sensitivity is as follows. Having a dataset X_1, the sensitivity of the predictor i is obtained from a new set X_2, where the i th-predictor has been eliminated. The characteristics that make up X_1 are used as a second step, resulting in the precision Y_1. The third step is to use the new feature set X_2 and obtain Y_2. Finally, the sensitivity for the i-th predictor is $Y_2 - Y_1$. A tool used to better visualize the sensitivity is the percentage change, which is calculated as

$$Percentage\ change = \frac{Y_2 - Y_1}{Y_1} \times 100 \quad (1)$$

4. Methodology

This section explains how the study was carried out, the procedures used, and how the results were analyzed.

4.1. EMG Data

The dataset used in this study was obtained from [14] and comprised muscle signals recorded from nine individuals aged between 23 and 27. The dataset included five men and four women without musculoskeletal or nervous system disorders, obesity problems, or amputations. The dataset captured muscle signals during five distinct arm and hand movements: arm flexion at the elbow joint, arm extension at the elbow joint, finger flexion, finger extension, and resting state. The acquisition utilized four bipolar channels and a reference electrode positioned on the dorsal region of the wrist of each participant. During the experimental procedure, the participants were instructed to perform each movement for 6 s, preceded by an initial relaxation period of 2 s. Each action was repeated 20 times to ensure adequate data for analysis. The data were sampled at a frequency of 1.5 kHz, allowing for detailed recordings of the muscle signals during the movements.

The database was divided into two sets. The first one (90%) was used to select the characteristics for the classification and hyperparameters. This first set was subdivided into the training and validation sets, which were used to calculate the objective functions of the metaheuristic algorithms. On the other hand, the second set (10%) was used for the final validation of the classifier. This second set was not presented to the network until the final validation stage, to check the level of generalization of the algorithm.

4.2. Signal Processing

This section explains the filtering process applied to the EMG signals before extracting the features needed for classification. Digital filtering was done using a fourth-order Butterworth filter with a passband ranging from 10 Hz to 500 Hz. This filtering aimed to remove unwanted noise and highlight relevant signals.

It is important to note that the database was subjected to analog filtering from 10 Hz to 500 Hz using a combination of a low-pass filter and a high-pass filter in series. These controllers used the second-order Sallen–Key topology. In addition, a second-order Bainter–Notch band-stop filter was produced to remove the 60 Hz interference generated by the power supply.

4.3. Feature Extraction

The characterization of EMG signals is required for their classification since individual signal values have no practical relevance for classification. Therefore, a feature extraction step is needed to find useful information before extracting the features of the signal. The features are based on the statistical method and are calculated in the time domain. Temporal features are widely used to classify EMG signals due to their low complexity and high computational speed. Moreover, they are calculated directly from the EMG time series. Table 1 illustrates the characteristics used [14,26].

Table 1. Most common time-domain indicators in the classification of EMG signals.

N°	Feature Extracted	Abbr.	N°	Feature Extracted	Abbr.
1	Average amplitude change	AAC	14	Variance	VAR
2	Average amplitude value	AAV	15	Wavelength	WL
3	Difference absolute standard deviation	DASDV	16	Zero crossings	ZC
4	Katz fractals	FC	17	Log detector	LOG
5	Entropy	SE	18	Mean absolute value	MAV
6	Kurtosis	K	19	Mean absolute value slope	MAVSLP
7	Skewness	SK	20	Modified mean absolute value type 1	MMAV1
8	Mean absolute deviation	MAD	21	Modified mean value type 2	MMAV2
9	Willson amplitude	WAMP	22	RMS value	RMS
10	Absolute value of the third moment	Y3	23	Slope changes	SSC
11	Absolute value of fourth moment	Y4	24	Simple square integral	SSI
12	Absolute value of the fifth moment	Y5	25	Standard deviation	STD
13	Myopulse percentage rate	MYOP	26	Integrated EMG	IEMG

Within the context of EMG signals, the features shown in Table 1 represent different quantitative aspects generated by muscle activity. The definition or conceptualization of each of these characteristics is presented below [17].

1. Average amplitude change: The average amplitude change in the EMG signal over a given time interval. It represents the average variation in the signal amplitude during this period.

$$AAC = \frac{1}{N}\sum_{k=1}^{N} |x_{k+1} - x_k| \qquad (2)$$

where x_k is the k-th voltage value that makes up the signal and N is the number of elements that constitute it.

2. Average amplitude value: This is the average of the amplitude values of the EMG signal. It indicates the average amplitude level of the signal during a specific time interval.

$$AAV = \frac{1}{N}\sum_{k=1}^{N} x_k \qquad (3)$$

3. Difference absolute standard deviation: This is the absolute difference between the standard deviations of two adjacent segments of the EMG signal. It measures extraction and abrupt changes in signal amplitude.

$$DASDV = \sqrt{\frac{1}{N-1}\sum_{k=1}^{N-1} (x_{k+1} - x_k)^2} \qquad (4)$$

4. Katz fractals: This refers to the fractal dimension of the EMG signal. It represents the self-similarity and structural complexity of the signal at different scales.

$$FD = \frac{\log_{10}(N)}{-\log_{10}(\frac{m}{L}) + \log_{10}(N)} \qquad (5)$$

where L is the total length of the curve or the sum of the Euclidean distances between successive points, m is the diameter of the curve, and N is the number of steps in the curve.

5. Entropy: This measures the randomness and complexity of the EMG signal. The higher the entropy, the greater the harvest and unpredictability of the signal.

$$SE(X) = -\sum_{k=1}^{n} P(x_k) \log_2 P(x_k) \qquad (6)$$

where $SE(X)$ is the entropy of the random variable X, $P(x_i)$ is the probability that X takes the value x_i, and n is the total number of possible values that X can take.

6. Kurtosis: This measures the shape of the amplitude distribution of the EMG signal. It indicates the number and concentration of extreme values relative to the mean.

$$K = \sum_{k=1}^{N} \frac{(x_k - \bar{x})^4}{Ns^4} \qquad (7)$$

where N is the size of the dataset, x_k is the k-th value of the signal, $\bar{x}$ is the mean of the data, and s is the standard deviation of the dataset.

7. Skewness: This is a measure of the asymmetry of the amplitude distribution of the EMG signal. It describes whether the distribution is skewed to the left or the right relative to the mean.

$$SK = \sum_{k=1}^{N} \frac{(x_k - \bar{x})^3}{Ns^3} \qquad (8)$$

8. Mean absolute deviation: This is the average of the absolute deviations of the amplitude values of the EMG signal concerning its mean. It indicates the mean spread of the data around the mean.

$$MAD = \frac{1}{N} \sum_{k=1}^{N} |x_k - \bar{x}| \quad (9)$$

9. Wilson amplitude: This measures the amplitude of the EMG signal to a specific threshold. It represents the muscle force or electrical activity generated by the muscle.

$$WAMP = \frac{1}{N} \sum_{k=1}^{N-1} f(|x_{k+1} - x_k|) \quad (10)$$

$$f(x) = \begin{cases} 1 & \text{if } x > L, \\ 0 & \text{otherwise} \end{cases} \quad (11)$$

In this study, a threshold L of 0.05 V is considered.

10. The absolute value of the third moment: This is the absolute value of the third statistical moment of the EMG signal. It is a proportion of information about the symmetry and shape of the amplitude distribution.

$$Y3 = \left| \frac{1}{N} \sum_{k=1}^{N} x_k^3 \right| \quad (12)$$

11. The absolute value of the fourth moment: This is the absolute value of the fourth statistical moment of the EMG signal. It describes the concentration and shape of the amplitude distribution.

$$Y4 = \left| \frac{1}{N} \sum_{k=1}^{N} x_k^4 \right| \quad (13)$$

12. The absolute value of the fifth moment: This is the absolute value of the fifth statistical moment of the EMG signal. It provides additional information about the shape and amplitude distribution of the signal.

$$Y5 = \left| \frac{1}{N} \sum_{k=1}^{N} x_k^5 \right| \quad (14)$$

13. Myopulse percentage rate: This is the average of a series of myopulse outputs, and the myopulse output is 1 if the myoelectric signal is greater than a pre-defined threshold.

$$MYOP = \frac{1}{N} \sum_{k=1}^{N} \phi(x_k) \quad (15)$$

where $\phi(x_k)$ is defined as

$$\phi(x) = \begin{cases} 1 & \text{if } x > L, \\ 0 & \text{otherwise} \end{cases} \quad (16)$$

In this work, L is defined as 0.016.

14. Variance: This measures the dispersion of the amplitude values of the EMG signal to its mean. It indicates the lack of signal around its average value.

$$VAR = \frac{1}{N-1} \sum_{k=1}^{N} x_k^2 \qquad (17)$$

15. Wavelength: This is the average distance between two consecutive zero crossings in the EMG signal. It is the information ratio regarding the frequency and period of the signal.

$$WL = \sum_{k=1}^{n} |x_k - x_{k-1}| \qquad (18)$$

16. Zero crossings: This refers to the number of times that the EMG signal crosses the zero value in each time interval. It indicates polarity changes and signal transitions.

$$ZC = \sum_{k=1}^{n-1} f(x) \qquad (19)$$

where

$$f(x) = \begin{cases} 1 & \text{if } x_k x_{k+1} < 0 \text{ and } |x_k - x_{k+1}| \geq L, \\ 0 & \text{otherwise} \end{cases} \qquad (20)$$

17. Log detector: An envelope detector is used to measure the amplitude of the EMG signal on a logarithmic scale. It helps to bring out the most subtle variations in the signal.

$$LOG = exp\left(\frac{1}{N} \sum_{k=1}^{N} log(|x_k|)\right) \qquad (21)$$

18. Mean absolute value: This is the average of the absolute values of the EMG signal. It represents the average amplitude level of the signal regardless of polarity.

$$MAV = \frac{\sum_{K=1}^{n} |x_K|}{N} \qquad (22)$$

19. Mean absolute value slope: The average slope of the EMG signal is calculated using the absolute values of the amplitude changes in a specific time interval. It indicates the average rate of change in the signal.

$$MAVSLP_k = MAV_{k+1} - MAV_k \qquad (23)$$

20. Modified mean absolute value type 1: This is a modified version of the average of the absolute values of the EMG signal. It is used to reduce the effect of higher-frequency components.

$$MMAV1 = \frac{1}{N} \sum_{k=1}^{N} w_k |x_k| \qquad (24)$$

where w_k is defined as

$$w_k = \begin{cases} 1 & 0.25N \leq k \leq 0.75N, \\ 0 & \text{otherwise} \end{cases} \qquad (25)$$

21. Modified mean value type 2: This is a modified version of the average of the amplitude values of the EMG signal. It is used to reduce the effect of higher-frequency components.

$$MMAV2 = \frac{1}{N} \sum_{k=1}^{N} w_k |x_k| \qquad (26)$$

where w_k is defined as

$$w_k = \begin{cases} 1 & 0.25N \leq k \leq 0.75N, \\ \frac{4k}{N} & k < 0.25N, \\ \frac{4(N-k)}{N} & otherwise \end{cases} \qquad (27)$$

22. Root mean square (RMS): This is the square root of the average of the squared values of the EMG signal. It represents a measure of the effective amplitude of the signal.

$$RMS = \sqrt{\frac{1}{N} \sum_{k=1}^{N} x_k^2} \qquad (28)$$

23. Slope changes: This refers to the number of slope changes in the EMG signal. It indicates inflection points and changes in the direction of the signal.

$$SSC = \sum_{k=1}^{n} f(x), \qquad (29)$$

where

$$f(x) = \begin{cases} 1 & if \quad x_k < x_{i+1} \text{ and } x_k < x_{k-1}, \\ 1 & if \quad x_k > x_{i+1} \text{ and } x_k > x_{k-1}, \\ 0 & otherwise \end{cases} \qquad (30)$$

24. Simple square integral: This is the integral value of the squares of the EMG signal in a specific time interval. It provides a measure of the energy contained in the signal.

$$SSI = \sum_{k=1}^{N} x_k^2 \qquad (31)$$

25. Standard deviation: This measures the dispersion of the amplitude values of the EMG signal for its average. It indicates the variability of the signal around its mean value.

$$STD = \sqrt{\frac{1}{N} \sum_{k=1}^{N} (x_k - \bar{x})^2} \qquad (32)$$

26. Integrated EMG: This is the integral value of the absolute amplitude of the EMG signal in each time interval. It provides a measure of total muscle activity.

$$IEMG = \sum_{k=1}^{N} |x_k| \qquad (33)$$

After extracting the characteristics, a matrix of arrangements was created with the features. This matrix comprised rows corresponding to the 20 tests carried out by eight people and for the different movements (five movements of the right arm). In contrast, the columns corresponded to the 26 predictors multiplied by the four channels.

4.4. Feature Selection

Figure 1 shows the methodology for the selection of characteristics. GA was used to select features to minimize the classification error of the validation data for a specific set of features used as input to a multilayer perceptron. The model hyperparameters were selected manually. The same input data from 9 of the 10 participants that comprised the database were used for the feature and hyperparameter selection.

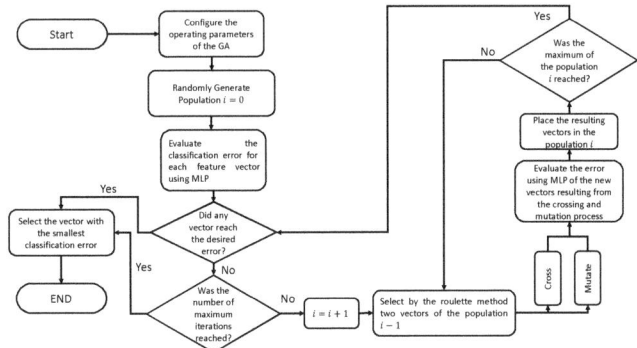

Figure 1. Methodology based on the proposal given by [14] for the selection of features by GA.

Table 2 shows the initial parameters used in GA for feature selection. These parameters include the initial population, the mutation rate, and the hyperparameters of the MLP, among others.

Table 2. Configuration used by GA for the selection of classification features.

Name	Configuration
Number of genes	104
Number of parents	100
Iteration number	25
Mutation percentage	2%
Selection operator	Roulette wheel
Crossover operator	Two-point
Mutation operator	Uniform mutation
Hidden layers	4
Number of hidden neurons per layer	150
Activation function of the hidden layers	Hyperbolic tangent
Activation function of the output layers	Sigmoid
Learning rate	0.0001
Epochs	10
Mini-batch size	20
Training data	60% of the data
Testing data	20% of the data
Validation data	20% of the data

4.5. Design and Integration of the Metaheuristic Algorithms and MLP

For the selection of the hyperparameters of the neural network, the PSO and GWO techniques were used. The cost criterion was the error of the validation stage. First, the completed data were divided into training, testing, and validation sets. The training set was used to train the neural network, the test set was used to fit the hyperparameters of the network, and the validation set was used to evaluate the final performance of the model.

Table 3 shows the initial parameters used in the PSO algorithm for the selection of the hyperparameters of the neural network. These parameters include the size of the particle population, the number of iterations, the range of values allowed for each hyperparameter (hidden neurons, epochs, mini-batch size, and learning rate), and the initial values for the coefficients of inertia, personal acceleration, and social acceleration. The Clerc and Kennedy method was used to calculate the coefficients in the PSO algorithm [27].

Table 3. Configuration of initial parameters used for the PSO algorithm, calculated using the Clerc and Kennedy method.

Name	Configuration
Coefficients of inertia	0.729
Personal accelerations	1.49
Global acceleration	1.49
Number of particles	12
Max iterations	35
Hidden neurons	[50 300]
Number of hidden layers	2
Epochs	[5 40]
Mini-batch size	[10 100]
Learning rate	[0.0001 0.01]
Activation function of the hidden layers	Hyperbolic tangent
Activation function of the output layers	Sigmoid
Training data	60% of the data
Testing data	20% of the data
Validation data	20% of the data

On the other hand, Table 4 shows the initial values for the hyperparameter selection process for GWO. Unlike PSO, only the initial number of individuals and the maximum number of iterations must be selected, in addition to the intervals for the MLP hyperparameters.

Table 4. Configuration of initial parameters used for the GWO algorithm.

Name	Configuration
Number of wolfs	25
Max iterations	35
Hidden neurons	[50 300]
Number of hidden layers	2
Epochs	[5 40]
Mini-batch size	[10 100]
Learning rate	[0.0001 0.01]
Activation function of the hidden layers	Hyperbolic tangent
Activation function of the output layers	Sigmoid
Training data	60% of the data
Testing data	20% of the data
Validation data	20% of the data

The different stages of the general methodology for the integration of the PSO and GWO algorithms with an MLP neural network for hyperparameter selection are shown in Figure 2.

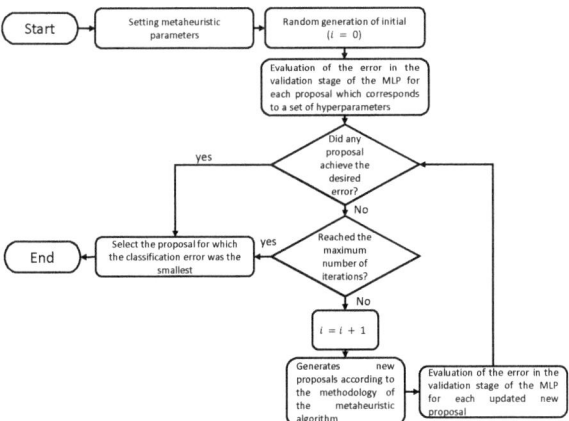

Figure 2. Proposed methodology for the selection of hyperparameters of MLP.

5. Results

This section presents and analyzes the results obtained from the multiple stages of the methodology.

5.1. Feature Selection

Table 5 shows the characteristics that GA selected from 104 predictors. In total, 55 features were selected and used as inputs in an MLP to classify the data and select the hyperparameters, representing a 47% reduction in features. A final classification percentage of 93% was achieved.

Table 5. Features selected as the best subset of characteristics for classification of signals.

Acronym	Channel
AAC	1 and 2
IEMG	All
MAV	1, 2 and 4
MAVSLP	1 and 4
MMAV1	All
VAR	1, 2 and 4
FC	1, 2 and 4
K	1,2 and 4
Y3	1
MYOP	1, 3 and 4
AAV	2 and 4
DASDV	2 and 4
LOG	2 and 3
MMAV2	2 and 3
SSC	2
SSI	2, 3 and 4
STD	2, 3 and 4
WL	2, 4
ZC	2, 3 and 4
MAD	2, 3 and 4
WAMP	2, 3 and 4
SE	3
SK	3 and 4
RMS	4
Y4	4
Y5	4

As shown in Figure 3, initially, the feature selection process had an error rate of 14%. GA improved the performance during the first iterations and reduced the errors to 11%. However, it stalled at a 10% error for eight iterations and an 8% error for 12 iterations. This deadlock occurred when existing candidate solutions had already explored most of the search space and new feature combinations that significantly improved the performance were not found. At this point, GA became stuck in a local minimum. This deadlock was overcome by implementing the mutate operation. In this case, it was possible that, during the 10% error plateau period, some mutation introduced in a later iteration led to the exploration of a new combination of features that improved the performance. This new solution could have been selected and propagated in the following generations, finally allowing it to reach a classification value of 93%.

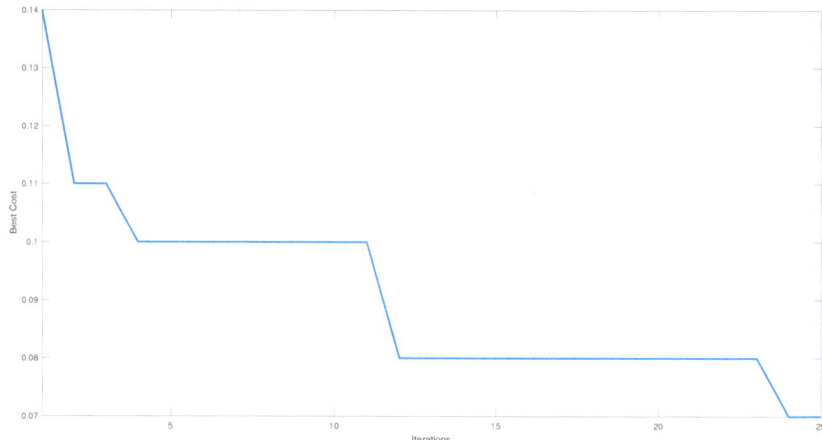

Figure 3. Reduction in the classification error due to the selection of features through GA.

In order to ensure that the feature selection process was carried out correctly and that only predictors that allowed high classification were selected, a sensitivity analysis was carried out. In Figure 4, the bar graph is shown, where the percentage decrease or increase in precision can be observed concerning the classification obtained at the end of the character selection stage, which was 93%.

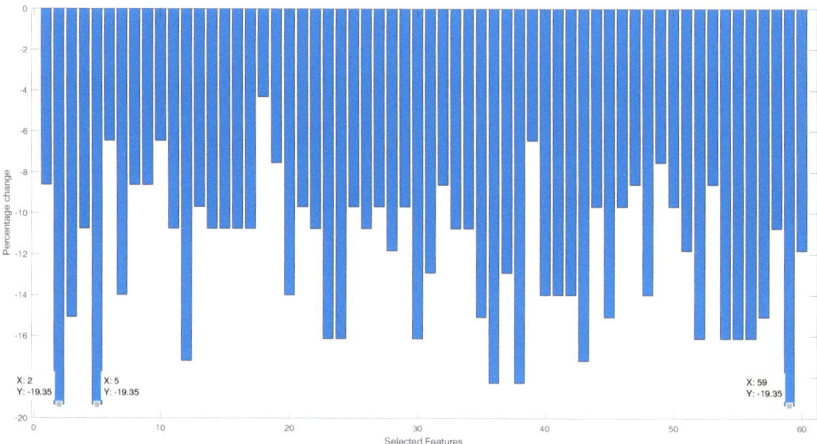

Figure 4. Sensitivity analysis of classification reduction percentages by predictor.

It is observed that feature number 18, which corresponds to the mean absolute value type 1 of channel 2, has the lowest percentage decrease in classification when eliminated. On the other hand, the characteristics with the most significant contributions are the absolute value of the fifth moment channel 4, integrated EMG channel 1, and modified mean value type 1 channel 1. When comparing the characteristics that present a more significant contribution against those of lesser contribution, it is seen that the type 1 modified mean value appears in both limits. The difference occurs in the channel from which the characteristic is extracted. Therefore, the exact predictor can have more or less importance in the classification depending on the muscle from which it is extracted.

5.2. Hyperparameter Selection

As shown in Figure 5, in the GWO implementation process, there is an error rate of 14% with the initial values proposed for the hyperparameters. This indicates that the initial solutions have yet to find the best set for the problem since, prior to the selection of the hyperparameters, there is a classification percentage of 93%, and it is found that the efficiency after the hyperparameter adjustment process is more significant than or equal to that of the previous phase.

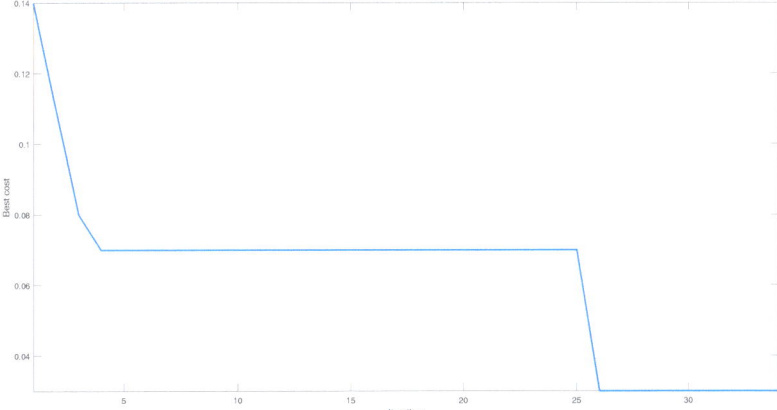

Figure 5. Reduction in the error due to the selection of hyperparameters by GWO.

In iteration 4, a reduction in error to 7% is observed. The proposed solutions have found a hyperparameter configuration that improves the model performance and reduces the error. During subsequent iterations, they continue to adjust their positions and explore the search space for better solutions. As observed during iterations 5 to 20, a deadlock is generated. However, later, it is observed that the error drops to 3%, which indicates that the GWO has managed to overcome this problem and find a solution that considerably improves the classification.

A possible reason that the GWO was able to exit the deadlock and reduce the error may be related to the intensification and diversification of the search. During the first few iterations, the GWO may have been in an intensification phase, focusing on exploiting promising regions of the search space based on the positions of the pack leaders. However, after a while, the GWO may have moved into a diversification phase, where the gray wolves explored new regions of the search space, allowing them to find a better solution and reduce the error to 3%.

Table 6 shows the values obtained for the MLP hyperparameters using GWO, achieving classification in the validation stage of 97%. When comparing the values implemented in the feature layer, it is noteworthy that the number of hidden layers was reduced from 4 to 2. On the other hand, the total number of neurons was reduced from 600 to 409. However, the epochs increased from 10 to 33 after hyperparameter selection. This indicates that the model required more opportunities to adjust the weights and improve its performance on the training dataset. Similarly, the mini-batch size is increased from 20 to 58, indicating that it needs more information during each training stage to adjust the weights.

Table 6. Hyperparameters selected as the best subset for classification of signals given by GWO.

Name	Value
Hidden neurons layer 1	204
Hidden neurons layer 2	205
Epochs	33
Mini-batch size	58
Learning rate	0.00223750

Finally, the learning rate increased from 0.0001 to 0.002237, which showed that the neural network learned faster during training. The results indicate that the selection of the hyperparameters improved the efficiency of the model by reducing its complexity, without compromising its classification ability.

Figure 6 shows the error reduction in selecting hyperparameters by PSO. The best initial proposal achieves a 13% error. After this, there is a stage where the error percentage is kept constant until iteration 6. From there, the error is reduced to 8%. Once this error is reached, it remains constant until iteration 27. Once iteration 28 begins, an error of 7% is achieved, representing only a 1% improvement. This 1% improvement is not a significant increase and could be attributed to slight variations in the MLP training weights.

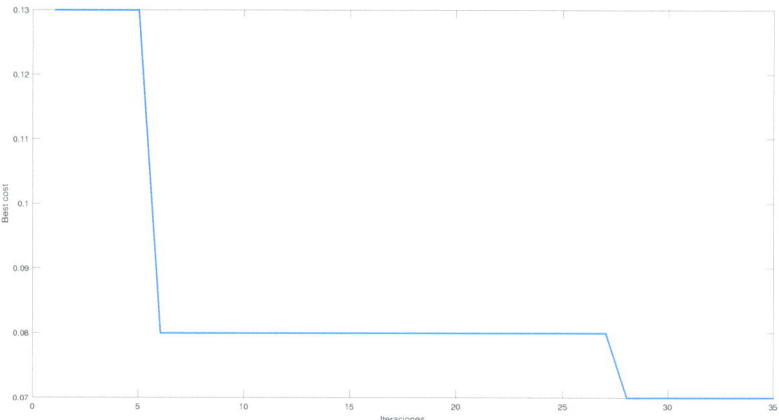

Figure 6. Reduction in the error due to the selection of hyperparameters by PSO.

On the other hand, Table 7 shows the calculated values of the MLP hyperparameters through PSO; the precision achieved is less than that achieved by GWO, being 93%. Despite this, a 50% reduction in hidden layers is also achieved, and it manages to maintain the precision percentage obtained in the feature selection stage with fewer neurons than achieved by GWO, being 359. However, similarly to the values obtained by GWO, the epochs increase to 38. Moreover, the mini-batch size is increased from 50. Finally, the learning rate increases from 0.0001 to 0.0010184. This smaller amount of information used for training, and the smaller learning steps and smaller number of neurons, justify the 4% decrease in classification.

When comparing Figures 5 and 6, it is observed that both start with error values close to 15%, and, after the first few iterations, there is an improvement close to 50%, achieving an error close to 8%. Hence, both algorithms have a period of stagnation, in which GWO is superior as it obtains a second improvement of 50%, achieving errors of 3%. On the other hand, although, visually, PSO managed to overcome the stagnation, it only managed to reduce the error to 1%, which does not represent a significant improvement and can be attributed to variations within the MLP parameters, such as the weights, and not to the selection of the hyperparameters.

Table 7. Hyperparameters selected as the best subset for classification of signals given by PSO.

Name	Value
Hidden neurons layer 1	155
Hidden neurons layer 2	204
Epochs	38
Mini-batch size	46
Learning rate	0.0010184

5.3. Validation

After selecting the characteristics and hyperparameters, the rest of the signals that comprised the database were used to validate the results obtained, since this information had never been used before. Figure 7 shows the graphs of the error in the training stage (60% of the data corresponding to 9 of 10 people, equivalent to 600 data to be classified), the test stage (40% of the data corresponding to 9 out of 10 people, equivalent to 200 data to classify), and the validation stage, which corresponded to data from the tenth person (equivalent to 100 data). It is noted that the data to be classified are formed from the number of people × the number of movements × the number of repetitions.

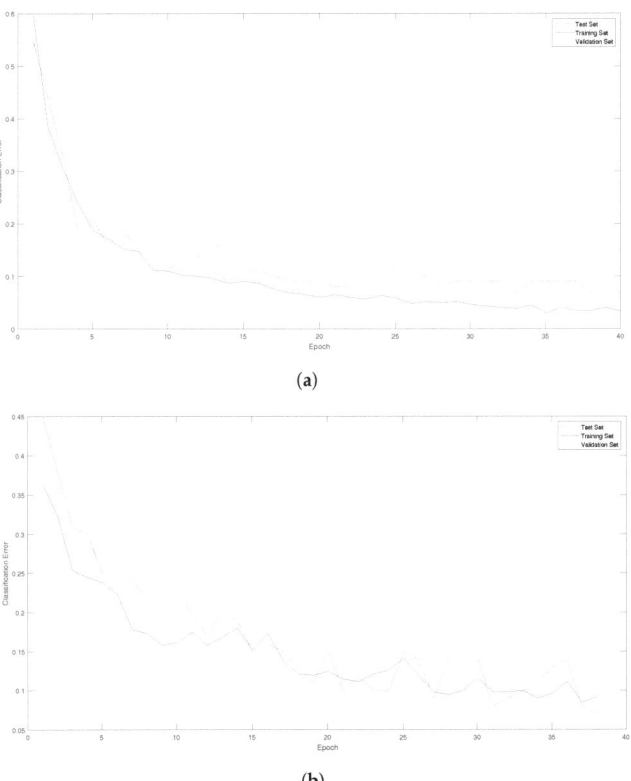

Figure 7. The error in training, testing, and validating a model using (**a**) GWO hyperparameters and (**b**) PSO hyperparameters.

Additionally, these graphs allow us to verify the overfitting in the model. The training, test, and validation errors were plotted in each epoch. If the training error decreases while the test and validation errors increase, this suggests the presence of overfitting. However, the results indicated that the errors decreased evenly across the three stages, suggesting that the model can generalize and classify accurately without overfitting. In addition, the percentage for the hyperparameter values given by GWO only decreased by approximately 4% for new input data, reaching 93% accuracy. Meanwhile, for PSO, 3% was lost in the classification, achieving a final average close to 90%.

6. Discussion

The following comparative Table 8 presents the classification results obtained in previews papers related to the subject of study, compared to the results obtained in this work.

Table 8. Comparative analysis of classification results.

Ref.	Classification Model	Accuracy
[14]	SVM	91%
[15]	SVM	90.92%
[13]	SVM	97.5%
[10]	Recurrent neuronal network	95.7%
[8]	SVM	88%
[28]	MLP	88.8%
[29]	MLP	94.10%
This work	MLP	93%

In this work, an approach based on hyperparameter optimization using PSO and GWO was used to improve the performance of a multilayer perceptron in the classification of EMG signals. This approach performed comparably to other previously studied methods.

However, during the experimentation, there were stages of stagnation. Several reasons explain this lack of success. First, the intrinsic limitations of PSO and GWO, such as their susceptibility to stagnation at local optima and their difficulty in exploring complex search spaces, might have made it challenging to obtain the best combination of hyperparameters [30]. Other factors that might have played a role include the size and quality of the dataset used, since the multilayer perceptron requires a more considerable amount of data to generalize [31].

Despite these limitations, the proposed approach has several advantages. On the one hand, it allows us to improve the performance of the multilayer perceptron by optimizing the key hyperparameters, which is crucial to obtain a more efficient model. Although the performance is comparable with that of other methods, the metaheuristics-based approach manages to reduce the complexity of the model, indicating its potential as an effective strategy for the classification of EMG signals.

Furthermore, the use of PSO and GWO for hyperparameter optimization offers a systematic and automated methodology, making it easy to apply to different datasets and similar problems. It avoids manually tuning hyperparameters, which is messy and error-prone.

It is important to note that each method has its advantages and limitations, and the appropriate approach may depend on factors such as the size and quality of the dataset, the complexity of the problem, and the available computational resources.

7. Conclusions

The proper selection of hyperparameters in MLPs is crucial to classify EMG signals correctly. Optimizing these hyperparameters is challenging due to the many possible combinations. This work uses the PSO and GWA algorithms to find the best combination of hyperparameters for the neural network. Although 93% accuracy has been achieved in classifying EMG signals, there is still room for improvement. Some possible factors that prevent higher accuracy may be the size of the EMG signal database. One way to overcome these problems is to obtain more extensive and robust databases. It is also possible to use data augmentation techniques to generate more variety in the signals. Another possible solution could be to use more advanced EMG signal preprocessing techniques to reduce noise and interference from unwanted signals. Different neural network architectures and optimization techniques can also be considered to improve the classification accuracy further. It is pointed out that the use of a reduced database in this work was part of an initial and exploratory approach to assessing the feasibility of the methodology. This strategy made it possible to obtain valuable information on the effectiveness of the approach before applying it to more extensive databases.

In addition, it is essential to point out that, in this work, no normalization of the data was performed, which might have further improved the performance of the MLP model. Therefore, it is recommended to consider this step in future work to achieve better performance in classifying EMG signals. It is essential to highlight that the cost function used in metaheuristics algorithms is crucial for its success. In this work, the error in the validation stage of the neural network was used as the cost function to be minimized. However, alternatives include sensitivity, efficiency, specificity, ROC, and AUC. A cost function that works well in one issue may not work well in another. Therefore, exploring different cost functions and evaluating their performance is advisable before making a final decision. Another factor that should be considered in this work is the initialization methodology of the network weights. Such considerations and initialization alternatives are subjects for future work that must be analyzed. In general, the selection of hyperparameters is a fundamental step in the construction and training of neural networks for the classification of EMG signals. With the proper optimization of these hyperparameters and the continuous exploration of new techniques and methods, significant advances can be made in this area of research.

Finally, although other algorithms are recognized for their robustness and ability to handle complex data, the MLP proved a suitable option due to the nature of EMG signals. The flexibility of the MLP to model nonlinear relationships was crucial since the interactions between the components were highly nonlinear and time-varying. Furthermore, the MLP has shown good performance even with small datasets, which was necessary considering the limited data availability.

Author Contributions: Conceptualization, M.A.; methodology, M.A.; software, M.A.; validation, M.A.; formal analysis, M.A. and D.I.; investigation, M.A.; resources, J.R.-R.; writing—original draft preparation, M.A., J.R.-R. and D.I.; writing—review and editing, M.A., J.R.-R. and D.I.; visualization, M.A.; supervision, J.R.-R. and D.I. All authors have read and agreed to the published version of the manuscript.

Funding: This research received no external funding.

Institutional Review Board Statement: Not applicable.

Informed Consent Statement: Not applicable.

Data Availability Statement: Access to the database used in this article can be obtained by emailing any of the authors. Please note that the authors reserve the right to decide whether to share the database and may have specific requirements or restrictions regarding its distribution.

Acknowledgments: We thank Consejo Nacional de Humanidades, Ciencia y Tecnología (CONAHCYT) for the national scholarship for doctoral students, which allowed us to carry out this research.

Conflicts of Interest: The authors declare no conflict of interest.

References

1. Jia, G.; Lam, H.K.; Ma, S.; Yang, Z.; Xu, Y.; Xiao, B. Classification of electromyographic hand gesture signals using modified fuzzy C-means clustering and two-step machine learning approach. *IEEE Trans. Neural Syst. Rehabil. Eng.* **2020**, *28*, 1428–1435. [CrossRef]
2. Albahli, S.; Alhassan, F.; Albattah, W.; Khan, R.U. Handwritten digit recognition: Hyperparameters-based analysis. *Appl. Sci.* **2020**, *10*, 5988. [CrossRef]
3. Yang, L.; Shami, A. On hyperparameter optimization of machine learning algorithms: Theory and practice. *Neurocomputing* **2020**, *415*, 295–316. [CrossRef]
4. Du, K.L.; Leung, C.S.; Mow, W.H.; Swamy, M.N.S. Perceptron: Learning, generalization, model selection, fault tolerance, and role in the deep learning era. *Mathematics* **2022**, *10*, 4730. [CrossRef]
5. Vincent, A.M.; Jidesh, P. An improved hyperparameter optimization framework for AutoML systems using evolutionary algorithms. *Sci. Rep.* **2023**, *13*, 4737. [CrossRef] [PubMed]
6. Mirjalili, S.; Mirjalili, S.M.; Lewis, A. Grey wolf optimizer. *Adv. Eng. Softw.* **2014**, *69*, 46–61. [CrossRef]
7. Purushothaman, G.; Vikas, R. Identification of a feature selection based pattern recognition scheme for finger movement recognition from multichannel EMG signals. *Australas. Phys. Eng. Sci. Med.* **2018**, *41*, 549–559. [CrossRef]
8. Too, J.; Abdullah, A.; Mohd Saad, N.; Tee, W. EMG feature selection and classification using a pbest-guide binary particle swarm optimization. *Computation* **2019**, *7*, 12. [CrossRef]
9. Sui, X.; Wan, K.; Zhang, Y. Pattern recognition of SEMG based on wavelet packet transform and improved SVM. *Optik* **2019**, *176*, 228–235. [CrossRef]
10. Xiu, K.; Xiafeng, Z.; Le, C.; Dan, Y.; Yixuan, F. EMG pattern recognition based on particle swarm optimization and recurrent neural network. *Int. J. Perform. Eng.* **2020**, *16*, 1404. [CrossRef]
11. Bittibssi, T.M.; Zekry, A.H.; Genedy, M.A.; Maged, S.A. sEMG pattern recognition based on recurrent neural network. *Biomed. Signal Process. Control* **2021**, *70*, 103048. [CrossRef]
12. Li, Q.; Zhang, A.; Li, Z.; Wu, Y. Improvement of EMG pattern recognition model performance in repeated uses by combining feature selection and incremental transfer learning. *Front. Neurorobot.* **2021**, *15*, 699174. [CrossRef]
13. Cao, L.; Zhang, W.; Kan, X.; Yao, W. A novel adaptive mutation PSO optimized SVM algorithm for sEMG-based gesture recognition. *Sci. Program.* **2021**, *2021*, 9988823. [CrossRef]
14. Aviles, M.; Sánchez-Reyes, L.M.; Fuentes-Aguilar, R.Q.; Toledo-Pérez, D.C.; Rodríguez-Reséndiz, J. A novel methodology for classifying EMG movements based on SVM and genetic algorithms. *Micromachines* **2022**, *13*, 2108. [CrossRef]
15. Dhindsa, I.S.; Gupta, R.; Agarwal, R. Binary particle swarm optimization-based feature selection for predicting the class of the knee angle from EMG signals in lower limb movements. *Neurophysiology* **2022**, *53*, 109–119. [CrossRef]
16. Li, X.; Yang, Y.; Chen, H.; Yao, Y. Lower limb motion pattern recognition based on IWOA-SVM. In Proceedings of the Third International Conference on Computer Science and Communication Technology (ICCSCT 2022), Beijing, China, 30–31 July 2022; Lu, Y., Cheng, C., Eds.; SPIE: Bellingham, WA, USA, 2022.
17. Toledo-Pérez, D.C.; Rodríguez-Reséndiz, J.; Gómez-Loenzo, R.A.; Jauregui-Correa, J.C. Support vector machine-based EMG signal classification techniques: A review. *Appl. Sci.* **2019**, *9*, 4402. [CrossRef]
18. Raez, M.B.I.; Hussain, M.S.; Mohd-Yasin, F. Techniques of EMG signal analysis: Detection, processing, classification and applications. *Biol. Proced. Online* **2006**, *8*, 11–35. [CrossRef]
19. Bi, L.; Feleke, A.g.; Guan, C. A review on EMG-based motor intention prediction of continuous human upper limb motion for human-robot collaboration. *Biomed. Signal Process. Control* **2019**, *51*, 113–127. [CrossRef]
20. Argatov, I. Artificial neural networks (ANNs) as a novel modeling technique in tribology. *Front. Mech. Eng.* **2019**, *5*, 30. [CrossRef]
21. Zemzami, M.; El Hami, N.; Itmi, M.; Hmina, N. A comparative study of three new parallel models based on the PSO algorithm. *Int. J. Simul. Multidiscip. Des. Optim.* **2020**, *11*, 5. [CrossRef]
22. Jain, M.; Saihjpal, V.; Singh, N.; Singh, S.B. An overview of variants and advancements of PSO algorithm. *Appl. Sci.* **2022**, *12*, 8392. [CrossRef]
23. Nematzadeh, S.; Kiani, F.; Torkamanian-Afshar, M.; Aydin, N. Tuning hyperparameters of machine learning algorithms and deep neural networks using metaheuristics: A bioinformatics study on biomedical and biological cases. *Comput. Biol. Chem.* **2022**, *97*, 107619. [CrossRef] [PubMed]
24. Nanda, S.J.; Panda, G. A survey on nature inspired metaheuristic algorithms for partitional clustering. *Swarm Evol. Comput.* **2014**, *16*, 1–18. [CrossRef]
25. Andonie, R. Hyperparameter optimization in learning systems. *J. Membr. Comput.* **2019**, *1*, 279–291. [CrossRef]
26. Asghari Oskoei, M.; Hu, H. Myoelectric control systems—A survey. *Biomed. Signal Process. Control* **2007**, *2*, 275–294. [CrossRef]
27. Clerc, M.; Kennedy, J. The particle swarm—Explosion, stability, and convergence in a multidimensional complex space. *IEEE Trans. Evol. Comput.* **2002**, *6*, 58–73. [CrossRef]
28. Fajardo, J.M.; Gomez, O.; Prieto, F. EMG hand gesture classification using handcrafted and deep features. *Biomed. Signal Process. Control* **2021**, *63*, 102210. [CrossRef]
29. Luo, R.; Sun, S.; Zhang, X.; Tang, Z.; Wang, W. A low-cost end-to-end sEMG-based gait sub-phase recognition system. *IEEE Trans. Neural Syst. Rehabil. Eng.* **2020**, *28*, 267–276. [CrossRef]

30. Tran, B.; Xue, B.; Zhang, M. Overview of particle swarm optimisation for feature selection in classification. In *Lecture Notes in Computer Science*; Lecture notes in computer science; Springer International Publishing: Cham, Switzerland, 2014; pp. 605–617.
31. Dargan, S.; Kumar, M.; Ayyagari, M.R.; Kumar, G. A survey of deep learning and its applications: A new paradigm to machine learning. *Arch. Comput. Methods Eng.* **2020**, *27*, 1071–1092. [CrossRef]

Disclaimer/Publisher's Note: The statements, opinions and data contained in all publications are solely those of the individual author(s) and contributor(s) and not of MDPI and/or the editor(s). MDPI and/or the editor(s) disclaim responsibility for any injury to people or property resulting from any ideas, methods, instructions or products referred to in the content.

Article

Radiation Dose Tracking in Computed Tomography Using Data Visualization

Reem Alotaibi [1,*] and Felwa Abukhodair [1,2]

1. Faculty of Computing and Information Technology, King Abdulaziz University, Jeddah 21589, Saudi Arabia
2. Center of Excellence in Smart Environment Research, King Abdulaziz University, Jeddah 21589, Saudi Arabia; fabukhodair@kau.edu.sa
* Correspondence: ralotibi@kau.edu.sa

Abstract: Radiation dose tracking is becoming very important due to the popularity of computerized tomography (CT) scans. One of the challenges of radiation dose tracking is that there are several variables that affect the dose from the patient side, machine side, and procedures side. Although some tracking software programs exists, they are based on static analysis and cause integration errors due to the heterogeneity of Hospital Information Systems (HISs) and prevent users from obtaining accurate answers to their questions. In this paper, a visual analytic approach is utilized to track radiation dose data from computed tomography (CT) through the use of Tableau data visualization software. The web solution is evaluated in real-life scenarios by domain experts. The results show that the visual analytics approach improves the tracking process, as users completed the tasks with a 100% success rate. The process increased user satisfaction and also provided invaluable insight into the analytical process.

Keywords: visual analytics; radiation dose; big data; tracking; computerized tomography (CT)

Citation: Alotaibi, R.; Abukhodair, F. Radiation Dose Tracking in Computed Tomography Using Data Visualization. *Technologies* 2023, 11, 74. https://doi.org/10.3390/technologies11030074

Academic Editors: Juvenal Rodriguez-Resendiz, Gerardo I. Pérez-Soto, Karla Anhel Camarillo-Gómez and Saul Tovar-Arriaga

Received: 16 May 2023
Revised: 4 June 2023
Accepted: 8 June 2023
Published: 10 June 2023

Copyright: © 2023 by the authors. Licensee MDPI, Basel, Switzerland. This article is an open access article distributed under the terms and conditions of the Creative Commons Attribution (CC BY) license (https://creativecommons.org/licenses/by/4.0/).

1. Introduction

Recent advances in technology have led to the evolution of data sources, forms, and structures in healthcare. This evolution has made healthcare a source of big data. One of the many available sources of data includes computerized tomography (CT) scans and other imaging modalities. However, the equipment used to generate such imagery exposes patients to harmful ionizing radiation [1], though it can produce numerous pieces of information about the patient's body parts, protocol, images, and radiation dosage. Even though a single dose may be small, radiation exposure is cumulative [2,3]; hence, there is a risk of overdose. The cohort under study in [4] had a higher baseline risk of cancer due to the cumulative doses of repeated CT scans.

Radiation dose tracking has, consequently, become increasingly important owing to the popularity of CT scans. Several variables affect the tracking dose, such as the patient's age, the size of the body part being scanned, the total size of the body, the scanner model and pitch, and the scanning protocol [5]. For example, for a chest CT scan, the patient lies on a flat bed, and then the machine circulates around him/her, emitting a beam that goes through the patient's body (in this case, the chest). When the machine completes a full cycle, it creates a slice that is a 2-D image of the lungs and the inside of the chest. The number of slices is determined based on the study protocol, and a computer processes these images to be displayed on the monitor, and they may even be rendered in a 3-D form [6]. In radiology, a study means all the procedures a patient underwent to obtain a specific scan protocol.

Tracking radiation dosages is an important and mandatory practice in healthcare facilities [7]. In some facilities, teams are assigned to carry out this task manually or by using commercial software, such as DoseTrack (version 1.0) by Sectra [8]. Manual tracking is unfeasible and has the potential for human error. However, commercial tools are based on static analysis, which limits the possibilities of exploring data due to problems such as predefined reports, no interwind filters, and complex interfaces for inexperienced users [9]. In addition, this commercial tracking software can create integration errors, such as errors while extracting data from the PACS. PACS is a technology used in healthcare facilities for storing, retrieving, presenting, and sharing images produced by medical hardware [10] thanks to the heterogeneity of Hospital Information Systems (HISs) [11].

Using visual analytics to track radiation dosages will allow users to track a single patient or cohort of patients, will include temporal information, and will enable one to observe patterns over time, find trends, and even optimize the practices in use through gained insights.

In this paper, we propose a visual analytical dashboard called VATrack to track radiation dose data for CT scans through employing visual analytics techniques. It monitors the radiation dose in patients undergoing CT scans. The dashboard processes multivariate temporal data and identifies any radiation overdose while reducing human intervention in the radiation dose tracking process and enhancing the performance in terms of accuracy and time. Domain experts have evaluated the proposed dashboard.

The remainder of this paper is organized as follows. Section 2 discusses some background information and sheds light on existing solutions for radiation dose tracking. Section 3 presents the proposed solution for applying visual analytics to monitor the radiation dosage for patients undergoing CT scans and the experimental design in more detail. The performance evaluation of the proposed solution is discussed in Section 4 and the results are presented in Section 5. Section 6 discusses the results. Finally, Section 7 concludes the thesis and provides suggestions for future research.

2. Related Work

Visual analytics is not the same as data or information visualization, even though they have a similar goal, which is to provide insight to the end-users so that they can make good and informed decisions. Visual analytics was first defined by Thomas and Cook as "The science of analytical reasoning facilitated by interactive visual interfaces" [12]. Data visualization is the representation of information or data using visual elements, with no regard to how to manage and process the data.

Therefore, when the problem is not simple or clear enough to be modelled using automatic analysis, the need for human cognitive abilities arises; hence, the use of visual analytics becomes necessary. Visual analytics has attracted considerable interest because it aids in gaining an understanding of problems and solutions; hence, it has been adopted in several application areas such as space, healthcare, social networks, deep learning, and many more.

There are available tools to track radiation doses. Kovacs et al. proposed an engine that calculates radiation exposure from a CT scan with several analysing schemes called the "RE3-Radiation Exposure Extraction Engine" [13]. The engine was integrated within the hospital's PACs to recover radiation dose data using image data and DICOM headers. This integration could, however, create several problems, such as errors in data extraction from different DICOM sources or other integration errors, whereas our solution is completely independent.

The authors in [14] used SharePoint (2013) to create a report generator to monitor the radiation dose in digital mammography and digital breast tomosynthesis. It supports different user roles and operates in a mixed imaging environment. In addition, it was integrated within DICOM and the PACS. However, it provides information as static visualizations to support decision making and quality assurance, whereas the proposed solution provides interactive visuals that support explanatory analysis in addition to the basic functions of monitoring.

Finally, the authors in [15] used the commercial software DoseWatch (version 1.3) (GE Healthcare). They confirmed that commercial dose tracking software is not sufficient on its own, even though it is true to what the developers claim in terms of tracking and monitoring, but inexperienced or busy users require additional time and effort because the tools are static and not interactive; thus, the user feels less motivated to search for answers.

3. VATrack Dashboard

To the best of our knowledge, there have been no applications of visual analytics in diagnostic imaging radiation dose tracking. Therefore, in this paper, a web solution is presented that provides tracking functionality and an exploratory environment for CT scan radiation dose data. The solution is based on visual analytics techniques and Tableau software. The proposed dashboard was evaluated using use case scenarios. To build the proposed dashboard, there were four phases, as follows [16].

- Phase 1: Conducting interviews with domain experts to collect and analyse the requirements.
- Phase 2: Abstraction of data and requirements; our data are multivariate temporal data.
- Phase 3: During this iterative phase, we conduct the following:
  - Generate multiple designs;
  - Develop and implement the selected design;
  - Evaluate prototypes;
  - Improve the design based on user-required improvements.
- Phase 4: Deploy the dashboard and interview users and domain experts for feedback [17].

3.1. Requirement Analysis

To collect the requirements, several meetings with domain experts were held. During this process, the required functions were defined as follows:

- Automate the tracking process and reduce human intervention as much as possible.
- Identify high-dose studies according to European DRLs (Diagnostic Reference Levels).
- Search by patient ID.
- Study comparisons.
- Grouping study protocols into three categories (Head, Chest, and Abdomen).
- Redefine age classes to be:
  - 0–3 months;
  - 4 months–1 year;
  - 2–5 years;
  - 6–16 years;
  - 16+ years.
- Reduce study attributes to provide good tracking and summary information—we chose only 13 relevant attributes. These attributes were chosen based on domain experts' opinions and reviews of what and how the radiology department performed the tracking process and what they used for their tracking process.

Additionally, we were able to ascertain our user characteristics:

- Educational level: Bachelor or higher;
- Experience: they are experienced in the field but not in the tracking process;
- Disabilities: No visual disabilities such as colour blindness or loss of vision;
- Technical expertise: Basic level of computer skills.

Finally, we determined the following general constraints:

- No integration with HIS is required;
- No data pre-processing by users;
- Platform-independent.

3.2. CT Radiation Tracking Dataset

Data records for 2019 were obtained from the Radiology Department at King Abdulaziz University Hospital, Jeddah, Saudi Arabia. They are multivariate data generated by Dosewatch software, not raw data from medical scanners, exported into an Excel file format. The dataset consisted of 98 columns. The number of rows varied depending on the patient count and the type of CT scan protocol conducted. It included patient information, such as Patient ID, date of birth, age class, sex, weight (kg), height (cm), and BMI. It also contained information pertaining to the study performed, such as study protocol name, study time, series type, mean CTDI, DLP, the operator, and the imaging machine.

The sample consisted of 2755 records that represented 946 patients. Data were adversely affected by user input errors, such as entering the wrong file number as a study ID or the wrong date of birth. Unfortunately, patient size, BMI, and weight data are missing, which has created an obstacle in tracking and identifying radiation overexposure cases. This will affect tracking of chest and abdominal CT scans, as head CT scan dosage limits are determined by age, as shown in Table 1.

Table 1. European diagnostic reference levels [18].

Exam	Age or Weight Group	EDRL	
		CTDI vol, mGy	DLP, mGycm
Head	[0 ≤ 3 months]	24	300
	[3 months ≤ 1 year]	28	385
	[1 ≤ 6 years]	40	505
	[≥6 years]	50	650
Thorax			
	<5 kg	1.4	35
	[5 ≤ 15 kg]	1.8	50
	[15 ≤ 30 kg]	2.7	70
	[30 ≤ 50 kg]	3.7	115
	[50 ≤ 80 kg]	5.4	200
Abdomen			
	< 5 kg		45
	[5 ≤ 15 kg]	3.5	120
	[15 ≤ 30 kg]	5.4	150
	[30 ≤ 50 kg]	7.3	210
	[50 ≤ 80 kg]	13	480

Table 1 represents the European Diagnostic Reference Levels (DRLs) followed by King Abdulaziz University Hospital and is used to identify any study that exceeded the radiation limits to be investigated in more detail. It covers three main areas: Head, Chest, and Abdomen. For head-related CT scans, it classifies the safe dosages by age, while for chest-related CT scans and abdomen-related CT scans, it uses weight for classification purposes.

Data pre-processing and cleaning was carried out by defining the conditions in Tableau that would be applied to the extracted data when uploaded. First, data were cleaned by removing records with missing patient IDs, invalid ID numbers, or invalid values in basic patient attributes (such as no gender given or an unrealistic date of birth provided). Subsequently, data aggregation occurred for studies with protocols that consisted of more than one step. Some CT scan protocols consisted of more than one step; hence, in the dataset, there were as many records for a study per patient as the number of protocol steps. This situation proved problematic because there were also some protocols with an unknown number of steps, meaning they could not be defined in the system because the physician decided how many steps there would be at the time of the procedure. To overcome this problem, VATrack aggregated each study using the patient ID, Study ID, and protocol name (local study description) to ensure that the protocol steps (series description) were covered.

3.3. Design

The dashboard must be easy to use and intuitive since our users will have only a basic level of computer skills and limited time to learn or be trained due to busy schedules, yet it still must be able to provide users with summary information at a glance. A dashboard that is capable of this is shown in Figure 1.

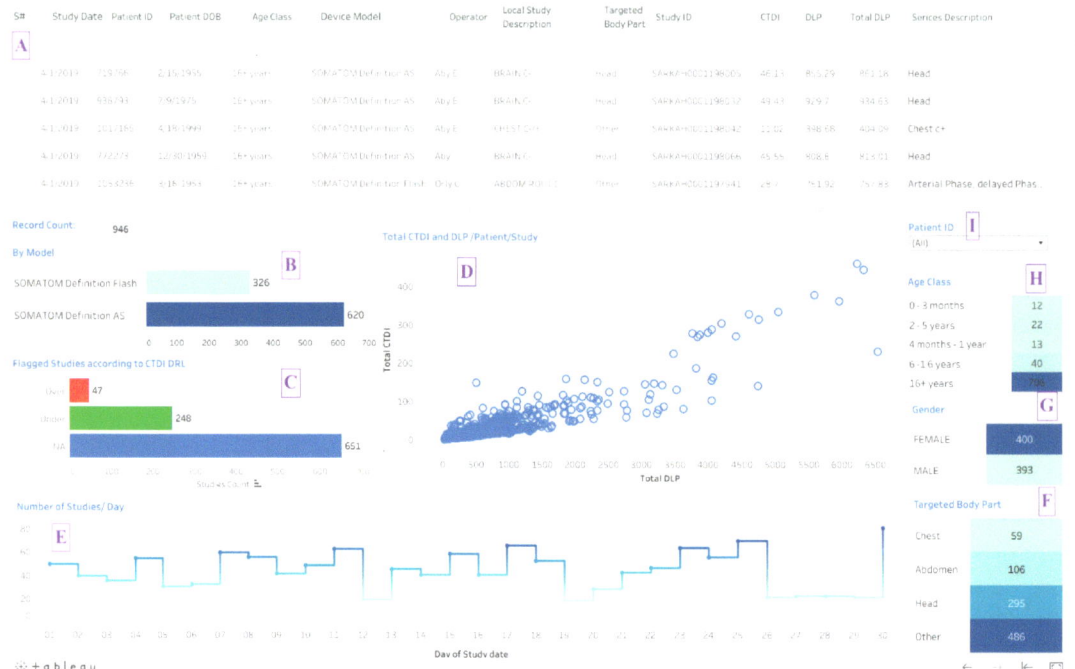

Figure 1. VATrack: Radiation dose tracking and monitoring interactive visual display. (**A**) main view, (**B**) imaging device model, (**C**) flagged studies (red indicates overdose studies and green indicates safe dose studies), (**D**) scatter plot of studies in relation to DLP and CTDI, (**E**) timeline to show the number of studies per day over the selected period, (**F**) study protocols categorized according to the targeted body part, (**G**) gender bar chart, (**H**) age classes using bar chart b, and (**I**) patient ID search bar.

The dashboard consists of:
- Nine coordinated views;
- Search bar;
- Several interaction techniques (filtering, linking and brushing, and details on demand).

Coordinated views are used, whereby clicking on or selecting an area in one view updates the remaining views. In addition, the search bar updates all nine views. The interaction techniques depend on the view, as shown below.

Figures 1A and 2 are the main view of the top third, which is a scrollable table structure wherein the 13 attributes are displayed at the beginning. The table is populated with complete data; each record represents a patient, clicking on a cell will highlight his/her record, and the remaining views will not be updated. In addition, under the table, there is a counter for the records that are currently available.

Figure 2. Dashboard main view.

Figures 1B and 3 are a horizontal bar graph that represents the imaging device model and acts as a filter by clicking on the bars, and the numbers at the outside end of each bar represent the number of studies using each model.

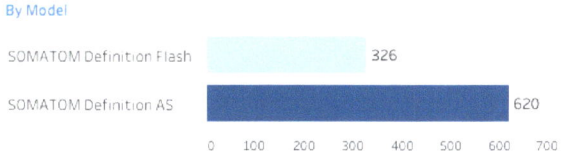

Figure 3. Imaging device model.

Figures 1C and 4 are also a horizontal bar graph that represents red-colour overdose studies, green-colour safe dose studies, and NA for studies that have yet to be categorized. The bar graph also acts as a filter by clicking on the bars, and the numbers at the outside end of each bar represent the number of studies within each category. We used bar graphs as filters to provide the necessary information and to visualize the comparison of data among categories.

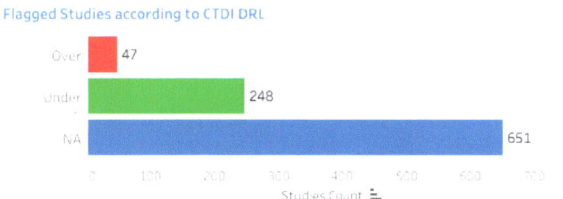

Figure 4. Flagged studies (red indicates overdose studies and green indicates safe-dose studies).

Figures 1D and 5 show a scatter plot that represents each study according to its DLP and CTDI. Users can click on an individual point or use 'sliding' by selecting an area on the plot that will update the rest of the dashboard; also, there are 'infotip' details-on-demand that appear when you rest the pointer on a point. We use a scatter plot to show the relationship between the two variables and allow the user to identify other patterns in the dataset.

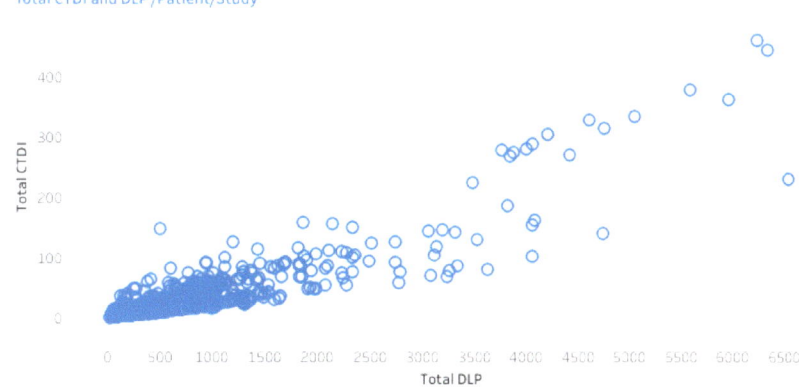

Figure 5. Scatter plot of studies in relation to DLP and CTDI.

Figure 1E is a dotted-line graph that represents the study date, and the dots represent the number of studies per day. It also has a brushing and linking interaction technique by selecting an area on the graph that updates the rest of the dashboard and 'infotip' details-on-demand, as seen in Figure 6.

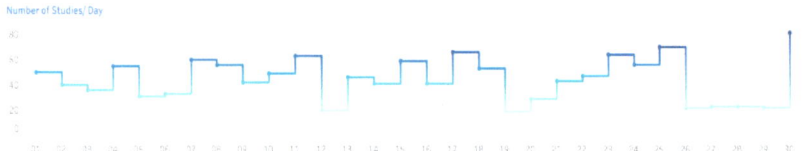

Figure 6. Total CTDI DLP per patient/study.

Figures 1F and 7 are a bar graph wherein each bar represents a group of CT scan protocols related to body area, which also acts as a filter, and the numbers inside each bar represent the number of patients in the category.

Targeted Body Part

Chest	59
Abdomen	106
Head	295
Other	486

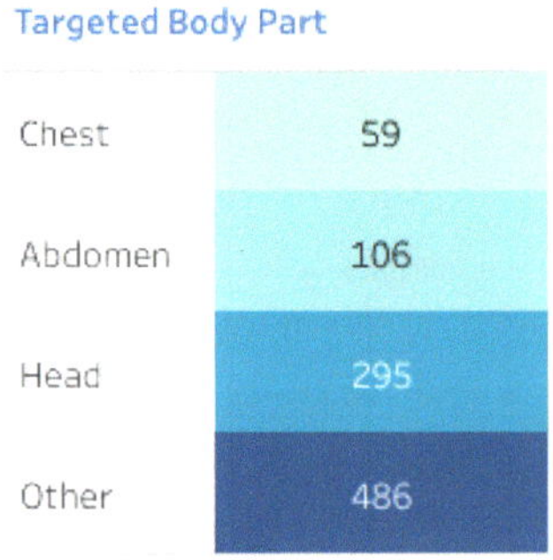

Figure 7. Study protocols categorized according to the targeted body part.

Figures 1G and 8 are a bar graph wherein each bar represents a gender, which also acts as a filter, and the numbers inside each bar represent the number of patients in the category.

Gender

Gender	Count
FEMALE	400
MALE	393

Figure 8. Gender bar chart.

Figures 1H and 9 are a bar graph wherein each bar represents an age class; it also acts as a filter, and the numbers inside each bar represent the number of patients in the class.

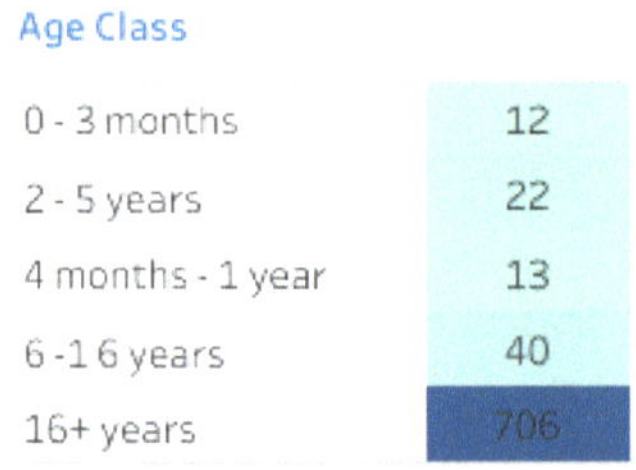

Age Class

0 - 3 months	12
2 - 5 years	22
4 months - 1 year	13
6 -1 6 years	40
16+ years	706

Figure 9. Age classes using the bar chart.

Figures 1I and 10 show a search bar wherein the user can filter the dashboard by patient ID, as shown in Figure 1.

Figure 10. Patient ID search bar.

Using these different visualization components as filters provides insights from an initial look and also conveys as much information as possible without overloading the user with multiple irrelevant records to scroll through. In addition, we used colour-coded views to indicate that the deeper the colour, the higher the number (that is, except for the "flagged studies" filter, which uses a traffic-light colour code to indicate high risk for red and safety for green).

3.4. Implementation

For testing and evaluation purposes, a proof of concept (PoC) implementation of the interactive visual display was developed. Tableau is a visual analytics software platform that provides several functions to individuals and enterprises. It helps with creating interactive and connected dashboards across an organization from pre-processing data, whether from a spreadsheet, SQL database, or in the cloud, to publishing compelling results to provide insights and answers acquired through the analytical process [19,20].

Tableau Desktop (version 2020.4) was used because data visualization and analysis functionalities are its main features, and the researcher(s) have greater familiarity with the software. Tableau was used to build the interactive visual display. The next section describes the steps taken to build the solution depicted in Figure 1.

Data Interaction Components

First, Tableau was connected to a data-source Excel (version 2019) file, as shown in Figure 11. The rules defined in data pre-processing are applied automatically whenever the user uploads new data.

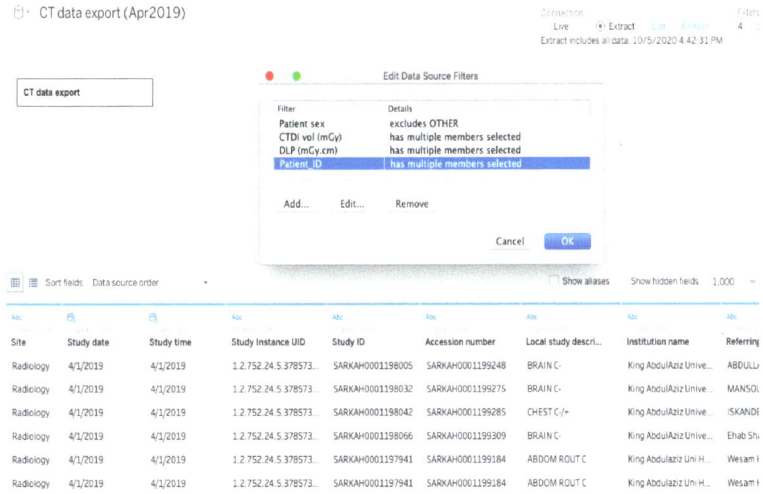

Figure 11. Connecting data source to Tableau software.

In Figure 1, only 13 different attributes are displayed in the data source rather than 98. The age-class interaction component is created in Tableau, as shown in Figure 12; according to user needs, age classes are split differently. To calculate the age month, which is not available in source data, Tableau subtracts the date of birth from the study date and converts that period to the number of months; then, the desired classes are applied using conditional statements. In this study, the linking and brushing interaction techniques were used.

```
Age Class
0 - 3 months          12
2 - 5 years           22         Age Class                                        X
4 months - 1 year     13         IF [Age months] <= 3
                                 THEN "0 - 3 months"
6 - 1 6 years         40         ELSEIF [Age months] > 3 and [Age months] < 13
                                 THEN "4 months - 1 year"
16+ years            706         ELSEIF [Age months] >12 AND [Age months] <= 60
                                 THEN "2 - 5 years"
                                 ELSEIF [Age months] >60 AND [Age months] <=192
                                 THEN "6 -1 6 years"
                                 ELSEIF [Age months]>192
                                 THEN "16+ years"
                                 ELSE "NA"
                                 END
```

Figure 12. Defining age-class filter complying to user specifications.

For the body part interaction component shown in Figure 1, four groups were created, each containing specific scanning protocols, as shown in Figure 13. In addition, there were eight other data interaction components:

- Model (B in Figure 1): a bar graph shows the number of studies performed using different devices. In addition, it acts as a cross-filter based on the medical scanners used.
- Flagged studies according to CTDI and DLP (C in Figure 1): a bar graph represents studies based on the total radiation dosage per study; hence, for each study per patient, the software sums the CTDI and DLP values then compares the total to the standard used in Figure 1. If it is over the standard, then it is in the red category, while if it is under the standard, then it is placed in the green category. In addition, it acts as a cross filter, which updates the dashboard based on whichever bar is clicked on.
- Number of studies/day (E in Figure 1): a dotted line graph represents the number of studies performed each day. In addition, it acts as a cross-filter when sliding and linking a group of days or selecting a single day. Further, it has an infotip feature.
- Gender (G in Figure 1): a bar graph shows the number of patients according to gender—female or male. In addition, it acts as a cross-filter when clicked on.
- Patient ID (I in Figure 1): a search bar filters data based on entering the patient ID in the search bar or by selecting a patient ID from a single-value drop-down list.
- Total CTDI and DLP/patient/study (D in Figure 1): this is a scatter plot wherein each point represents a study showing the radiation dose in CTDI and DLP measures. In addition, it acts as a cross-filter when sliding and linking a group of points or by selecting a single point. Further, it has an infotip feature.

To remove what the user may think of as redundancies, a field called "series description" was created, as shown in Figure 14. Each scanning protocol consists of a different number of steps, and each step has its own radiation dose value in the DLP field. For example, in Figure 14, a protocol with four steps actually appears four times in the data source, but in Figure 14, it appears as a single record, and the aggregate value of the radiation dose is in the DLP field.

All charts and input boxes in Figure 14 are cross-filters that can be applied simultaneously. The table at the top displays the records available after applying the data interaction components. Users can click on several interaction components and immediately inspect the dashboard. On the scatter plot and the timeline at the bottom, the user can either click on a single value/day or slide a group of dots or days.

Figure 13. Defining body-part filter based on scanning protocols.

Figure 14. Listing all protocol steps in a single record.

All charts and input boxes in Figure 1 are cross-filters that can be applied simultaneously. The table at the top displays the records available after applying the data interaction components. Users can click on several interaction components and immediately reflect on the dashboard. On the scatter plot and the timeline at the bottom, the user can either click on a single value/day or brush a group of dots or days.

4. Evaluation

The evaluation focused on use-case scenarios, completion accuracy, and time. We also used a questionnaire-based evaluation. A usability test was conducted to evaluate, measure, and improve the usability of the proposed solution. The usability test examined the user-friendliness of the solution. According to Jakob Nielsen, a renowned usability expert, usability consists of five components: learnability, efficiency, memorability, error minimization, and satisfaction [21]. A "User and Task Observation" test method was chosen.

4.1. Study Design

An on-site usability test was conducted at King Abdulaziz University Hospital. It took place at participants' offices on Sunday 17 October 2021 from 8:30 a.m. to 2:45 p.m. There were some waiting periods between the test sessions since a number of participants were working. The test started by greeting the participants and explaining the goal of the study, which was to assess the usability of VATrack interface design and interactivity. Four test users who worked as medical physicists in the department participated; they were responsible for monitoring radiation doses and exposure in all forms at the hospital, including tracking radiological procedure doses.

It was difficult to include additional participants since the remainder were technicians who were operating the scanning machines, and it would have been very difficult to suspend the workflow. Then, a consent form was signed by each participant to obtain his/her agreement to conduct the study. Then, the observer introduced VAtrack as a dashboard web-based solution that served as a medium to track and analyse radiation doses of CT scans retrospectively.

The dashboard also performed data pre-processing and provided users with easy access to tracking information and efficient interactivity that allowed users to answer related questions. Subsequently, scenarios were provided to test users one by one. Each individual session lasted approximately one hour, during which test users performed the scenarios both ways. The first participant started by using the current existing process and then VATrack, while the remaining participant started with VATrack then the current existing process. We made the switch in order to ensure the completion of the test because the first test participant became reluctant, tired, and kept checking the time. All participants had the same order of scenarios.

4.2. User and Task Observation

In this usability testing method, a tester observed a test user performing a given task while documenting and timing the process. Any success or failure was recorded. In addition, the test users were not coached. User success was measured by the percentage of tasks completed correctly and the time taken to complete the task.

4.2.1. Identifying Users

This study illustrates how visual analytics enhances the radiation dose tracking process in the radiology department at King Abdulaziz University Hospital. To identify the user group, we started by looking at who was going to use the dashboard: the radiologists at the hospital. We then had to identify their characteristics, so we interviewed radiologists at the hospital to get their thoughts about the current tracking method deployed there, the problems they faced, their needs, and what they felt should be changed. During the conversations, we also learned about their technical skill levels. Four users working in the radiology department volunteered to participate in the testing phase. One of them who was responsible for the tracking task was involved in VATrack's data collection phase. Two of them were female and the other two were male. Three of them were in their late twenties to mid-thirties and one was in his/her early forties.

4.2.2. Task Scenarios

To create the test scenarios, we first listed three basic tasks that needed to be completed regularly by the end users plus two additional related tasks. We determined these tasks by reviewing department tracking files and discussing their contents with domain experts in the Radiology Department at the hospital. These scenarios were completed both ways—the current process and using the proposed dashboard—to compare the two solutions fairly. The task scenario activities were as follows:

- Participating users needed to provide a quick summary presentation of studies and patients from April 2019 that included patients' total counts and distributions within the three main categories of CT scan studies (head, chest, and abdomen) by age classes using both the VATrack dashboard and the process currently used to obtain that information.
- Participating users needed to create a report that showed gender and age distributions within chest CT scan studies using both the VATrack dashboard and the process currently used to obtain that information.
- Participating users needed to know why the patient with ID 1054533 had received a high dosage that was over the limits of the European Diagnostic Reference Levels using both the VATrack dashboard and the process currently used to obtain that information.
- Participating users needed to create a report of adult head CT scan studies performed with a device model "Somatom Definition AS" using both the VATrack dashboard and the process currently used to obtain that information.
- Participating users needed to create a report of overdose studies in children aged 0–3 months for the month of April 2019 using both the VATrack dashboard and the process currently used to obtain that information.

To complete Activities 1 and 2, test users had to explore the data and complete Activities 3 and 5. Further, the data had to be analysed. Finally, to complete Activity 4, participants had to compare the data. In addition, any age criterion was based on the standard used by the department. Age classes were as follows:

- 0–3 months;
- 4 months–1 year;
- 2–5 years ;
- 6–16 years;
- 16+ years.

5. Results

Using the VATrack dashboard, all participants successfully completed the five task scenarios shown in Table 2. However, while following the current procedure used in the department of radiology, 10% of the tasks were completed correctly, 45% were completed with errors, and 45% were not completed (Figure 15).

Table 2. Task completion time. NA means the task hasn't been completed.

VATrack	Test User 1	Test User 2	Test User 3	Test User 4
Activity 1	00:47.23	00:07.50	00:11.60	00:19.00
Activity 2	00:21.17	00:39.50	00:04.52	00:19.95
Activity 3	00:28.47	00:39.51	01:24.91	00:21.90
Activity 4	03:16.84	01:22.95	00:17.50	00:25.10
Activity 5	02:19.99	01:54.65	00:01.00	00:01.00
Total	07:13.70	04:44.11	01:59.53	01:26.95
Current Process	Test User 1	Test User 2	Test User 3	Test User 4
Activity 1	09:32.83	01:20.00	02:14.64	NA
Activity 2	01:37.60	03:46.32	02:04.42	NA
Activity 3	03:00.47	02:15.91	00:00.00	NA
Activity 4	03:55.97	04:03.25	01:13.85	NA
Activity 5	19:35.28	03:30.87	04:30.05	NA
Total	37:42.13	14:56.35	10:02.96	00:00.00

For completion time, Figure 16 compares the total times the test users took to perform each activity both ways. As depicted in Figure 16, the activities performed using VATrack took less time to complete with a 100% success rate. On the other hand, being familiar with the current process did not have a significant impact on the time required to complete each activity. Activity 1 represented a basic task the users performed in the department every month. This activity had the longest completion time of 27 min and 36 s, and it was one of the activities that users either did not complete or completed with errors. The test user who already performed this task monthly stopped after approximately 20 min because it would take the entire morning to perform correctly.

In contrast, performing the same Activity 1 using VATrack took around 4 min and 17 s for this particular user, and it took two test users a second to complete, with them noting: "no need to do anything, it is clear by looking at the dashboard".

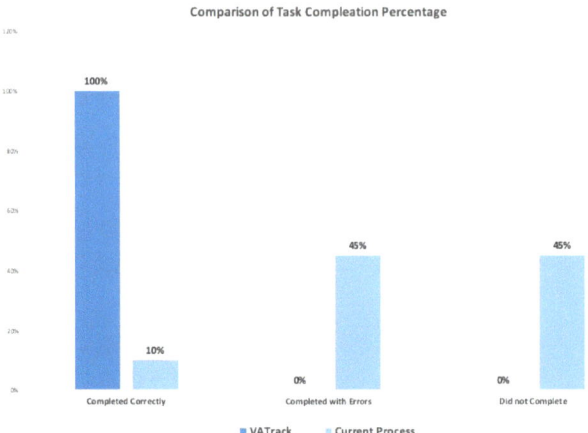

Figure 15. Comparison of task completion percentage.

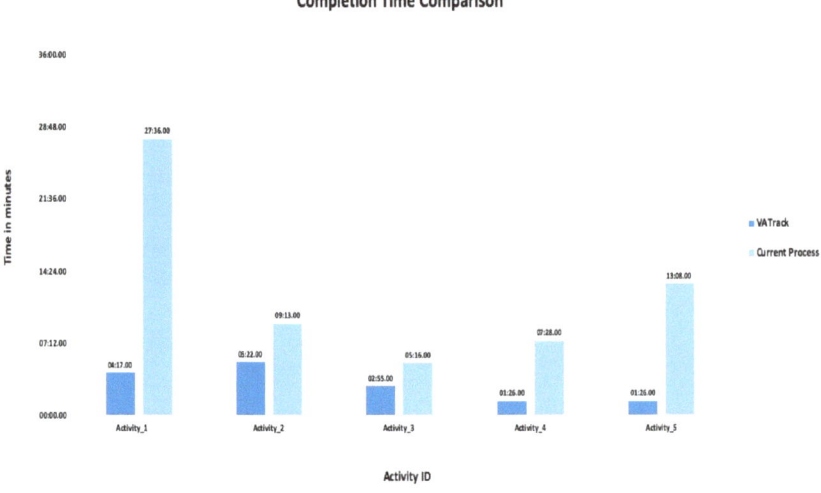

Figure 16. Comparison of task completion time.

Activity 2 took five minutes and 16 s using VATrack and had a 100% success rate, while following the current process took nine minutes and 13 s with a 0% success rate, putting it into the categories of 'completed with errors' and 'did not complete'. Some of the user errors in this activity occurred during data filtration in Excel; for example, a user chose scanning protocols with "chest" in the name but did not know that "thorax" was an alternative name for the chest. Another user did not complete the task because he was confused by the amount of data.

Activity 3 took two minutes and fifty-five seconds using VATrack with a 100% success rate. Following the current process, it took five minutes 16 s while falling into the three categories 'completed correctly', 'completed with errors', and 'did not complete'. Some of the user errors in this activity while following the current process involved guessing the answer based on his/her knowledge and experience. A user who did not complete this activity knew that he/she needed to compare information against certain standards but did not know what they were.

Activity 4 took one minute and twenty-six seconds using VATrack with a 100% success rate. Using the current process, it took seven minutes and twenty-six seconds and also fell into the three categories 'completed correctly', 'completed with errors', and 'did not complete'. Some of the user errors in this activity while following the current process involved included any scanning protocol with the words 'head' and 'adult' in the name. Additionally, one user used "Series description", not "Study Protocol Name", which was usually used.

Finally, Activity 5 using VATrack took one minute and twenty-six seconds with a 100% success rate. The current process took 13 min and 8 s with a 0% success rate.

Many human errors occurred while following the current process, such as missing records while aggregating the dosages of a single study or adding a CT scan protocol that was not included in a given category; in addition, there were mistakes when recalculating age groups or not calculating them at all. The same went for the use of diagnostic reference levels. In general, looking at the raw data while working with the current process led to the test users becoming reluctant and confused.

Questionnaire

After the test users completed the required tasks, they were asked to contribute and answer an electronic questionnaire that summarised their experience and the usability of the new solution. The first question, as shown in Figure 17, asked users to choose, after trying both methods, which they felt performed each activity better, and 100% chose VATrack.

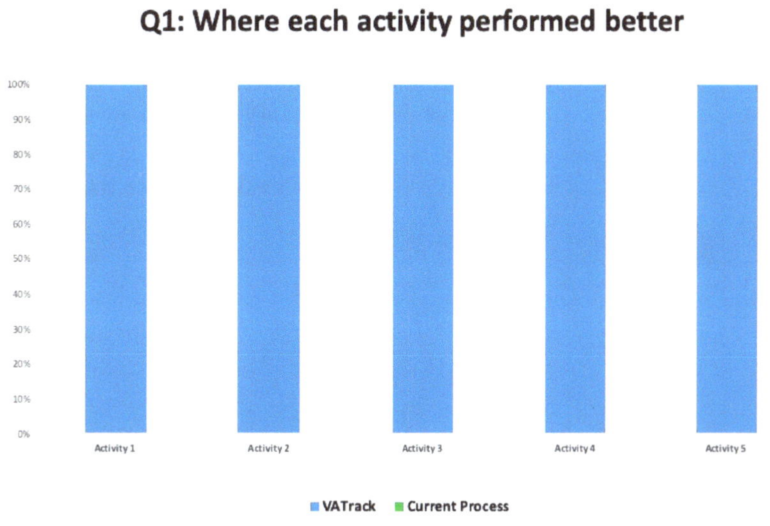

Q1: Where each activity performed better

Figure 17. Test users' responses to the first question of the questionnaire.

The second question was about the overall satisfaction with the VATrack design according to the clarity of information, comfort of usage, visual appeal, and usefulness. In general, the responses fell into two categories, as seen in Figure 18: 62% of participants were very satisfied, and 38% were satisfied.

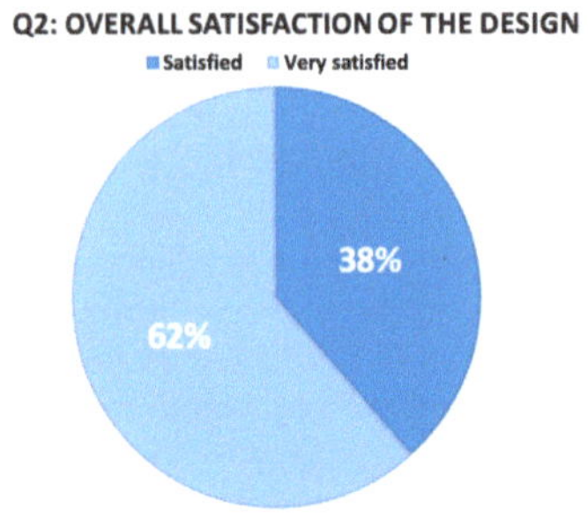

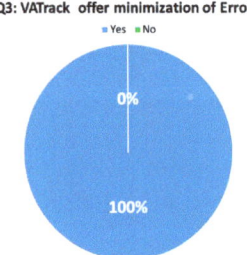

Figure 18. Test users' responses to the second question of the questionnaire.

The third question was about VATrack's ability to minimise errors while performing tasks or in data pre-processing. All test users, as shown in Figure 19, agreed that VATrack minimised errors. In addition, all test users, as shown in Figure 20, agreed with this.

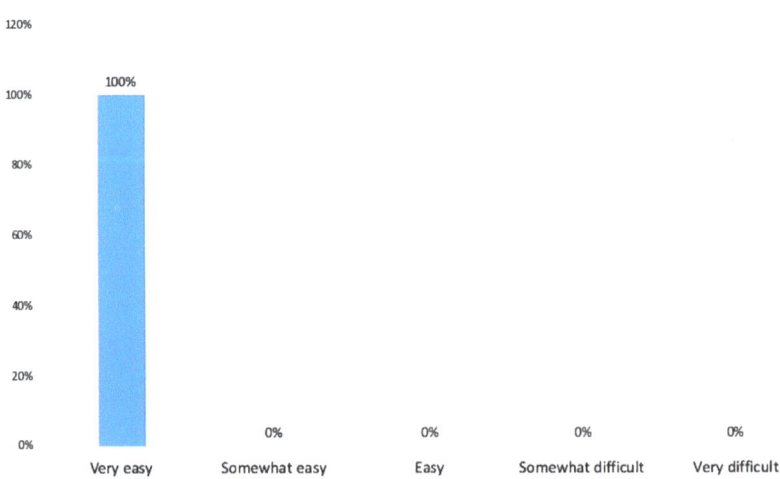

Figure 19. Test users' responses to the third question of the questionnaire.

Figure 20. Test users' responses to the fourth question of the questionnaire.

The remaining question was about what particular aspect(s) VATrack users liked or disliked and what they thought would improve VATrack. In general, users liked the speed of obtaining the required answers and that the ease of use—interacting with it—was intuitive. In addition, they liked not having to constantly check the DRLs since they were built-in.

In addition, the users liked several simultaneous filters. One complaint was that even though VATrack was much faster than the current process, it needed a better internet connection to perform smoothly. This could be solved using a desktop version of VATrack. Second, if it is possible to change the used Diagnostic Reference Levels (DRLs).

For the suggested improvements in general, users were very happy with their experience with VATrack. One user suggestion was to allow multiple DRLs such as American DRLs and Saudi DRLs in addition to the European DRLs. This point will be added in our future work to support different DRLs.

Users needed more flexibility than commercial radiation dose-tracking software offered, and age classes had to be redefined in accordance with standards used in a user environment, which cannot be done with Dosewatch software. In addition, users needed to be able to customize the view to display CT scan protocols as groups of the targeted body part and to aggregate multi-step protocols into a single record. Plus, users needed to be able to dig deeper once they had identified which patient had been exposed to high radiation doses and answer some important questions such as: 'Why?', 'Which device?', 'Who was the operator?', 'Was it a cumulative or single-dose?', and many more.

6. Discussion

Due to the current increased use of CT scans, patients are increasingly prone to radiation overdose hazards, such as an increased risk of developing cancer [2]; hence, there is a need to keep track of every possible venue where a person might be exposed to radiation, and one of these venues is the hospital.

For example, in [22], the cumulative radiation dose from diagnostic imaging such as a CT scan measured over a partial period—from diagnosis until radiotherapy planning—was 71.5 mSv (range: 11.9–131.9 mSv) for breast cancer patients; this is not negligible. It is also important to note that monitoring is not only to track the cumulative dose but also to support the justification and optimization of procedures.

Before proceeding in detail, we must differentiate between radiation exposure tracking and radiation dose tracking. According to Rehani [2], exposure tracking involves tracking all radiological examinations a patient has undertaken regardless of the dose, whereas radiation dose tracking takes into consideration the dose of each examination.

Healthcare providers, policymakers, regulators, researchers, industries, and, most importantly, patients can benefit from radiation exposure and dose tracking. For instance, in terms of patient safety to avoid/reduce instances of radiation exposure overdose, healthcare providers can benefit from such tools in decision support, resource savings, and better-quality radiation datasets for researchers.

When tracking, patients may be tracked as individuals or groups. In addition, one might track radiological procedure types and counts if a patient has undergone several procedures, the radiation dose data of those procedures, or both. Tracking diagnostic imaging approaches is not an easy task: one of the main difficulties is that radiation dosage varies for the same examination type depending on the scanner, protocol, part of the body to be imaged, and total body size, even if it is the same scanner.

Using the proposed solution shown in Figure 1, users can start by filtering cases using the "flagged studies" filter on the left, examining data more closely on the scatter plot, or viewing detailed patient and study information from the table at the top. In addition, each filter is colour-coded such that the darker/more intense the colour, the higher the radiation count.

7. Concluding Remarks

In this study, we have addressed several problems associated with radiation dose tracking software solutions, including lack of flexibility, ineffective analysis support, and poor interactivity, which tend to decrease user satisfaction with such solutions. The main contributions of this work include:

- Design and develop the proposed visual analytics dashboard to track radiation doses to which patients are exposed.
- Clean the data and organize it automatically in an interactive dashboard.
- Apply several filters simultaneously to the data using several interactive styles. Additionally, the versatility of different filters allows users to explore and manipulate data to answer specific questions.

The aim of this study was to provide a generic, flexible, and easy-to-use visual analytics display tool for radiation dose data from medical scanners and to implement a proof-of-concept prototype that proves our design. The results show that the visual analytic approach improves the tracking process, as users completed the tasks with a 100% success rate.

In future work, we will utilize VATrack exploratory data analysis to allow users to create their own radiation dose benchmarks. They can then include them with other diagnostic reference level standards by providing patient cohort comparison features to emphasize trends and patterns. In addition, to overcome the limitation of a retrospective analysis style, VATrack will have a built-in predictive model that can predict future scenarios based on user-defined criteria and generate alerts based on the radiation dose. Furthermore, VATrack can be used with other diagnostic imaging types where tracking is required, such as nuclear imaging. More, we will make VATrack customizable in a way that allows the user to determine protocols within each targeted body category because this cannot be automated based on protocol names considering there is no standard naming convention in the department. Finally, we will create a desktop version of VATrack to ensure reliability and availability, thus avoiding internet connection issues.

Author Contributions: R.A. and F.A.; Conceptualization, methodology, investigation, analysis, resources, writing—original draft, preparation, writing—review and editing final draft, project administration and funding acquisition. All authors have read and agreed to the published version of the manuscript.

Funding: The Deanship of Scientific Research (DSR) at King Abdulaziz University, Jeddah, funded this project under grant No. KEP-PhD-29-612-1443.

Informed Consent Statement: Informed consent was obtained from all subjects involved in the study.

Data Availability Statement: Not applicable.

Conflicts of Interest: The authors declare no conflict of interest.

Abbreviations

BMI	Body Mass Index
CTDI	Computed Tomography Dose Index
CT Scan	Computerized Tomography
DICOM	Digital Imaging and Communication in Medicine
DLP	Dose Length Product
EMR	Electronic Medical Record
HIS	Hospital Information System
PACS	Picture Archiving Communication System

References

1. Chuah, L.F.; Klemeš, J.J.; Bokhari, A.; Asif, S.; Cheng, Y.W.; Chong, C.C.; Show, P.L. Chapter 3—A review of intensification technologies for biodiesel production. In *Biofuels and Biorefining*; Gutiérrez-Antoni, C., Gómez Castro, F.I., Eds.; Elsevier: Amsterdam, The Netherlands, 2022; pp. 87–116. [CrossRef]
2. Rehani, M.M. Patient radiation exposure and dose tracking: A perspective. *J. Med. Imaging* **2017**, *4*, 031206. [CrossRef] [PubMed]
3. Crowley, C.; Ekpo, E.; Carey, B.; Joyce, S.; Kennedy, C.; Grey, T.; Duffy, B.; Kavanagh, R.; James, K.; Moloney, F.; et al. Radiation dose tracking in computed tomography: Red alerts and feedback. Implementing a radiation dose alert system in CT. *Radiography* **2021**, *27*, 67–74. [CrossRef] [PubMed]

4. Sodickson, A.; Baeyens, P.F.; Andriole, K.P.; Prevedello, L.M.; Nawfel, R.D.; Hanson, R.; Khorasani, R. Recurrent CT, cumulative radiation exposure, and associated radiation-induced cancer risks from CT of adults. *Radiology* **2009**, *251*, 175–184. [CrossRef] [PubMed]
5. Smith-Bindman, R.; Wang, Y.; Yellen-Nelson, T.R.; Moghadassi, M.; Wilson, N.; Gould, R.; Seibert, A.; Boone, J.M.; Krishnam, M.; Lamba, R.; et al. Predictors of CT Radiation Dose and Their Effect on Patient Care: A Comprehensive Analysis Using Automated Data. *Radiology* **2016**, *282*, 182–193. [CrossRef] [PubMed]
6. Buzug, T.M. Computed Tomography. In *Springer Handbook of Medical Technology*; Kramme, R., Hoffmann, K.P., Pozos, R.S., Eds.; Springer: Berlin/Heidelberg, Germany, 2011; pp. 311–342. [CrossRef]
7. Frane, N.; Bitterman, A. Radiation Safety and Protection. 2020. Available online: https://www.ncbi.nlm.nih.gov/books/NBK557499/ (accessed on 7 June 2023).
8. Åhlström Riklund, K.; Andersson, J.; Lundman, J.; Granberg, C. Experiences with large-scale radiation exposure monitoring in Västerbotten County, Sweden. In *Proceedings of the EuroSafe Imaging 2016*; European Society of Radiology: Vienna, Austria, 2016.
9. Hintenlang, D.E. *Patient Dose Tracking for Imaging Studies*; University of Florida: Gainesville, FL, USA, 2014; p. 66.
10. Ashworth, J. What Is PACS? Library Catalog. Section: Uncategorized. 2018. Available online: www.jpihealthcare.com (accessed on 7 June 2023).
11. Maddalo, M.; Trombetta, L.; Torresin, A.; Colombo, P.; Vanzulli, A.; Vighi, G.; Righini, A.; Vismara, L.; Pola, A. *A Feasibility Study on the Installation of a Multicentre System for Patient Organ Dose Monitoring and Reporting in Computed Tomography*; European Congress of Radiology: Vienna, Austria, 2014.
12. Cook, K.; Thomas, J. *Illuminating the Path: The Research and Development Agenda for Visual Analytics*; U.S. Department of Energy: Washington, DC, USA, 2005; Volume 54.
13. Kovacs, W.; Weisenthal, S.; Folio, L.; Li, Q.; Summers, R.M.; Yao, J. Retrieval, visualization, and mining of large radiation dosage data. *Inf. Retr.* **2016**, *19*, 38–58. [CrossRef]
14. Barufaldi, B.; Schiabel, H.; Maidment, A.D.A. Design and implementation of a radiation dose tracking and reporting system for mammography and digital breast tomosynthesis. *Phys. Med.* **2019**, *58*, 131–140. [CrossRef] [PubMed]
15. De Bondt, T.; Mulkens, T.; Zanca, F.; Pyfferoen, L.; Casselman, J.W.; Parizel, P.M. Benchmarking pediatric cranial CT protocols using a dose tracking software system: A multicenter study. *Eur. Radiol.* **2017**, *27*, 841–850. [CrossRef] [PubMed]
16. Sedlmair, M.; Meyer, M.; Munzner, T. Design Study Methodology: Reflections from the Trenches and the Stacks. *IEEE Trans. Vis. Comput. Graph.* **2012**, *18*, 2431–2440. [CrossRef] [PubMed]
17. Carpendale, S. *Evaluating Information Visualizations*; Springer: Berlin/Heidelberg, Germany, 1970; pp. 19–45. [CrossRef]
18. Vano, E.; Miller, D.; Martin, C.; Rehani, M.; Kang, K.; Rosenstein, M.; Ortiz-Lopez, P.; Mattsson, S.; Padovani, R.; Rogers, A. ICRP Publication 135: Diagnostic Reference Levels in Medical Imaging. *Ann. ICRP* **2017**, *46*, 1–144. [CrossRef] [PubMed]
19. Tableau. What is Tableau. 2022. Available online: https://www.tableau.com/why-tableau/what-is-tableau (accessed on 7 June 2023).
20. Murray, D. *Tableau Your Data! Fast and Easy Visual Analysis with Tableau Software*, 1st ed.; Wiley Publishing: Hoboken, NJ, USA, 2013.
21. Nielsen, J. *Usability Engineering*; Morgan Kaufmann Publishers Inc.: San Francisco, CA, USA, 1994.
22. Choi, J.S.; Rim, C.H.; Kim, Y.B.; Yang, D.S. Cumulative radiation exposure dose of diagnostic imaging studies in breast cancer patients. *Int. J. Radiat. Res.* **2019**, *17*, 275–281. [CrossRef]

Disclaimer/Publisher's Note: The statements, opinions and data contained in all publications are solely those of the individual author(s) and contributor(s) and not of MDPI and/or the editor(s). MDPI and/or the editor(s) disclaim responsibility for any injury to people or property resulting from any ideas, methods, instructions or products referred to in the content.

MDPI AG
Grosspeteranlage 5
4052 Basel
Switzerland
Tel.: +41 61 683 77 34

Technologies Editorial Office
E-mail: technologies@mdpi.com
www.mdpi.com/journal/technologies

Disclaimer/Publisher's Note: The title and front matter of this reprint are at the discretion of the Guest Editors. The publisher is not responsible for their content or any associated concerns. The statements, opinions and data contained in all individual articles are solely those of the individual Editors and contributors and not of MDPI. MDPI disclaims responsibility for any injury to people or property resulting from any ideas, methods, instructions or products referred to in the content.

www.ingramcontent.com/pod-product-compliance
Lightning Source LLC
LaVergne TN
LVHW072336090526
838202LV00019B/2429